STP 1194

Application of Accelerated Corrosion Tests to Service Life Prediction of Materials

Gustavo Cragnolino and Narasi Sridhar, editors

ASTM Publication Code Number (PCN)
04-011940-27

ASTM
1916 Race Street
Philadelphia, PA 19103

Library of Congress Cataloging-in-Publication Data

Application of accelerated corrosion tests to service life prediction of materials / Gustavo Cragnolino and Narasi Sridhar, editor.
(STP ; 1194)
"The ASTM Symposium on Application of Accelerated Corrosion Tests to Service Life Prediction of Materials was held during 16-17 Nov., 1992 in Miami, Fla. ... sponsored by ASTM Committee G-1 on Corrosion of Metals"--Foreword
"ASTM publication code number (PCN) 04-011940-27
Includes bibliographical references and index.
ISBN 0-8031-1853-8
1. Corrosion and anti-corrosives--Testing--Congresses. 2. Accelerated life testing--Congresses. 3. Stress corrosion--Testing--Congresses. I. Cragnolino, Gustavo, 1940- . II. Sridhar, Narasi, 1964- . III. ASTM Committee G-1 on Corrosion of Metals. IV. ASTM Symposium on Application of Accelerated Corrosion Tests to Service Life Prediction of Materials (1992 : Miami, Fla.) V. Series: ASTM special technical publication ; 1194.
TA418.74.A67 1994
620.1'1223--dc20 93-47376
CIP

Photocopy Rights

Peer Review Policy

Each paper published in this volume was evaluated by three peer reviewers. The authors addressed all of the reviewers' comments to the satisfaction of both the technical editor(s) and the ASTM Committee on Publications.

To make technical information available as quickly as possible, the peer-reviewed papers in this publication were printed "camera-ready" as submitted by the authors.

The quality of the papers in this publication reflects not only the obvious efforts of the authors and the technical editor(s), but also the work of these peer reviewers. The ASTM Committee on Publications acknowledges with appreciation their dedication and contribution to time and effort on behalf of ASTM.

Printed in Ann Arbor, MI
February 1994

Foreword

The ASTM symposium on Application of Accelerated Corrosion Tests to Service Life Prediction of Materials was held during 16–17 Nov. 1992 in Miami, FL. It was sponsored by ASTM Committee G-1 on Corrosion of Metals. Gustavo Cragnolino and Narasi Sridhar, both of the Center for Nuclear Waste Regulatory Analyses in the Southwest Research Institute served as symposium cochairmen and editors of this publication.

Contents

Overview

The initial impetus for this symposium was the long-term (hundreds to thousands of years) performance requirements in the disposal of high-level nuclear wastes that have driven the development of various life-prediction approaches for waste containers. Life prediction has also gained increasing importance in other applications due to aging facilities and infrastructure (for example, nuclear power plants, aircraft, concrete structures), heightened concerns regarding environmental impact (for example, hazardous waste disposal, oil and gas production, and transportation), and economic pressures that force systems to be used for extended periods of time without appropriate maintenance. Life prediction in the context of this publication pertains essentially to structures or systems that are undergoing corrosive processes. For systems subjected to purely mechanical failure processes such as fatigue and creep, life prediction techniques have advanced to a greater degree. Accelerated laboratory corrosion tests, which in the past have focused on screening tests for materials ranking and selection and quality control tests for materials certification, also have to be re-evaluated for their usefulness to life prediction. A major objective of this symposium was to provide a forum for discussing the approaches to life prediction used by various industries. A second objective, especially relevant to the mission of ASTM, was to discuss the appropriateness of various accelerated corrosion tests to life-prediction. The papers in this volume cover many industries, although some areas, such as nuclear waste disposal, are more heavily represented than others. However, the goal of being able to compare life prediction versus actual performance is yet to be achieved in many of the industries represented in this volume. The papers in this volume are classified into three sections: Laboratory and Field Analysis Techniques, Life Prediction Techniques in Various Applications, and Experimental Techniques.

Laboratory and Field Data Analysis Techniques

All the papers in this section describe various ways in which laboratory or field data can be extrapolated to predict service life. A systematic approach to design and life prediction, as described by Staehle, would entail a definition of environmental conditions, material conditions, and failure modes, all combined in a probabilistic framework. Several examples from the literature for the definition of failure modes are provided in this paper. One important aspect of this paper is the organization of experimental data in a potential-pH framework so that failure modes can be defined for a given material under a given set of environmental conditions. While the examples cited in this paper use the potential-pH diagrams as the basis for failure mode definition, other methods such as the definition of corrosion potential as a function of time can also be used to define failure modes. The latter approach is described in other papers in this volume (for example, Macdonald et al., Sridhar et al.). The use of Weibull statistics in defining the probability of failure by stress corrosion cracking is also illustrated by Staehle. Finally, the approach to predicting the overall probability of failure by a combination of failure modes, each with its own probability distribution is discussed. It must be noted that this approach can be combined with other methods of combining failure modes such as fault tree analysis. This method also differs from some of the other performance techniques that rely on developing an overall probability of failure through Monte-Carlo techniques whereby individual probabilities of various failure modes are not calculated separately. Nyborg and Lunde discuss an empirical crack-growth model along with a probabilistic assessment based on the uncertainties in the input parameters for the case of ammonia cracking of carbon steel storage tanks. While this model carries many uncertainties regarding the extrapolation of short-term crack-growth data to long-term prediction, the methodology illustrates the importance of early and periodic inspection in reducing failure probability. It also illustrates the sensitivity of failure probability to various material and design parameters, such that corrective actions can be pursued more effectively. The use of artificial neural network to synthesize both short-term laboratory data and longer-term field experience in similar environments to make expected service-life predictions in a rapid manner is discussed by Silverman. This technique is combined with an expert system to provide qualitative guidelines regarding the applicability of a specific material in a given environment for which only short-term data can be generated. McCuen and Albrecht discuss the use of various curve-fitting approaches in extrapolating atmospheric corrosion data collected for time periods ranging up to 23 years to predict end-of-service corrosion penetration at 75 years. They suggest that a composite model that combines a power-law behavior of corrosion penetration at short-times with a linear behavior at long-times is the most robust of the curve-fitting schemes. The uncertainties in predicted penetrations at 75 years due to the uncertainties in the assumed model are highlighted. Duncan et al. describe the use of empirically measured corrosion rates and Poisson distribution for failure to predict the cumulative probability of failure of transuranic waste drums stored at Hanford.

These papers also highlight the need for greater mechanistic (or deterministic as some prefer to call it) understanding of the various corrosion processes since extrapolations performed on the basis of parametric or statistical fitting of present data result in considerable variations in the predicted behavior, depending on the selection of the fitting method.

Life Prediction Techniques in Various Applications

The importance of an engineered barrier system in high-level nuclear waste disposal, as well as the interdependency of the engineering design and environmental conditions, is highlighted by Verink. The next six papers deal with various life-prediction techniques related to high-level waste disposal containers. The approach used by Ikeda et al. in predicting the performance of Ti containers involves the assumption that crevice corrosion initiation is inevitable under the Canadian vault conditions, but that propagation is limited by the availability of oxidants to the open surface. Hence, as time progresses, a deceleration of crevice corrosion propagation and eventual repassivation is predicted to occur. For the same repository and container design, Macdonald et al. use a variety of approaches to predict long-term performance. The mechanistic modeling of corrosion potential is of special importance because it can be used to determine the corrosion modes as a function of environmental factors. Macdonald et al. also calculate the upper bound in corrosion rate by assuming that rate of transport of oxygen or other radiolytic species determines the dissolution rate of Ti and show that the calculated corrosion rates are rather low. These models are further useful because they indicate the areas in which expenditures of experimental effort will be most fruitful. Beavers et al. also emphasize the need for mechanistic modeling and suggest that corrosion allowance materials whose corrosion rate can be well-defined coupled to a multibarrier system be given greater consideration for high-level nuclear waste packages in the U.S. program. Park et al. use fracture mechanics-based tests under a variety of loading conditions to measure stress corrosion crack growth rate of types 304L and 316L stainless steels and alloy 825 in repository groundwater environments concluding that no significant environmentally assisted crack growth was observed in these alloys. The limited environmental conditions examined makes it difficult to use this negative finding for long-term prediction. The approach suggested by Sridhar et al. is essentially similar to that of Ikeda et al., albeit for a different class of materials, namely Ni-based alloys and stainless steels. The long-term prediction is attempted by considering the evolution of corrosion potential and critical potentials (initiation and repassivation potentials) for localized corrosion. A crevice initiation model is presented and the use of repassivation potential as a bounding parameter for container performance assessment is examined. The need for detailed mechanistic justification of crevice repassivation potential and the shortcomings of the current models are pointed out. Sjöblom reviews a variety of scenarios to be considered for assessing the safety of copper containers in the Swedish high-level waste program.

The use and limitations of accelerated laboratory tests to predict service performance of nuclear reactor components are examined in the next two papers. Perkins and Shann show that a higher temperature laboratory test of various types of zircaloy fuel cladding can be used to distinguish the performance of these claddings in service. In contrast, Palumbo et al. warn that certain well-known accelerated laboratory tests may not be able to distinguish subtle variations in alloy 400 samples from two different lots that however, result in significant differences in service performance.

The last two papers in this section cover two widely different industries. The corrosion of oil and gas production components occurs under complex environmental and flow conditions. The paper by Kolts and Buck reviews various empirical correlations between corrosion or erosion rates and environmental and flow parameters. The corrosion and erosion of the infrastructure has become a topic of great concern both in the U.S. and elsewhere. Andrade and Alonso review the factors affecting the service performance of reinforced concrete structures, although the example they cite is mainly related to low-level radioactive waste vaults or bunkers. The importance of preventing or delaying the onset of active corrosion of steel is pointed out. It is also noted that unreinforced concrete structures have lasted for many centuries.

Experimental Techniques

A new index for crevice corrosion susceptibility, based on the concept of change from passive to active behavior due to IR potential drop is presented by Xu and Pickering. The advantage of this technique, in addition to being consistent with some of the observed crevice corrosion phenomena, is the ease with which it can be modeled on a mechanistic basis. A possible limitation may be its inapplicability to stainless steels and other highly passivating alloys. The use of graphite fiber wool as a crevice forming device to accelerate stress corrosion cracking of type 304 stainless steel and alloy 600 is examined by Akashi. The salient feature of this work is the use of exponential probability distribution in comparing the accelerated laboratory test to documented service life. Aaltonen et al. propose a multipotential test technique whereby a number of stress corrosion cracking specimens under a range of applied potentials can be exposed to a given environment simultaneously to determine critical potentials for stress corrosion cracking. The results of this type of test can be used to evaluate some of the proposed methodologies in papers on life prediction mentioned previously. Tsujikawa et al. examine the use of spot welded specimen, which simulates both the effects of crevices and residual stresses for predicting stress corrosion cracking. The important result from this paper is that the critical potential for stress corrosion cracking is the same as the repassivation potential for crevice corrosion. This simplifies the task of performance assessment considerably because one critical potential can be used to evaluate several failure modes. However, these results need further scrutiny. From the mechanical aspect, Louthan and Porr suggest that specimen geometry has a significant effect on stress corrosion cracking susceptibility as measured in slow strain rate tests. The use of electrochemical potentiokinetic reactivation (EPR) test method to characterize the extent of sensitization of some austenitic stainless steels has been relatively well-established. Some semi-empirical models exist that use the EPR values to predict the susceptibility to stress corrosion cracking of certain nuclear reactor components. The EPR technique is extended to the case of a duplex stainless steel by Verneau et al.

A novel method to evaluate atmospheric corrosion beneath a thin electrolyte layer under heat transfer conditions is presented by Muto et al. The interesting feature of this paper is the comparison of accelerated laboratory test results using a rating number derived from Weibull distribution parameters to rating number from field exposure tests. The applicability of this test technique to studying corrosion under repeated wet and dry cycles and to moist environments needs to be examined.

The need for long-term life prediction of components exposed to corrosive conditions necessitates a re-evaluation of many of the accelerated corrosion test methods that are being used at present. As many of the papers in this volume suggest, the comparison between accelerated laboratory tests and service life data must be made in a statistical framework. The evaluation of appropriate test methods will also be aided by the simultaneous development of predictive models. It is hoped that the papers contained in this volume will stimulate further examination of the present-day corrosion test methods, many of which are contained in ASTM standards. We wish to thank the authors for their efforts in the publication of this volume. We also wish to thank the reviewers for their assistance in improving the quality of this publication, the ASTM staff and Mr. Arturo Ramos for their timely assistance in organizing the symposium and assembling this publication. Finally, we wish to thank Dr. Michael Streicher and Mr. Jefferey Kearns who provided the initial encouragement in organizing this symposium.

Gustavo Cragnolino

Narasi Sridhar

Center for Nuclear Waste Regulatory Analyses, Southwest Research Institute, San Antonio, TX; symposium cochairmen and editors.

Laboratory and Field Data Analysis Techniques

Roger W. Staehle[1]

RELATIONSHIP AMONG STATISTICAL DISTRIBUTIONS, ACCELERATED TESTING AND FUTURE ENVIRONMENTS

REFERENCE: Staehle, R. W., "**Relationship Among Statistical Distributions, Accelerated Testing, and Future Environments,**" Application of Accelerated Corrosion Tests to Service Life Prediction of Materials, ASTM STP 1194, Gustavo Cragnolino and Narasi Sridhar, Eds., American Society for Testing and Materials, Philadelphia, 1994.

ABSTRACT: The relationships among future environments, accelerated testing and statistical distributions are discussed. A stepwise procedure for predicting performance is described which starts first with defining environments. Next steps include definition of modes and submodes of corrosion, the definition of materials, and superposition of environmental and mode definitions. Incorporating the dependencies of environmental variables in distribution parameters is discussed.

KEYWORDS: Prediction, corrosion, performance, environmental definition, mode and submode definition, Weibull, design.

INTRODUCTION

The purpose of this paper is to discuss important elements of the relationships among environments in which components must perform and the combination of statistical distributions and accelerated testing which can be used to predict performance in these environments. I also suggest a procedure whereby predictions can be made which incorporate considerations of different environments. The framework for this discussion is the Corrosion Based Design Approach (CBDA) which I have discussed in previous reviews[1, 2]. The principal steps of the CBDA include the following:

1. Environmental definition.
2. Material definition.
3. Mode and submode definition.
4. Superposition of environmental and mode/submode definitions.
5. Failure definition.
6. Statistical definition.
7. Accelerated testing.
8. Prediction.

[1]Adjunct Professor, Chemical Engineering and Materials Science, University of Minnesota, Minneapolis, Minnesota, 55455

9. Feedback and correction.
10. Modification and optimization of design, materials, environments, operations.

This discussion considers primarily steps 1, 3, 4, 6, 7 and 8.

ENVIRONMENTAL DEFINITION

Defining environments expected in service is the most crucial step in predicting performance. While it is often assumed that environments cannot be easily defined and therefore that corrosion cannot be so easily predicted, this is erroneous. This section describes an approach which is general and, if properly implemented, can be rigorous. Most important, until the environments are defined at the levels described in this section, the development of accelerated tests and the application of statistical distributions will not be useful.

The main ideas associated with the step of environmental definition are the following:

Multiple Environments on a Single Component

On a single component, for example a tube in a nuclear steam generator, several environments may activate modes of corrosion which will perforate the wall. Corrosive environments may occur on the inside surface and on the outside. Corrosive environments may vary along the length as there are tube supports, sludge, and flow differentials.

Several different modes of corrosion can occur associated with any one or all of these different environments. Thus, there is no single corrosive environment; rather there are distinguishable separate environments which need to be considered. Such distinguishable environments need to be identified first and their properties determined by some combination of experiments and analyses.

Nominal Classes

Environments are often thought of as nominal environments by designers when considering the performance of materials. Such nominal environments are usually too narrowly considered. Examples of nominal environments usually considered are pure water, drinking water, moist air, or sea water as they may apply to various components. The concept of "nominal" needs to be more inclusive since there is much more that can be known *a priori* than is usually accepted. "Nominal" needs to be considered in four classes by designers in determining what materials and designs are optimum and in developing predictions:

1. Major nominal—The major nominal environment is the general environment in which the component operates. Such an environment may be the pure water, drinking water, moist air or sea water. Unfortunately, this is often the environment used in qualifying components and materials.

2. Minor nominal—The minor nominal includes low concentrations of species which are usually well known but not always considered. In moist air these may be the acidic gases. In drinking water, these may be bacteria.

3. Local nominal—Local nominal environments are those that occur in crevices, under deposits and associated with long range cells. Local nominal environments are those which are affected by heat transfer and wetting and drying as these may be exacerbated by occluding geometries. The possibility of such local environments and the general range of chemistries which they may engender are knowable. These local environments need to be considered implicitly in design.

4. Accidental nominal—Certain environmental incursions are knowable. Condenser tubes are often perforated and cooling water enters steam systems. Prevailing winds bring H_2S gases from refineries to nearby industrial equipment as well as to automobiles. Oxygen enters deaerated systems. Most if not all of such environments are knowable and need to be considered in design, monitoring and inspection.

These four parts of "nominal" can serve as a useful check list for designers and materials engineers alike in determining the possibility of the various modes of corrosion which might occur.

Changes in Time

Environments on surfaces change in time as deposits build up, crevices are clogged and heat transfer persists. Environments change in time as the equipment sustains different circumstances such as :

1. Manufacturing.
2. Shipping.
3. Storage.
4. Construction.
5. Testing.
6. Start up.
7. Steady state.
8. Shutdown.

Failures may occur in any or all of these environments and there is extensive anecdotal information on failures which occur at all of these stages. For each of these, the range of the four nominal conditions exists and needs to be accounted for.

Intrinsic and Extrinsic Modes of Corrosion

Historically, mainly due to the writings of Fontana[3], the morphology of corrosion has been considered in terms of general corrosion, pitting, intergranular, parting, galvanic cell, crevice, stress corrosion cracking and erosion corrosion. Others have been variously added to this list or combined such as liquid metal embrittlement and splitting SCC into constant load and cyclic or corrosion fatigue. However, some of these "forms" of corrosion are really definitions of environments. For example, galvanic cells, crevices, cyclic stresses, and erosion are really descriptions of environments. Within these environments, general corrosion, pitting, intergranular corrosion, parting and SCC may occur. Thus, the latter are "intrinsic" to the material and any or all of them may occur in a crevice environment or in a galvanic cell.

Thus, in considering corrosion, these forms of corrosion which are in fact environments are not considered part of the intrinsic modes of corrosion. Crevices, cells, cyclic stressing and erosion are taken as defining environments rather than forms of corrosion.

Environmental Definition Diagrams

It is possible to define environments in terms of key parameters such as potential and pH. Whether the environment is a major nominal, minor nominal, local nominal or accidental nominal, the roles of environments on the surface can be specified primarily in terms of potential and pH. Sometimes, other features must be specified such as the species and their concentrations which affect passivity to the extent that these are not specified by potential and pH. This suggests that all the environments to which a given material or component is exposed can be summarized on a single diagram mainly with potential and pH as coordinates and possibly with a third coordinate which considers some kind of passivating or non-passivating parameter. Such a third coordinate may also be incorporated as contours on a two dimensional diagram.

Once the range of potential and pH for a given environment is specified, it can be compared with the domain for a mode of corrosion. Such an overlay is shown schematically in Fig. 1. Here, the domains for a single environment and a single mode of corrosion are shown in adjacent diagrams; their superposition is shown in Fig. 1c. The region where the environmental definition and the mode definition intersect is the region where the corrosion mode may be expected to operate. It is this region of intersection which is of interest for accelerated testing and statistical definition. In general, components may be exposed to several different environments simultaneously and several different modes or submodes of corrosion may occur in any or all of the environments.

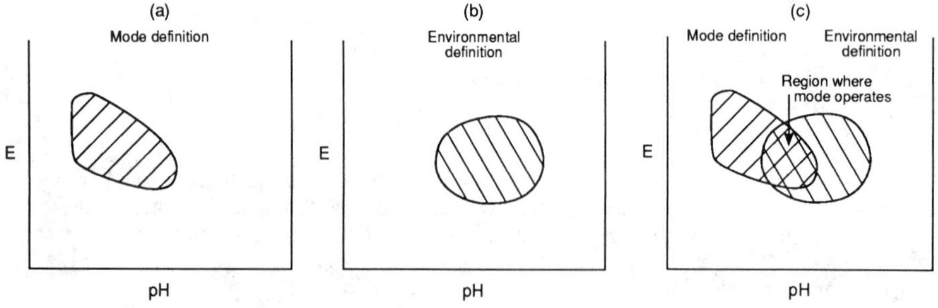

Figure 1 Schematic view of a corrosion mode diagram (a), an environmental definition diagram (b) and their superposition (c). From Staehle[2].

This procedure for overlaying properties of environments and materials with respect to corrosion to determine what reactions might occur was originally developed by Pourbaix[4, 5, 6]. Fig. 2 shows the superposition of environmental definition in 2b on a corrosion mode definition in 2a to obtain a superimposed result in 2c. Figs. 1 and 2 differ in that the former correlates kinetic data and the latter correlates thermodynamic data. In the former, potential and pH are useful since they are principal variables in kinetic processes. In the latter, potential and pH are useful since they are principal thermodynamic influences on the stability of metals. However, the kinetic and thermodynamic influences are complementary in that the latter provide the boundaries for the former. This is the reason that the superposition shown in Fig. 1 is so useful.

An approach to defining environmental conditions is shown in Fig. 3[1,7]. Here, environmental conditions are plotted with reference to the equilibrium potential-pH diagrams for Inconel 600 as it is used in the range of 300°C for tubing in nuclear steam generators. The various environmental conditions in Fig. 3 show patterns which may be expected but none have been based precisely on direct experimental measurements. This diagram is reassuring in that broad patterns can be suggested by inspection with some general knowledge of thermodynamics and kinetics.

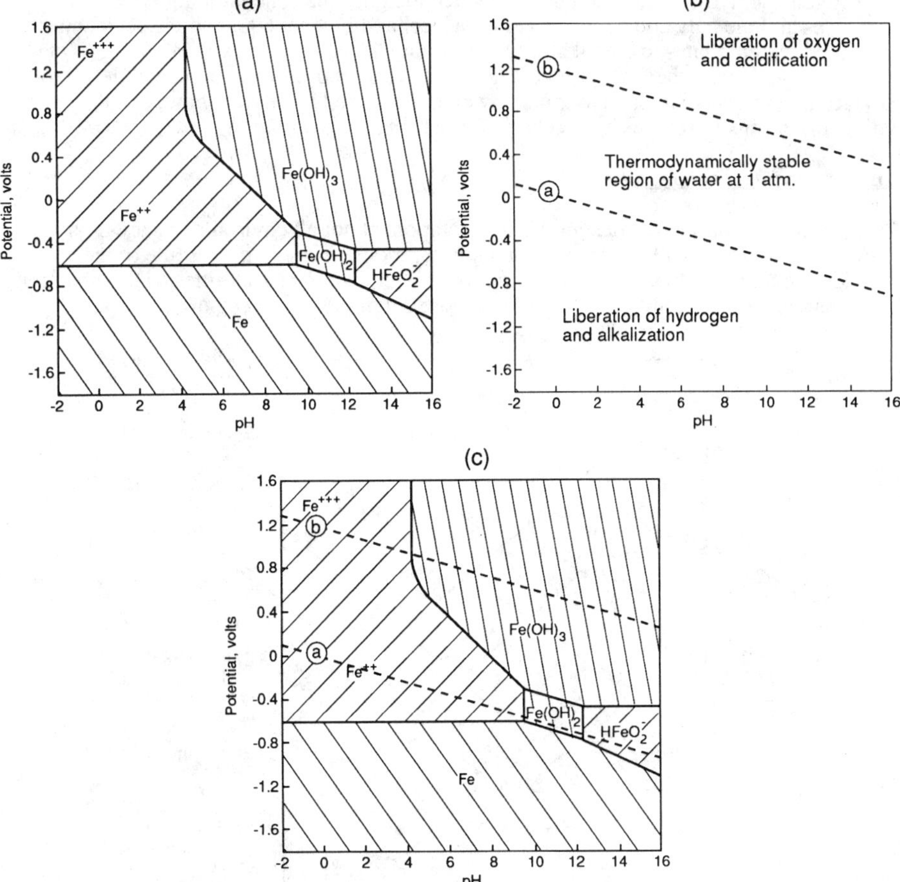

Figure 2 Superposition of potential-pH diagrams for iron and water. (a) Potential-pH diagram for iron. (b) Potential-pH diagram for water. (c) Superposition of diagrams to give integrated view of relationship between metal and environment. Adapted from Pourbaix[5].

Fig. 3 starts with an initially deaerated environment on a free surface in environment 1a. These free surface environments change their effects as oxygen, hydrogen and hydrazine are added. After considering the free surface, the effects of alkaline and acidic crevices are added as these might arise where heat flux acts to concentrate impurities which arise respectively from fresh water and salt water used in the cooling water circuit for the condensers.

Fig. 3 also shows that different conditions on both sides of the tube, primary and secondary water, in the free spans and in heat transfer crevices can be described on a single diagram. This is an important feature since the failure of a tube is still a failure regardless of where it occurs.

In terms of the four conditions of nominal Fig. 3 shows: the major nominal condition which is the deaerated condition; the minor nominal condition which includes hydrogen and hydrazine; local nominal which includes the concentrating effects of crevices and the galvanic coupling between steel and Inconel 600; and the accidental nominal which relates to the alkalinity and acidity as well as the oxygen.

Fig. 3 is not the final product. The final version of an environmental definition diagram would have to be determined by experiments and analyses; however, the diagram in Fig. 3 provides a good basis for reasonable expectations.

Environmental Probability

In considering the design life of a component not all environments are highly probable and some will occur only for short durations. For example, some of the accidental nominal conditions may occur only with certain probabilities. The subject of environmental probability is discussed in more detail in another review[8].

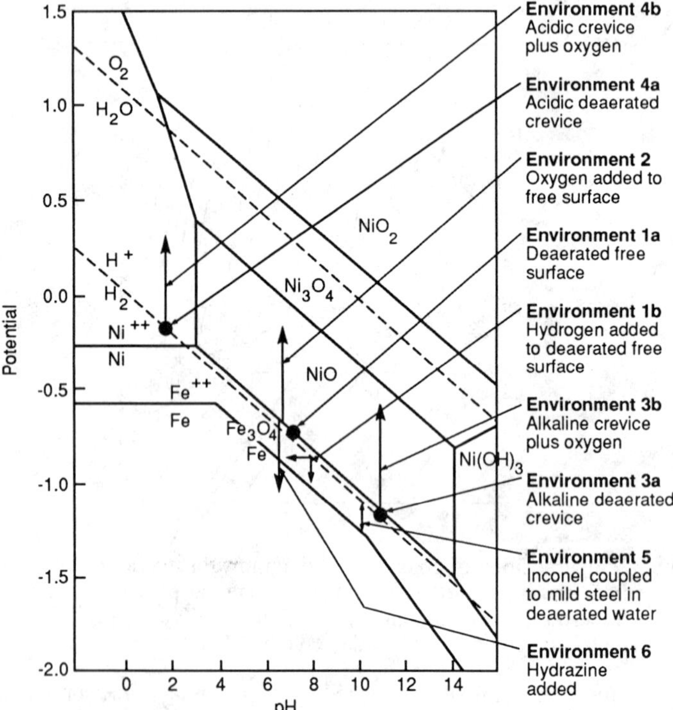

Figure 3 General environmental definition diagram for nickel and iron base alloys where crevices and galvanic connections are illustrated. Diagram taken for 300°C for nickel and iron in water. Environmental conditions are defined at the right and point to locations or ranges where the effects occur. Thermodynamic diagram based on the work of Chen[9].

The domain of potential and pH for environments which need to be treated probabilistically can be determined early since the associated chemical conditions can be readily known and investigated. However, when and for how long such conditions may exist may be known with less certainty and may have to be treated probabilistically. The integration of environmental probability into prediction is considered later in this discussion.

Stress Environments

This discussion is focused mainly on chemical environments. However, the stress environment is equally important especially as it affects SCC and corrosion fatigue. Of all the stress conditions which affect SCC, residual stresses are the most important. Such contributions need to be quantified and accounted for. The applied stresses usually play a minor role in corrosion processes because they are a relatively small fraction of the residual stresses and stress affects SCC according to a more or less fourth power relationship.

The intensity of SCC is also increased when stress is cycled as a ripple on a mean stress and as the cyclic frequency is reduced. These subjects have been treated elsewhere[10].

MATERIAL DEFINITION

Defining materials is not dealt with here but has been discussed in other reviews[1, 2, 7]. Briefly, in order to predict corrosion performance, it is necessary to define the following for metallic materials:

1. Major alloying elements.
2. Minor alloying elements.
3. Impurity elements.
4. Grain size and anisotropy.
5. Composition, distribution of elements and structure of grain boundaries.
6. Composition and distribution of second phases.
7. Cold work.
8. Yield strength.

Defining such factors attends to differences in performance which are often ascribed to "heat to heat variation." The role of some of these material factors also is fixed by heat treatments and cold work; once these are defined, the behavior of the material may be more or less fixed. However, it is not the heat treatment that affects corrosion, it is the distribution of elements at the grain boundaries, the distribution of second phases and the dislocation density.

When the overall failure probability is developed, material definition can be quantified and considered as a random variable[9]. This is not done here.

MODE AND SUBMODE DEFINITION

Modes of corrosion refer to the general morphologies of corrosion. These morphologies are essentially the intrinsic modes of corrosion: general corrosion, pitting, intergranular corrosion, parting and SCC. In addition to these modes there are variations. For example, for the alloy Inconel 600, in high temperature water in the range of 300°C there appear to be five submodes of SCC. The term, "submode," in this discussion is defined as an occurrence of a mode which can be distinguished by generally different dependencies upon alloy heat treatment, temperature, stress, and environmental composition.

Defining modes and submodes of corrosion is important to prediction because it is the occurrence of the modes and submodes which produces failures. It is important to know how many of these modes can occur in the several environments which may occur simultaneously on the component. It is the modes and submodes which are modeled statistically and which are the basis for accelerated testing. The question of which modes occur is answered by the environmental definition.

Distinguishing among submodes of corrosion, e.g. for SCC, is crucial when predicting performance. For example, the results of accelerated testing of one submode would not be applicable to another. Also, it might be assumed that one submode is more aggressive than another only to realize later that such an assumption is erroneous.

The principal implication of this discussion is that these modes and submodes are not surprises and occur in relatively regular patterns contrary to the older ideas of specific ions and magic circumstances of "susceptibility." Since the occurrence of the various modes and submodes is so regular, it is possible to determine which ones operate by superposition of the various environments in some format as described in Fig. 1.

The occurrences of different submodes of corrosion follow generally patterns of existence of protective films, solubility and the occurrence of hydrogen. Reasons for such dependencies and patterns have been discussed[2, 7]. In general, the locations where the various modes and submodes of corrosion can occur is knowable; Fig. 4 shows where SCC should be expected with reference to a polarization curve[7, 11]. The locations where SCC is shown in Fig. 4 are knowable generally from the potential-pH diagrams such as those shown for iron in Fig. 2. In Fig. 4 zone 1 corresponds to the region of potential where hydrogen-related processes generally occur. Zone 2 corresponds to the instability between the active peak and passivity. Zone 3 corresponds to the instability between passivity and transpassivity. If SCC would occur in all three zones in the same range of environments, there would be three "submodes."

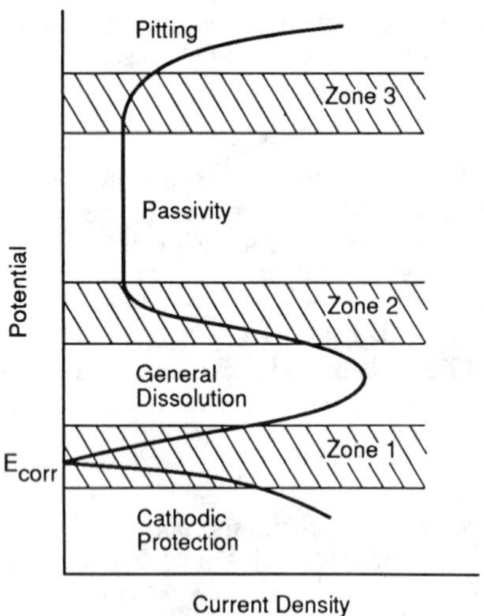

Figure 4 Schematic view of regions of potential in which SCC is reasonably expected to occur relative to the polarization characteristics. After Staehle[2].

An illustration of a mode diagram which describes pitting is shown in Fig. 5 where the occurrence of pitting is determined experimentally by polarization experiments. At the left are the results of polarization experiments and at the right is a corresponding potential-pH diagram with the pitting mode identified. Thus, when environments are defined as intersecting this region of pitting as for the superposition in Fig. 1c, pitting will occur.

Relative to general suggestions provided in Fig. 4 which applies only to a single pH, a broader view of modes of SCC for steels is shown in Fig. 6. Separate studies by Congleton et al.[12] and Parkins[13] have produced the data in Figs. 6a and 6b respectively. These figures are combined in Fig. 6c to produce an integrated mode diagram for the SCC of low alloy steel[2]. Fig. 6c is based on extensive experiments with a variety of materials and environments. Fig. 6d shows the data for pitting from Fig. 5 superimposed on the SCC diagram of Fig. 6c. Fig. 6d, thus, provides an integrated view of SCC and pitting of low alloy steels although there are some differences in the environments which are relevant to pitting and SCC.

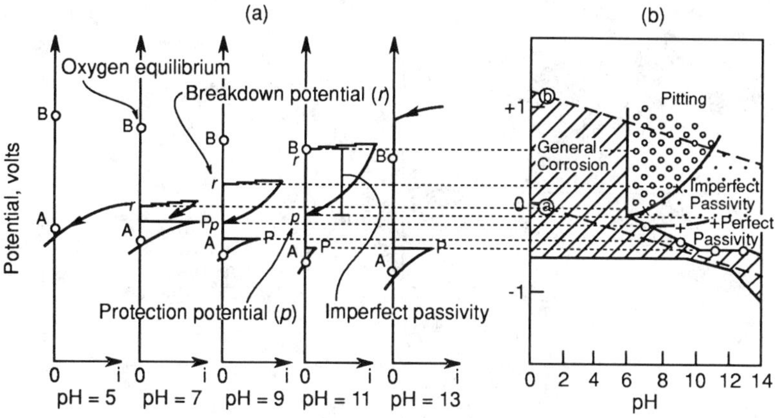

Figure 5 Superposition of kinetically determined conditions for pitting on the potential-pH diagram for iron. (a) Polarization curves at different values of pH. (b) Potential-pH diagram for iron with the domains of pitting and imperfect passivity also shown. Adapted from Pourbaix[6].

The patterns observed for SCC in Fig. 6c can be related to expectations which are derived from the potential-pH diagram for iron shown in Fig. 2. For example, the region in which hydrogen-related SCC is defined by the hydrogen equilibrium line is part of the water equilibrium. Secondly, the anodic modes above the hydrogen line show no SCC where iron has its minimum solubility. This suggests that the solubility of iron plays an important role in SCC. Third, in both the alkaline and acidic directions, SCC occurs where regions of passivity or metastable passivity occur adjacent to regions of great solubility. Note, in the acidic region the SCC mode is symmetric about the Fe_3O_4/Fe_2O_3 equilibrium and its metastable extension. This corresponds approximately to the Flade potential which identifies a transition from stable to unstable behavior of protective films. Finally, it is important to note that the acidic submode of SCC occurs regardless of the chemistry of the solutions studied.

Figs. 4, 5 and 6 show that the various modes including pitting and SCC and submodes of SCC follow dependencies upon the electrochemical potential. This pattern is illustrated also from the work of Subramanyam in Fig. 7 where specimens of stainless steel were exposed at 100% of their yield strength in a boiling 70% caustic solution. Plotted in Fig. 7 are the polarization curve, the time to failure, the mode of corrosion, and the regions of stable oxides. Several features are of particular interest. When the potential is in the passive range, corrosion does not occur and the time to fail was "infinite" in the framework of these experiments. Note that the specimens were small diameter wires. Second, the morphologies of SCC which occur follow patterns much like those suggested in Fig. 6. SCC occurs where there are transitions in stability. Pitting occurs at high potentials, and general dissolution occurs where it should in the active region. Fig. 7 is significant because it was conducted in a single environment with material from the same source; only the potential was changed whereas the data for Fig. 6 was collected from several sources and in several environments. The general patterns of both figures are quite consistent.

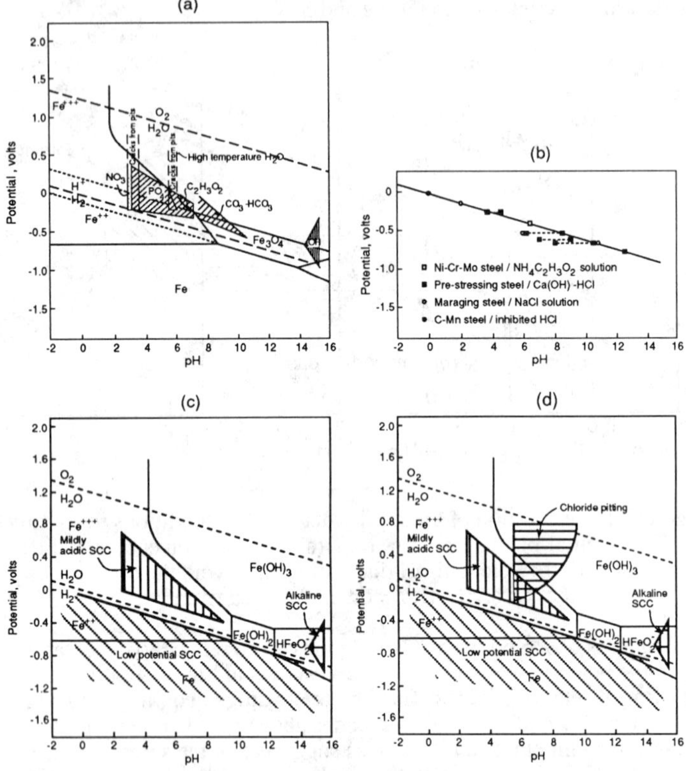

Figure 6 Regions of occurrence for different modes of SCC for low alloy steels exposed to aqueous solutions. Work shown from studies in different environments and for different alloys. (a) Shaded areas show regions where SCC occurs. (b) Line shows the potential below which hydrogen related SCC occurs. (c) Schematic synthesis showing three submodes of SCC based on (a) and (b). (d) Integration of mode definition diagrams for SCC and pitting of low alloy steel[2].

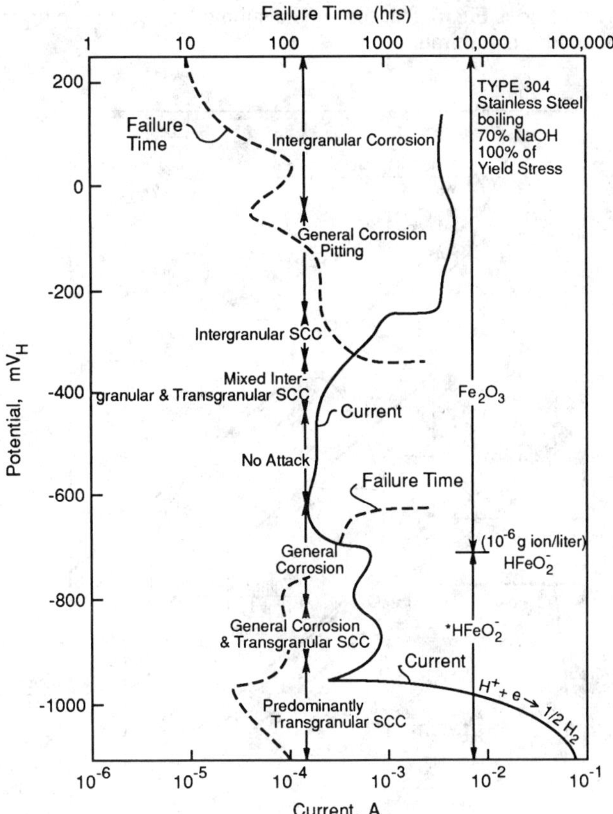

Figure 7 Effect of electrochemical potential on polarization current, time-to-failure and mode of corrosion for Type 304 stainless steel stressed at 100% of the yield stress in boiling 70% NaOH. Adapted from the work of Subrahmanyam et al.[14]

Another view of mode diagrams is shown in Fig. 8 for Inconel 600 in high temperature water in the range of 300°C. The development of this diagram has been already described[2, 7, 15]. This diagram was developed by analyzing data from many authors where separate submodes of SCC were investigated. Agreement among the many authors is quite good in terms of the individual submodes although their respective versions of the intensity of SCC varies since they all used different testing techniques. The pattern which results is quite similar to that for low alloy steels in Fig. 6: There is a submode of SCC in the alkaline region for Inconel 600 similar to that in Fig. 6; similarly, there is an hydrogen related submode; in the acidic region there is also an SCC submode. In addition, there is a submode at high potentials in the acidic region. This corresponds to SCC which is also observed in stainless steels in highly oxidizing media. Finally, there is at least one submode which includes transgranular SCC associated with lead impurities added to these solutions. For the present it is not known whether such SCC is unique to the lead or results from the lead fixing the local potential and pH.

Fig. 8 also shows, as does Fig. 6, that the anodic submodes become negligible when the solubility of the oxides is minimum.

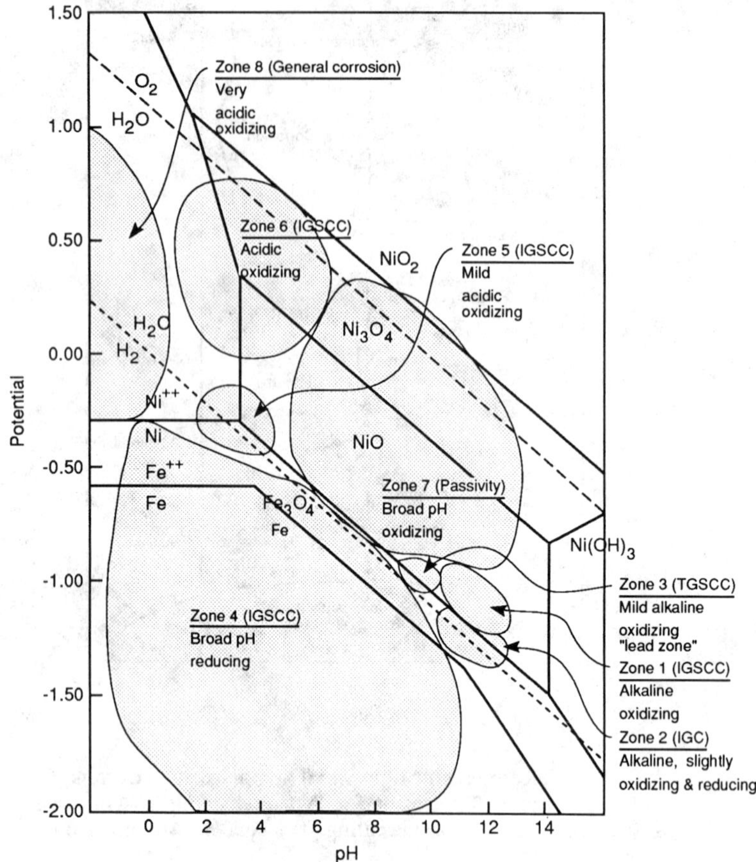

Figure 8 Modes of corrosion for Inconel 600 in the range of 300°C plotted on a potential-pH diagram for iron and nickel determined for 300°C. Oxygen/water and water/hydrogen equilibria at 300°C are shown. In addition to the submodes of stress corrosion cracking (zones 1, 3, 4, 5 and 6), a passivity mode (zone 7), an intergranular corrosion mode (zone 2) and a general dissolution mode (zone 8) are shown. From Staehle[7].

The distribution of modes and submodes shown in Fig. 8 may be taken one more step to indicate the intensity of SCC as affected by the environment. Such a three-dimensional plot is shown in Fig. 9. Unfortunately, the experimental procedures and resulting data upon which this figure is based are not perfectly coherent and Fig. 9 is based mostly on judgment. However, it does indicate the great value of such three-dimensional plots; and, accepting the uncertainties inherent in combining data from many investigators, the results seem quite regular and credible.

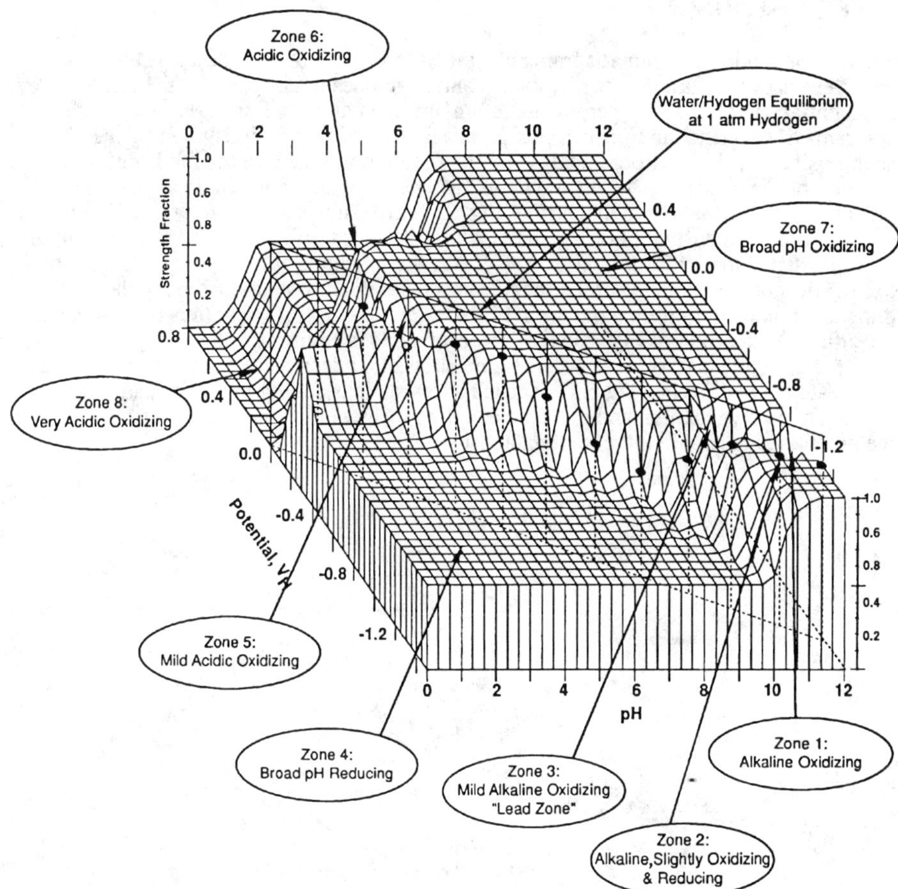

Figure 9 Three dimensional corrosion mode diagram for Inconel 600 in water at about 300°C. The useful strength is plotted versus the potential and pH. Vertical plane shows the location of the water/hydrogen equilibrium at standard conditions. Zones of corrosion are noted. From Staehle[7, 15].

The corrosion mode diagrams show that the corrosion modes can be summarized neatly in the framework of potential and pH. Further, there are many similarities among low alloy steels, stainless steels, and high nickel alloys. These diagrams show that the occurrence of the various corrosion modes and submodes follows regular patterns which adhere to patterns dominated by the relative stability of protective films and to the occurrence of molecular hydrogen.

These corrosion mode diagrams are sufficiently regular that they can be readily used in superpositions with environmental definition diagrams to define the intersections where various corrosion modes may occur. Once the intersections, according to Fig. 1, are determined, it is possible to determine which modes in which environments need to be correlated with statistical distributions and should be the subject of accelerated testing.

SUPERPOSITION

The mode definition diagrams illustrated in Figs. 5, 6 and 8 describe the occurrences of the various intrinsic modes and submodes of corrosion in the framework of potential and pH. The superposition of environmental definition and mode definition diagrams as suggested in Fig. 1 from a practical point of view is illustrated by the superposition of the environmental definition diagram in Fig. 3 on the mode definition diagram of Fig. 9 with the result shown in Fig. 10. Fig. 10a shows an isometric view of the three dimensional mode diagram of Fig. 9. It also shows a plane cut through it which corresponds to a constant pH. The plane, thus cut, plots useful strength versus potential as illustrated at the left of Fig. 10b. Fig. 10b at the left starts at high potentials and goes to low; the corresponding useful strength shows at the left a section from the three dimension mode definition diagram in Fig. 9. Here, the useful strength is plotted versus potential. At the right of Fig. 10b the effects of various environmental changes as shown in Fig. 3 are plotted. Essentially, this figure shows that decreasing the potential decreases the useful strength of Inconel 600. Those factors such as adding hydrazine, adding hydrogen, deoxygenation, and galvanic coupling of Inconel 600 to carbon steel all lower the potential and should increase the intensity of SCC.

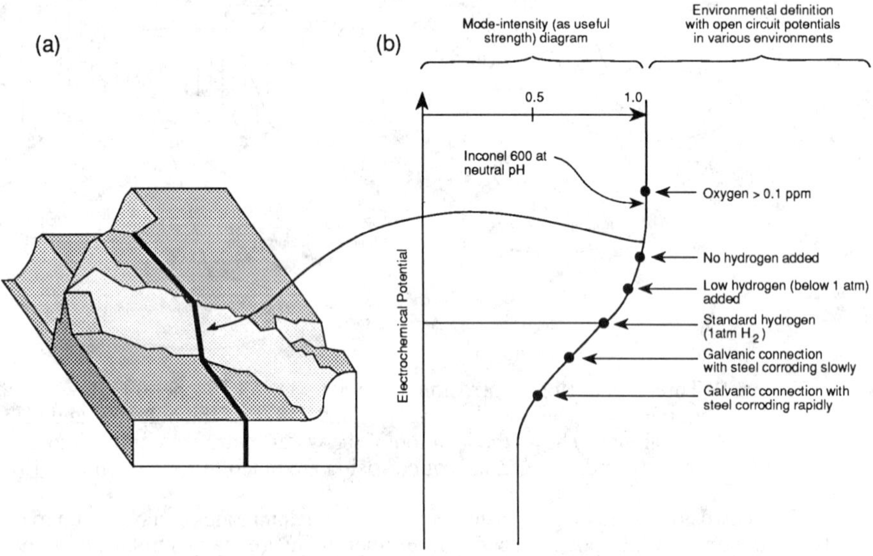

Figure 10 (a) Schematic view of three dimensional corrosion mode diagram for Inconel 600 at about 300°C with a slice taken at constant pH. (b) Useful strength versus potential from (a) and potentials shown which correspond to specific modifications of the environment based on the environmental definition diagram for Fe-Ni base alloys in high temperature water. After Staehle [2].

The superposition in Figure 10 is more complex than in Fig. 1. A simpler superposition could have been produced by superimposing Fig. 8 on Fig. 3 or Fig. 3 after it is embellished by experimental data.

STATISTICAL DEFINITION

Once the region or regions of intersection are identified based on superposition of mode and environmental definitions, it is appropriate to seek methods for determining how to predict performance of the modes and submodes as they may occur in the various possible environments. In order to make such predictions it has been useful to correlate data with statistical distributions.

There are many approaches to correlating data but they are not considered here. Only the Weibull distribution is used to illustrate how statistical distributions might be used to predict performance. It should be recognized that statistical distributions in themselves are not the data nor do they necessarily or intrinsically model the data. They simply provide useful means for correlating data.

The probability density function for the Weibull distribution is:

$$f(t) = \left[\frac{b}{\theta}\left(\frac{t - t_o}{\theta}\right)^{b-1}\right]\left\{\exp\left[-\left(\frac{t - t_o}{\theta}\right)^{b}\right]\right\} \qquad (1)$$

where:

$f(t)$ = probability density function
t = time
t_o = initiation time
b = Weibull slope
θ = characteristic value

The probability density function is integrated to give the cumulative failure probability:

$$F(t) = \int_{t_o}^{t_n} f(t)dx \qquad (2)$$

The result of this integration gives equation 3 which is the cumulative distribution.

$$F(t_n) = 1 - \exp\left\{-\left(\frac{t_n - t_o}{\theta}\right)^{b}\right\} \qquad (3)$$

Equation 3 when linearized, becomes equations 4 and 5:

$$\log\log\frac{1}{1-F(t)} = A + b\log(t_n - t_o) \qquad (4)$$

$$\text{Where: } b\log\frac{1}{\theta} + \log 2.303 = A \qquad (5)$$

Of special interest to this discussion is the significance of the distribution parameters, t_o, b, and θ which are respectively the initiation time, the Weibull slope, and the Weibull characteristic. The initiation time, t_o, in these equations is not a physical constant but rather a fitting parameter; however, determining its value is often useful because it gives a sense of when the failure process might have initiated. From a physical point of view the initiation time is actually a random variable and would have its own distribution and distribution parameters. The Weibull slope, b, is a measure of the dispersion of the data. "b" is referred to as the slope since it serves this function in equation 4 when the cumulative distribution is plotted. The Weibull characteristic, θ, is a measure of the central tendency of the data and here has units of time.

A principal value of the Weibull function is its flexibility in modeling a wide range of data. Figs. 11 and 12 illustrate the relationship between the probability density function and the cumulative distribution for selected values. Fig. 11a illustrates the effects of changing θ while holding the slope constant; Fig. 11b illustrates the effect of changing the slope while holding θ constant. Fig. 11b shows the interesting result that slopes of b of unity or less produce curves that have a more exponential pattern and thereby gives the Weibull function more flexibility. Fig. 12a and 12b show the cumulative distributions corresponding to the probability density functions in Figs. 11a and 11b. Here, the role of the Weibull slope, b, becomes more clear.

The application of Weibull correlations to data for the failure of steam generator tubes is illustrated in Fig. 13. Here the data for SCC in primary water as it affects the failure of U bends in U tube steam generators is summarized[16]. Fig. 13 shows a cumulative distribution of the type in Fig. 12; the slope is 4.5 and θ is 9.1 years with t_o as zero. In this plot 10% of the tubes have failed and the Weibull distribution fits the data well. Such data permit predicting when larger fractions of tubes would fail.

What makes these distributions useful at this point is the possibility that they can incorporate dependencies upon principal variables which affect the performance of materials. Such dependencies would be incorporated into the three distribution parameters of θ, b, and t_o. Such variables include temperature, stress, environmental composition and properties of the materials. These dependencies can be used in two ways:

1. The distribution parameters can be determined under aggressive conditions and extrapolated to nominal conditions. This would provide bases for predicting cumulative failure performance at relatively long times using relative short time data.

2. If the distribution parameters have been determined in terms of functional dependencies upon test variables, the resulting equations may be used to determine cumulative failure distributions when the variables change with time. Thus, if the temperature changes over time, the cumulative distribution can be determined as the temperature changes.

Modeling corrosion failure phenomena would be approached by determining the effect of environmental variables on θ, b, and t_o. To illustrate how this might be done consider the effect of stress, temperature and hydrogen ion concentration on time to failure. In general, these variables influence t_f as shown in equations 6-8:

$$t_f = \beta\sigma^{-n} \tag{6}$$

$$t_f = \gamma e^{Q/RT} \tag{7}$$

$$t_f = \alpha[H^+]^{-m} \tag{8}$$

where:

t_f = time to failure
n = exponent of stress
Q = thermal activation energy
T = absolute temperature
R = gas constant
$[H^+]$ = hydrogen ion activity
m = exponent of hydrogen ion effect
$\beta, \gamma, \alpha, \delta, \theta_o$ = constants

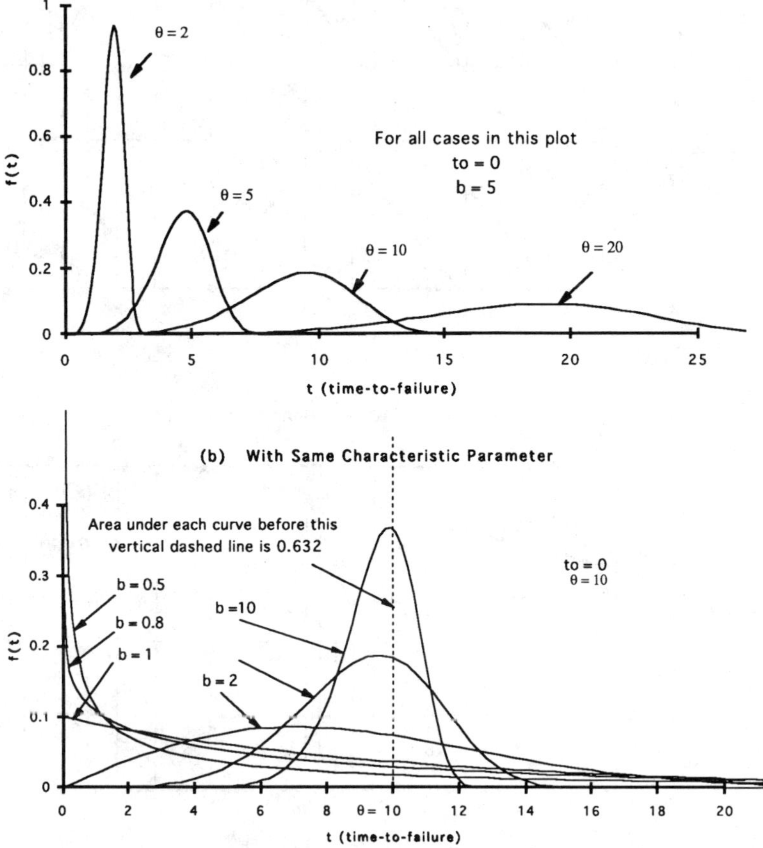

Figure 11 Probability density function for Weibull distribution. (a) b = 5 with θ = 2, 5, 10, 20; (b) θ = 10 with b = 0.5, 0.8, 1.0, 2, 5, 10. The initiation time, t_o, is taken as zero in both cases and the time to failure is taken as in arbitrary units.

These dependencies suggest that the parameters of the statistical distributions might be readily modeled. For example, the characteristic value of the Weibull distribution might be modeled by the product of equations 6-8 as shown in equation 9.

$$t_f = \delta[H^+]^{-m}e^{-Q/RT}\sigma^{-n} \tag{9}$$

$$\theta = \theta_o[H+]^{-m\theta}e^{Q_\theta/RT}\sigma^{-n_\theta} \tag{10}$$

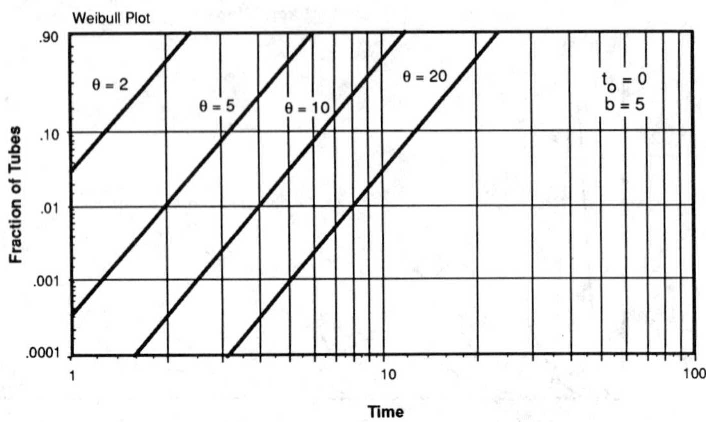

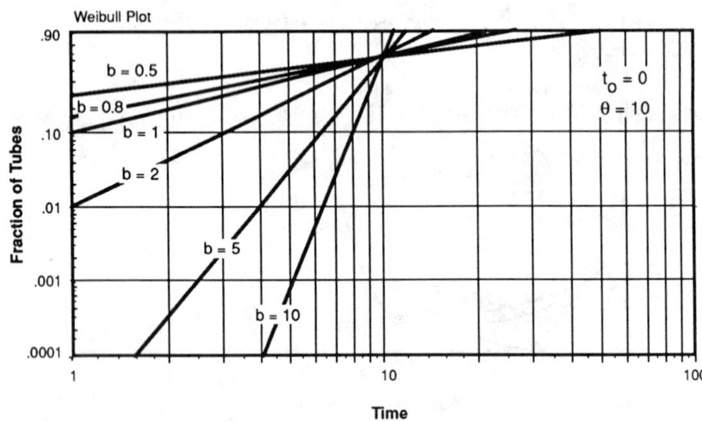

Figure 12 Linearized cumulative distribution for Weibull distribution. (a) $b = 5$ with $\theta = 2, 5, 10, 20$; (b) $\theta = 10$ with $b = 0.5, 0.8, 1.0, 2, 5, 10$. The initiation time, t_o, is taken as zero in both cases and the time to failure is taken as in arbitrary units.

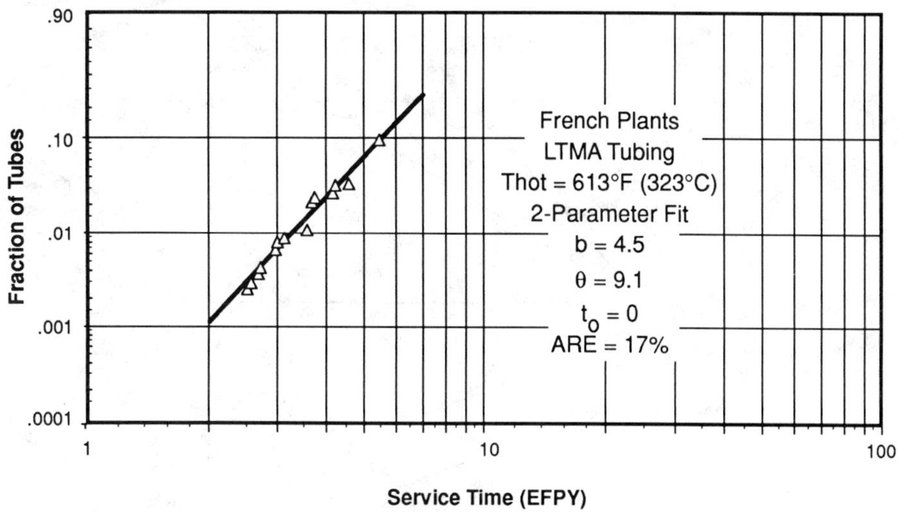

Figure 13 Fraction of tubes failed versus time in Weibull coordinates for SCC on the primary side for US made row 1 U-bends in French plants[16].

The initiation time and the Weibull slope might be modeled similarly. Once the distribution parameters are modeled, then:

$$f(t) = f(t, \sigma, T, H^+) \qquad (11)$$

In the form of equation (11) the probability density function and the cumulative distribution can be used for extrapolating accelerated data and for integrating over time when the environmental variables change.

It should be noted parenthetically here that there is a controversy concerning how corrosion processes should be modeled. This argument suggests that one must know the mechanism completely before it can be modeled. While such an objective is desirable, it is not achievable in practical times. Statistical correlations are quite adequate and widely utilized for characterizing corrosion data for use in design.

The fact that distribution parameters depend regularly on environmental variables is illustrated in Fig. 14 based on work by Shimada and Nagai[17]. Here, the time to failure of zircaloy-2 specimens when exposed to iodine at 350°C has been measured as a function of stress. Fig. 14a shows the cumulative distributions and Fig. 14b shows how the Weibull parameters depend on stress. These dependencies seem regular and their tendencies are in directions that seem intuitively reasonable.

The data of Fig. 14 show the dependencies when data can be obtained in relatively short times. When accelerated tests need to be extrapolated, it is necessary to obtain data at relatively shorter times and extrapolate them to longer times. Such an approach with the temperature variable is shown in Fig. 15[18]. Here, distributions are determined at relatively short times at high temperatures; these distributions are extrapolated to nominal conditions.

Conducting accelerated tests and integrating such work with statistical frameworks is discussed in a text by Nelson[18].

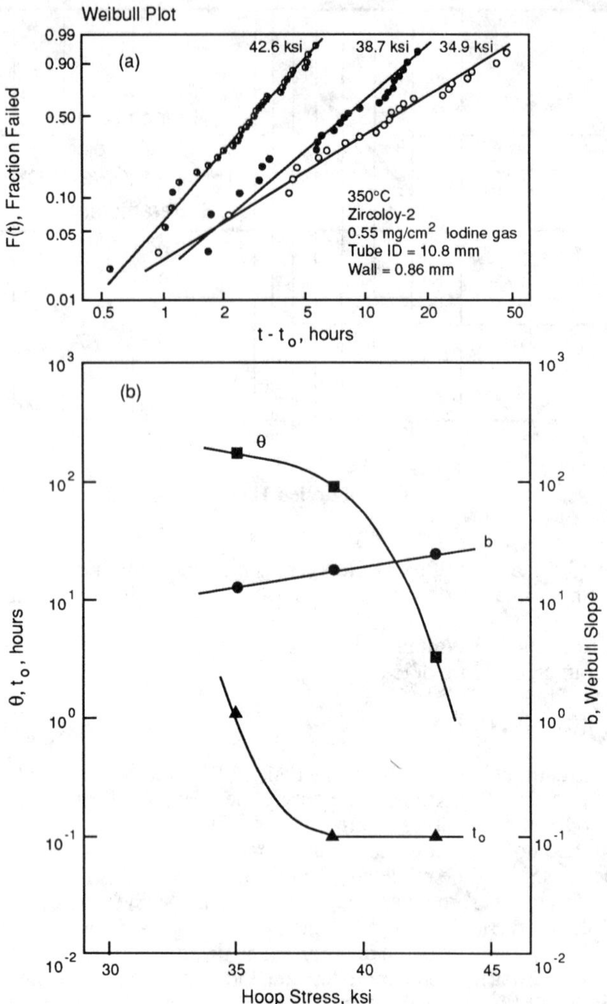

Figure 14 Effect of stress on the cumulative failure rate of zircaloy-2 in iodine gas at 350°C. (a) Cumulative distributions at three stresses. (b) Effect of stress on characteristic time, slope and initiation time. Adapted from Shimada and Nagai[17].

PREDICTING PERFORMANCE

For the present discussion, "predicting performance," means predicting the time to failure. Based on this discussion, predicting performance includes the following essential components:

1. A component is likely to be exposed to more than one explicitly different environment. In the simplest terms there may be a free surface environment and one associated with deposits in crevices. In general, there may be several more. Such environments need to be explicitly recognized and defined in the sense of Figs. 1 and 3.

2. Within the ranges of each of these different environments there may be several explicitly different modes and submodes of corrosion in the sense of the mode diagrams shown in Figs. 5, 6, and 8.

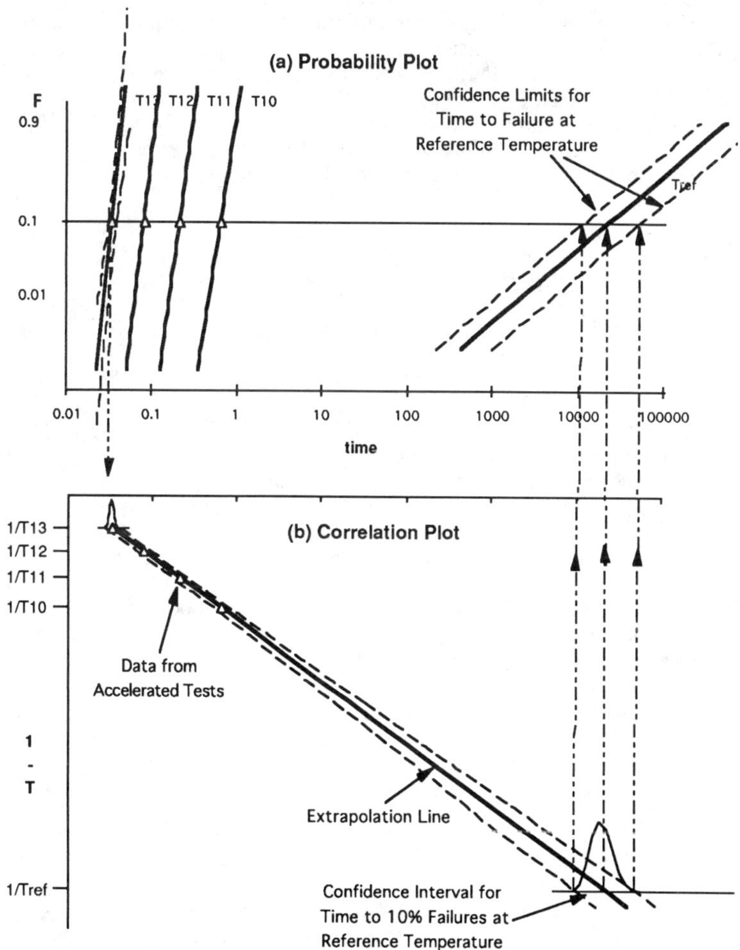

Figure 15 Probability plot (a) and correlation plot (b) for the effect of temperature on corrosion failure processes. T_{13}, T_{12}, . . . are temperatures of accelerated tests. From Staehle and Stavropoulos[8].

3. As a result of the environmental and mode definitions of points 1 and 2, there may be several explicitly different mode-environment processes, possibly three to six. For this discussion, let us designate these by their cumulative distributions F1(t), F2(t)... Fn(t). Thus, the component may be perforated by any or all of these mode-environment processes.

4. For the purpose of this discussion, I have omitted consideration of material definition as a part of the prediction process; however, it needs to be included. The material variables may be lumped together in some way depending on the desire for simplicity and convenience. This problem is not considered here.

5. Also, the problem of separating initiation and propagation is not considered here, but likewise needs to be considered.

6. The cumulative distributions need to be quantified by accelerated testing as suggested in Figs. 14 and 15.

7. At this point cumulative distributions are available for each of the mode-environment cases. The total failure probability, F_T is then determined from the product of success probabilities, assuming the individual probabilities are independent:

$$(1 - F_T) = (1 - F_1)(1 - F_2) \ldots (1 - Fn) \qquad (12)$$

and

$$F_T = 1 - (1 - F_1)(1 - F_2) \ldots (1 - F_n) \qquad (13)$$

8. F_T gives the total cumulative probability of failure for all of the mode-environment processes to which the component is exposed.

9. As suggested in Fig. 15 there would be corresponding confidence limits associated with the final F_T. Interestingly, it can be shown that the confidence band for F_T decreases as the number of possible failure processes increases.

10. In the framework suggested by equation 13, failure might be defined as failure at the 95% confidence limit at the design life. Such a case is illustrated in Fig. 16 where the cumulative probability of failure is plotted versus time with associated 95% confidence limits and for a change in temperature from an initial 100°C to 20°C after 100 years[8]. The objective here is to stay below 1% failure in 1000 years and 10% failure in 10,000 years. The plot of Fig. 16 is schematic and is not based on direct experimental data; however, it is indicative of how such a plot might appear.

The approach summarized in Fig. 16 is one such approach which might be taken for predicting performance. Certainly, less elaborate approaches might be taken where failure envelopes, safety factors, boundaries, or thresholds are used. Whether these use the statistical approach suggested here may be less important; however, it is necessary to use the steps involving environmental definition, mode definition, material definition, and superposition as bases for prediction.

ACKNOWLEDGMENTS

Much of the work upon which this discussion is based has resulted from discussions with Drs. J. A. Gorman and K. D. Stavropoulos of Dominion Engineering. The work which led to this paper has been supported and continues to be supported by the Department of Energy at Yucca Mountain, the Power Reactor and Nuclear Fuel Corporation in Japan, IHI Corporation of Japan, the Electric Power Research Institute, and several electric utilities. In particular I would like to thank Dr. J. Boak of DOE, Dr. N. Sasaki of PNC, Mr. H. Wakamatsu and Dr. M. Akashi of IHI, and Mr. C. Welty of EPRI for their support and encouragement. I would also like to acknowledge extensive discussions with Prof. T. Pigford, Prof. R. N. Parkins, and Dr. E. N. Pugh over the past several years. I appreciate very much being asked to participate in this meeting and I especially appreciate the leadership of Drs. G. Cragnolino and N. Sridhar for setting up the program.

I would also like to acknowledge the very great help of my associates who have helped in preparing the materials for this manuscript and for their continued help on all my projects: Mary Berg, Nancy Clasen, John Ilg, Barbara Skon, and Carolyn Swanson.

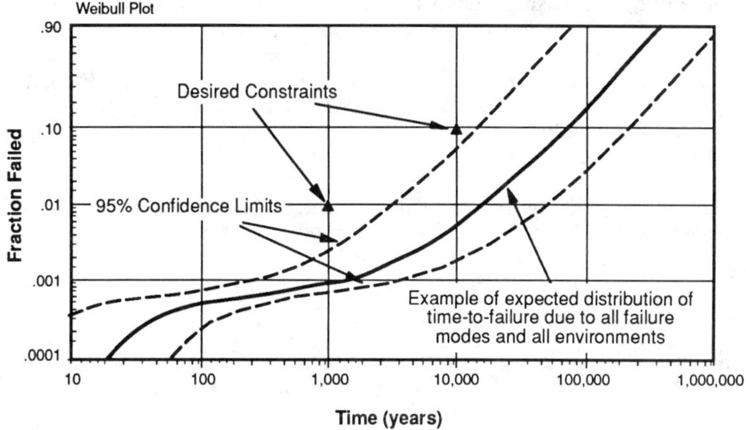

Figure 16 Schematic illustration of cumulative fraction failed versus time showing constraints, 95% confidence limits and a central curve including several modes of failure and a decreasing surface temperature. For this purpose failure is defined as exceeding 95% confidence limits for 1% failures in 1000 years and 10% failure in 10,000. All parameters here are arbitrary. The times here are relevant to containers for radioactive waste but are not based on experiment.

REFERENCES

[1] Staehle, R. W., "Environmental Definition," *Materials Performance Maintenance*, R. W. Revie,V. S. Sastri, M. Elboujdaini, E. Ghali, D. L. Piron, P. R. Roberge and P. Mayer, eds., Pergamon Press, Ottawa, Ontario, 1991.

[2] Staehle, R. W., "Development and Application of Corrosion Mode Diagrams," *Parkins Symposium on Fundamental Aspects of Stress Corrosion Cracking*, S. M. Bruemmer, E.I. Meletis, R. H. Jones, W. W. Gerberich, E. P. Ford, and R. W. Staehle, eds., TMS, Warrendale, Pennsylvania, 1992.

[3] Fontana, M. G., *Corrosion Engineering*, 3rd Ed., McGraw Hill, New York, 1986.

[4] Pourbaix, M, *Thermodynamics of Dilute Aqueous Solutions*, Edward Arnold & Co. 1949.

[5] Pourbaix, M, *Atlas of Electrochemical Equilibria in Aqueous Solutions*, CEBELCOR, NACE, Houston, 1974.

[6] Pourbaix, M, *Lectures on Electrochemical Corrosion*, Plenum Press, London 1973.

[7] Staehle, R. W., "Understanding'Situation-Dependent Strength': A Fundamental Objective in Assessing the History of Stress Corrosion Cracking," *Environment-Induced Cracking of Metals*, NACE, Houston, 1989.

[8] Staehle, R. W. and Stavropoulos, K. D. "Elements and Issues in Predicting the Life of Containers for Radioactive Waste," Report to the Department of Energy, Yucca Mountain Project Office, Corrosion Center, University of Minnesota, March 30, 1992.

[9] Chen, C. M., "Computer-Calculated Potential pH Diagrams to 300°C, Volume 2: Handbook and Diagrams," EPRI Report NP-3137, 1983.

[10] Parkins, R.N. Ed., *Life Prediction of Corrodible Structures*, to be published by NACE.

[11] Staehle, R.W., "Stress Corrosion Cracking of the Fe-Cr-Ni Alloy System," *The Theory of Stress Corrosion Cracking in Alloys*, Ed. J.C. Scully, NATO, 1971.

[12] Congleton, T. Shoji and R. N. Parkins, "Stress Corrosion Cracking of Reactor Pressure Vessel Steel in High Temperature Water," *Corrosion Science, 35*, 1985, pp. 633-650.

[13] Parkins, R.N., "The Use of Synthetic Environments for Corrosion Testing," *ASTM STP 970*, P.E. Francis, T. S. Lee, eds, American Society for Testing and Materials, Philadelphia, 1988.

[14] Subrahmanyam, D.V., Agrawal, A.K., and Staehle, R. W. "The Stress Corrosion Cracking Behavior of Type 304 Stainless Steel in Boiling 50-% NaOH Solution," *Proceedings of the 7th International Congress on Metallic Corrosion*, NACE, Houston, 1978.

[15] Staehle, R.W., and Gorman, J.A.,"Development and Application of Intensity and Operating Diagrams for Predicting the Occurrence and Extent of Stress Corrosion Cracking," Proccedings of an International Symposium, *Corrosion Science and Engineering*, In Honour of Marcel Pourbaix's 85th birthday, Rapports Techniques CEBELCOR, Brussels, vol. 157-158, RT 297-298, November 1989.

[16] Staehle, R.W., Gorman, G. A., Stavropoulos, K.D., Welty Jr., C.S., "Application of Statistical Distributions to Characterizing and Predicting Corrosion of Tubing in Steam Generators of Pressurized Water Reactors." *Life Prediction of Corrodible Structures*, To be published, NACE, Houston.

[17] Shimada, S., and Nagai, M., "Variation of Initiation Time for Stress Corrosion Cracking in Zircaloy-2 Cladding Tube," *Reliability Engineering*, 9, 1984.

[18] Nelson, W., *Accelerated Testing*, John Wiley & Sons, New York, 1990.

Rolf Nyborg[1] and Liv Lunde[1]

LIFE PREDICTION OF AMMONIA STORAGE TANKS BASED ON LABORATORY STRESS CORROSION CRACK DATA

REFERENCE: Nyborg, R. and Lunde, L., "Life Prediction of Ammonia Storage Tanks Based on Laboratory Stress Corrosion Crack Data," Application of Accelerated Corrosion Tests to Service Life Prediction of Materials, ASTM STP 1194, Gustavo Cragnolino and Narasi Sridhar, Eds., American Society for Testing and Materials, Philadelphia, 1994.

ABSTRACT: Stress corrosion crack growth of carbon steel in anhydrous ammonia has been studied at both ambient temperature and at -33 °C. This has resulted in the development of a crack growth model where the crack depth is predicted to be proportional to the square of the stress intensity factor and the square root of the exposure time. The crack growth is much slower at a low temperature of -33 °C than at ambient temperature. The crack growth model has been used together with field inspection results and probabilistic methods to obtain a quantitative measure for the probability of tank failure as function of time. This approach can be used by the ammonia storage tank operators to optimize the inspection intervals and procedures while assuring the safety of the tank.

KEYWORDS: stress corrosion cracking, carbon steel, ammonia, storage tanks, crack growth rate, probabilistic methods

Storage tanks for anhydrous ammonia suffer stress corrosion cracking in the welds. Spherical pressure vessels operating at ambient temperature are inspected with a few years' interval, and in many cases several stress corrosion cracks from 1 to 5 mm deep are found. The cracks are ground down, and repair welding is applied if the remaining wall thickness becomes too small. Many spheres have been repaired and inspected several times, and often new cracks develop between each inspection. Before the experiments described here were started, there was no information available about how quickly the

[1]Senior Research Scientist and Department Head, respectively, Institutt for energiteknikk, P.O.Box 40, N-2007 Kjeller, Norway

cracks actually had grown. Cracks could have grown slowly during the period from one inspection to the other, or could have been formed shortly after filling of the tank. In this period the oxygen content in the ammonia is high, and this gives a high susceptibility to stress corrosion cracking [1,2,3]. Information about the crack growth rate of carbon steel in ammonia is of vital importance from a safety point of view.

Until a few years ago it was believed that SCC did not occur in low temperature storage tanks operating at -33 °C, where the vapour pressure of ammonia is at atmospheric pressure. During the last years, stress corrosion cracks have been found also in several low temperature storage tanks [4,5,6,7]. The present experiments provide information about crack growth rates for both ambient temperature and low temperature storage tanks.

The tendency for stress corrosion cracking of ammonia storage tanks can be reduced by using low strength steel and soft welding electrodes, minimizing residual welding stresses and local hardness peaks, minimizing oxygen contamination and inhibiting with water additions [3]. However, this cannot guarantee against cracking in every case, and the ammonia storage tank operators have to assume the presence of cracks in their tanks. Fracture mechanics calculations can give critical crack sizes for each tank, and inspection can ensure that no significant cracks are present. Inspections must be repeated before any undetected or new cracks can grow to a critical size. Some information about crack growth rates is necessary in order to establish safe inspection intervals. The crack growth model presented here can provide a useful tool for this. However, model predictions must be compared with actual inspection results, and the model adjusted if necessary. Uncertainties in model predictions, complex stress distributions and lack of physical data on specific materials can give results with too high of an uncertainty to be of practical use. Probabilistic methods can provide tools for dealing with these various uncertainties in a rational manner. Such an approach can ensure safety of ammonia storage tanks while optimizing inspection intervals and procedures.

CRACK GROWTH EXPERIMENTS

Experimental Procedure

The experiments were performed with 25 mm wide compact tension (CT) specimens with 1.5 mm deep side grooves. The specimens were made from the normalized carbon steel St 52-3N (DIN 17100) with yield point 380 MPa and tensile strength 550 MPa. This steel corresponds to ASTM A537 Grade 1 or BS 4360 Grade 50D, and represents construction steels typically used for large ammonia storage tanks. The specimens were mounted in a stainless steel test container, which was then half filled with liquid ammonia. The oxygen content in the liquid ammonia was controlled by adjusting the air pressure in the test container before filling with ammonia, and the resulting air partial pressure was measured continuously during the experiment. The temperature of the liquid ammonia was kept constant at either 18 °C or -33 °C. The experiments were performed with 1 to 10 ppm oxygen and 50 ppm water in the liquid ammonia. Previous investigations have shown that this composition range gives the highest SCC susceptibility [1,2,3].

The load was applied to the specimen after stable environmental conditions had been obtained by means of a hydraulic cylinder mounted on top of the test container. Stress intensity factors between 30 and 120 MPa $m^{1/2}$ and exposure times between 20 and 900 hours were used. After the experiment, the specimens were broken apart and the fracture surface examined in a scanning electron microscope (SEM).

The first experiments were performed with CT specimens with fatigue precracks. Severe crevice corrosion attack was found inside the fatigue crack, and this attack seemed to prevent SCC. No stress corrosion cracks were found in these specimens. CT specimens with a sharp notch without fatigue crack obtained SCC readily, and specimens of this type were used for the crack growth studies.

Crack Growth At Ambient Temperature

A series of experiments was performed with constant load and stress intensity factor 80-85 MPa $m^{1/2}$ at 18 °C. Exposure times varied from 24 to 900 hours. Fig. 1 shows the maximum crack depth in each specimen as function of exposure time. There is a considerable spread in results, but evidently the crack growth slows down with time. The full-drawn line in the figure corresponds to an SCC crack depth proportional to the square root of the exposure time. This represents an upper bound for the crack depth at ambient temperature in these experiments.

SEM examination of the fracture surfaces showed a crevice corrosion attack in the outer part of the stress corrosion crack. This crevice corrosion attack seems to slow down the stress corrosion crack growth possibly by reducing the oxygen content in the ammonia near the crack tip or by changing the electrochemical conditions in the crack. Findings of deposits inside the cracks have been reported in both published [7] and unpublished tank inspection results. The presence of oxygen in the ammonia is a vital requirement for SCC of carbon steel in ammonia [1,8]. Oxygen in liquid ammonia does not produce a passive film of the same character as the one formed in aqueous systems, but forms an adsorbed film on the steel surface which maintains the corrosion potential at noble values. Iron dissolution can take place when this film is disrupted by plastic deformation, accompanied by oxygen reduction in oxygen filmed areas [8]. Pure ammonia has much lower conductivity than pure water, and for a deep crack to grow it is

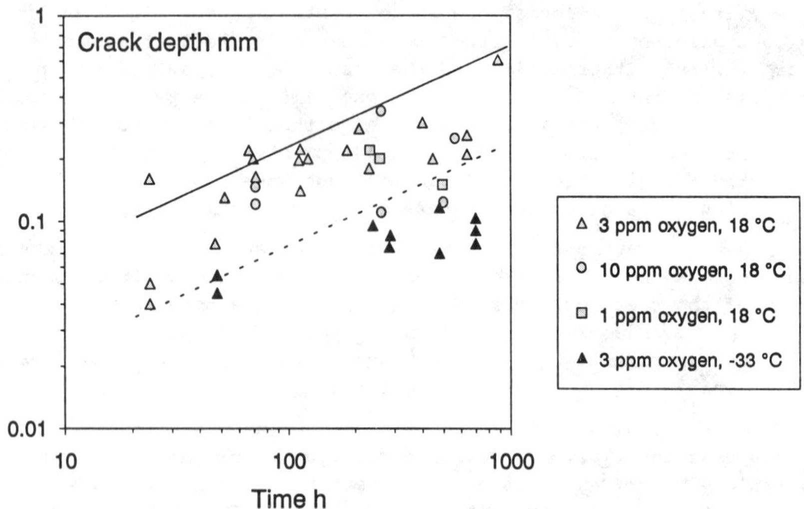

FIG. 1--Maximum stress corrosion crack depth for experiments with stress intensity factor 80 to 85 MPa $m^{1/2}$.

necessary to maintain a galvanic cell between an unfilmed surface at the crack tip and an oxygen filmed surface at the crack walls or sufficient migration of ionic species through the narrow crack to the surface.

Crack Growth At Low Temperature

Experiments with CT specimens at a low temperature of -33 °C showed that SCC initiation is very difficult at this temperature [9]. Several experiments with constant stress intensity factor in the range 80 to 120 MPa $m^{1/2}$ at -33 °C resulted in no stress corrosion cracks. Similar experiments at 18 °C resulted always in extensive SCC [9]. Experiments with slowly increasing stress intensity factor from 60 to 80 MPa $m^{1/2}$ resulted in very little SCC at -33 °C and extensive cracking at 18 °C. Previous low temperature experiments with thin-walled tubular specimens stressed by internal gas pressure showed that substantial stress corrosion cracking can be obtained in the laboratory at -33 °C when the plastic deformation and the strain rate is high enough [1]. Stress corrosion crack initiation is certainly possible at -33 °C, but it seems that the stress/strain and environmental conditions for SCC initiation are much more narrow at -33 °C than at ambient temperature. The plane strain situation in the CT specimens does not represent the worst case conditions for SCC initiation. The authors are currently running a research project focused on the conditions for SCC initiation in low temperature ammonia storage tanks.

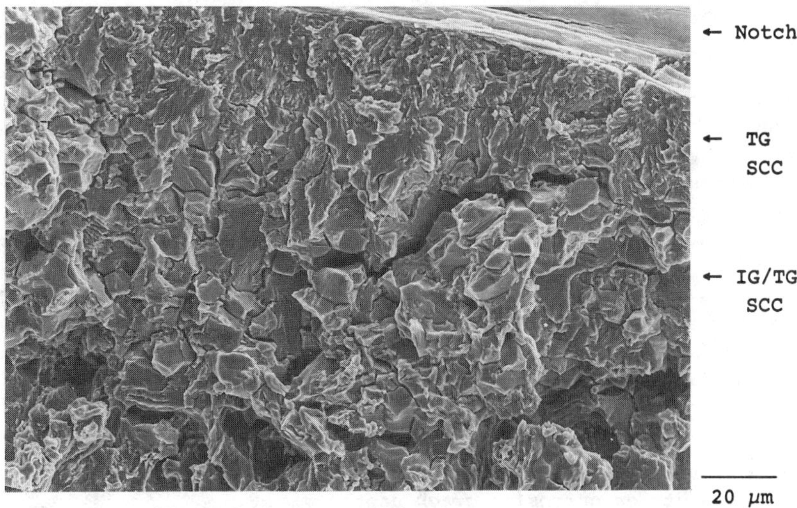

FIG. 2--Transgranular SCC at 18 °C (top) followed by intergranular SCC at -33 °C.

In order to be able to study crack growth at -33 °C it was necessary to initiate stress corrosion cracks at ambient temperature first. This was accomplished by a short exposure to ammonia at 18 °C with increasing stress intensity factor from 60 to 70 MPa $m^{1/2}$ during 3 hours. After this the specimen was exposed at -33 °C to a constant stress intensity factor of 80 MPa $m^{1/2}$ for a period varying from 48 to 700 hours. This procedure resulted in a purely transgranular crack with depth 20-40 μm resulting from the exposure at 18 °C, followed by a stress corrosion crack with a mixture of transgranular and intergranular cracking formed during the exposure at -33 °C, where the crack growth was much slower. The transition from one fracture mode to the other was easily recognizable, as shown in Fig. 2. Transgranular cracking is often associated with high crack growth rates for carbon steel in ammonia [3,10].

The results from the crack growth experiments at low temperature are included in Fig. 1, where the maximum depths of the stress corrosion cracks formed at -33 °C are indicated as filled points. The figure shows clearly that the stress corrosion crack growth rate is lower at -33 °C than at 18 °C. The deepest cracks at -33 °C were about one third of the largest crack depths at ambient temperature. The low temperature crack depths are too small and too few to determine the time correlation at -33 °C. However, if a similar time dependence is assumed at ambient and low temperature, the dashed line in Fig. 1 may be taken as an upper bound for the low temperature experiments. The dashed and solid lines in the figure differ by a factor 3.

DEVELOPMENT OF CRACK GROWTH MODEL

The results from crack growth experiments with stress intensity factors in the range 30 to 120 MPa $m^{1/2}$ are shown in Fig. 3. In this figure the crack depth divided by the square root of the exposure time is shown as function of stress intensity factor. According to Fig. 1 the maximum crack depth should be proportional to the square root of the exposure time for a given stress intensity factor. There is obviously a strong dependence of the stress intensity factor on the stress corrosion crack growth. The full-drawn line in the figure has a slope of 2, and gives an upper bound for stress corrosion cracking of this type of carbon steel in ammonia at ambient temperature. This line also represents a model for stress corrosion crack growth of carbon steel in ammonia:

$$a = 3x10^{-4}\ K^2 t^{1/2} \tag{1}$$

where

a = stress corrosion crack depth, mm

K = stress intensity factor, MPa $m^{1/2}$

t = time, years.

According to this model the maximum stress corrosion crack depth is proportional to the square of the stress intensity factor and the square root of the exposure time.

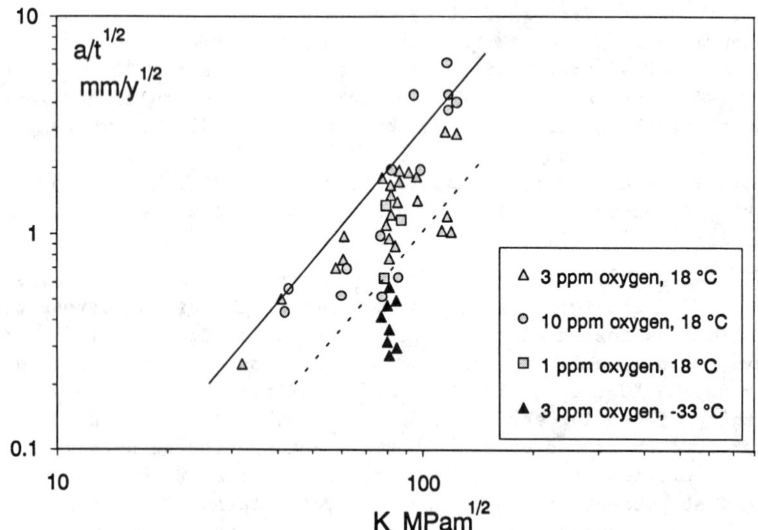

FIG. 3--Stress corrosion crack depth divided by square root of exposure time as function stress intensity factor.

The full-drawn line in Fig. 3 shows the model for crack growth at ambient temperature. Experiments at -33 °C have only been performed at a stress intensity factor of 80 MPa $m^{1/2}$, so there are not available data to make a similar model at low temperature. However, if a similar time and stress intensity factor dependence is assumed at ambient and low temperature, the dashed line in Fig. 3 can give an indication of crack growth in low temperature ammonia storage tanks. This line differs from the solid line by a factor 3.

The crack growth results can also be presented as in Fig. 4, where a/K^2 is shown as function of time. This figure shows the time dependence in the same way as Fig. 1, but Fig. 4 includes all experimental results with stress intensity factors ranging from 30 to 120 MPa $m^{1/2}$. The full-drawn line in the figure represents the model given in Equation 1. This line gives a reasonable upper limit for the data also when stress intensity values other than 80 MPa $m^{1/2}$ are included.

The line in Fig. 4 gives the crack depth after a certain time at a specific stress intensity factor. However, in order to make crack depth estimates, it is necessary to integrate Equation 1, taking into account the increase in stress intensity factor as the crack grows deeper. Typical stress intensity factors in an ammonia storage sphere can be in the range of 50 MPa $m^{1/2}$ with a 3 mm deep crack in an 11 mm thick wall [10]. With the present model, such a crack can grow further to 5 mm in four years and to 6 mm in 8 years. Inspection results show that most cracks are between 1 and 4 mm deep, even after several

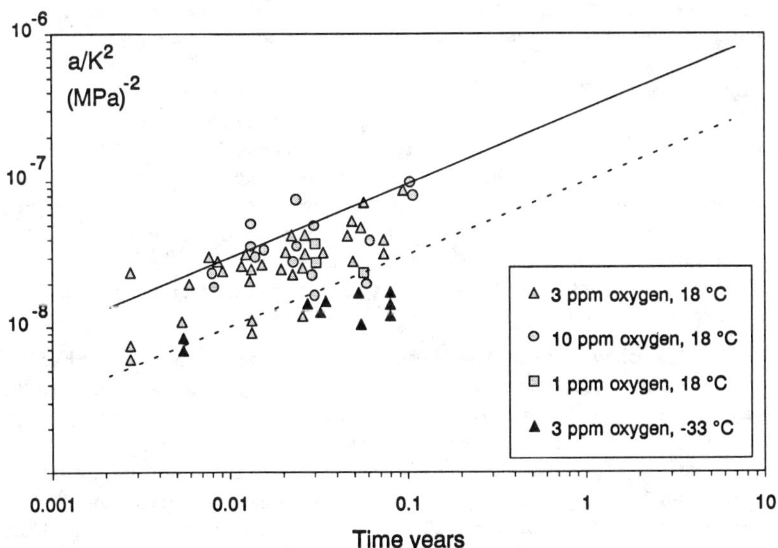

FIG. 4--Stress corrosion crack depth divided by square of stress intensity factor as function of time.

years' exposure, with a few instances of 7 to 10 mm deep cracks [10]. Comparison of inspection results with the model can be difficult because inspection results often report detected crack depths without any estimation of the actual stress intensity factor. Inspection results gathered by the group of ammonia producers sponsoring the present work indicate that the crack growth predicted by the model is in the same range as experienced by inspection of ammonia storage tanks. The model seems to give somewhat low values compared to a few very deep cracks found in storage tanks in the past. However, in the early days of finding SCC in ammonia spheres, little, if any, distinction was made between SCC and fabrication defects. Hence, cracks could be a combination of a fabrication defect with stress corrosion cracking propagating from it, the total depth being significantly greater than the stress corrosion cracking element.

Fig. 4 illustrates that model calculations for operating periods of several years obviously will involve considerable uncertainties, as the model is based on experimental data with duration up to 0.1 year and considerable spread in results. The data in Fig. 1 can be fitted to a line with a slope in the range from 1/3 to 1/2. The lines in the figure represent the steepest slope giving a reasonable correlation to the data. This should give conservative values for crack growth.

A square root of time dependence in the stress corrosion crack depth might suggest that a diffusion-controlled process is the rate-controlling factor in the stress corrosion crack growth process. The oxide layer observed on the crack surfaces might reduce crack growth by consuming the oxygen dissolved in the ammonia inside the crack. As discussed above, the presence of oxygen in the ammonia is a vital requirement for the propagation of stress corrosion cracks. The cracks are very narrow, and if the oxide layer on the crack surfaces fills most of the space in the crack, transport of dissolved oxygen or other species from the bulk liquid through the narrow, oxide-filled crack to the crack tip may be very slow. This may account for the reduction in crack growth rate as the crack grows deeper, and may eventually result in a cease in crack growth. A continued crack growth is then dependent on some mechanism for reactivation of the crack, for instance as a result of load variations during emptying and filling of the tank or temperature variations.

Fig. 4 includes also the dashed line a factor 3 lower than the full-drawn line, representing crack growth at low temperature in the same way as in Figs. 1 and 3. It must be emphasized that this only gives an indication of crack growth behaviour at low temperature. There are not enough experimental data to justify model calculations of crack growth in low temperature storage tanks. As discussed above, SCC initiation is much more difficult at low temperature, and other variables like oxygen content in the ammonia, residual welding stresses and shrinkage stresses during filling are probably important for SCC in low temperature ammonia storage tanks. Although stress corrosion cracking has been found in several low temperature storage tanks over the last few years, it should be noted that there are other low temperature storage tanks operating satisfactorily without any evidence of SCC, even after careful examination and NDT. Institutt for energiteknikk is currently conducting a new ammonia SCC research project aiming at determining the conditions for SCC initiation in welds in low temperature ammonia storage tanks.

The crack growth model described here was developed from experimental results from a carbon steel with yield point 380 MPa, which is quite representative for steels used in ammonia storage tanks. The model is not directly applicable for steels with different strength. It has been shown previously for carbon steel in ammonia that SCC susceptibility increases with the strength of the metal [3,10].

APPLICATION OF CRACK GROWTH MODEL

Practical use of the crack growth model is associated with considerable uncertainties. The most significant uncertainty in the present crack growth model is probably the uncertainty in the stress intensity dependence, as expressed in the exponent 2 in Equation 1 and the slope of the line in Fig. 3 [11]. Another uncertainty in the model is in the time dependence, as described above. Also important is the geometry of the CT specimens used in the experiments, where the absence of a fatigue precrack means that an initiation stage and early crack growth in the notch bottom must take place before stress intensity controlled crack growth takes over. This is probably a main reason for the spread in crack growth results in the experiments [12].

Other uncertainties arise when the crack growth model is applied for a specific storage tank. For old tanks all material properties may not be known, especially concerning the welds where the cracking occurs in practice. The stress distribution is usually quite complex, and involves also large residual welding stresses. Procedures for inspection, repair, recommissioning and operation of the tanks are important both for the size of cracks left undetected and the probability for initiating new cracks.

The observed reduction in crack growth has been explained by crevice corrosion for ammonia SCC, as discussed above, and by work hardening of the crack tip in carbonate-bicarbonate SCC [13]. Inspection results with shallow cracks even after several years' exposure indicate a reduction in crack growth with time as predicted by the present model. The residual welding stresses are usually tensile near the plate surface, but shift to compressive in the middle of the plate. Thus, the stress corrosion cracks grow into a region with low or compressive residual stresses, and this may also slow down crack growth.

Probabilistic Analysis

The uncertainties and complexities discussed above makes it difficult to obtain quantitative assessments of structural integrity of ammonia storage tanks. An evaluation of a crack of uncertain size growing at an uncertain rate towards some uncertain critical size would seem to give results with such uncertainty that they would be of little practical use. Applying conservative estimates all the way may then give a final result which is unrealistically conservative, with no indication of the probability of that result actually occurring. If one instead assigns a statistical distribution to each parameter that has some uncertainty incorporated, probabilistic techniques can supply a tool to treat all these uncertainties together in a more realistic manner. Probabilistics uses a statistical distribution in an equation instead of a fixed value, as used in a deterministic approach. Instead of calculating a single result based on fixed inputs, the probability of a result is calculated based on the statistical distribution of the input. The result will in this case be a probability of failure versus time. With this type of approach it is possible to determine how the different parameters affect the probability of failure. It is also possible to evaluate the probability of detection of a crack using different inspection techniques with different sensitivities, and how this affects the continued operation of the tank [11,14].

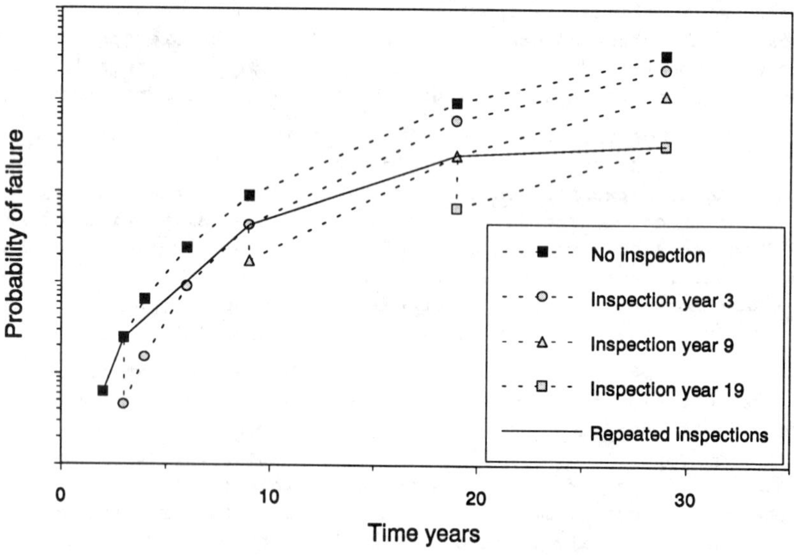

FIG. 5--Probability of failure for a new stress-relieved ammonia storage tank with inspection in years 3, 9 and 19 without detection of cracks.

A probabilistic analysis has been performed for some selected ammonia storage tanks [11]. This analysis uses the crack growth equation and fracture mechanics calculations with the tank geometry, wall thickness, probable size of undetected fabrication defects and residual and operating stress values for each tank. A statistical distribution has been assigned to each parameter with some uncertainty involved, and the probability of fracture of the tank up to any point in time has been calculated.

The result of a probabilistic analysis of a new ammonia storage tank is shown in Fig. 5 [11]. The probability of failure is shown on a logarithmic scale, but without actual values for the probability as requested by the tank owner. This tank was made from high toughness steel and was stress relieved, 100 % radiographed and acoustic emission tested for fabrication defects. SCC problems would not be expected in such a tank. The calculated probability of failure is indeed very low and remains low even without inspection during a 30 year period. Without periodical inspection the probability of failure increases over time as a result of the crack growth predicted by the crack growth model. The probability of failure will decrease every time the tank is inspected without detecting any cracks. Even with regular inspections the probability of failure will increase during the 30 year period, as indicated by the full-drawn line in the figure. This is because the predicted crack sizes are so small that they would probably not be detected even with a sensitive test technique.

The result of a probabilistic analysis of an old ammonia storage tank is shown in Fig. 6. This tank was not stress relieved, and was inspected after nearly 30 years of operation. At this inspection a

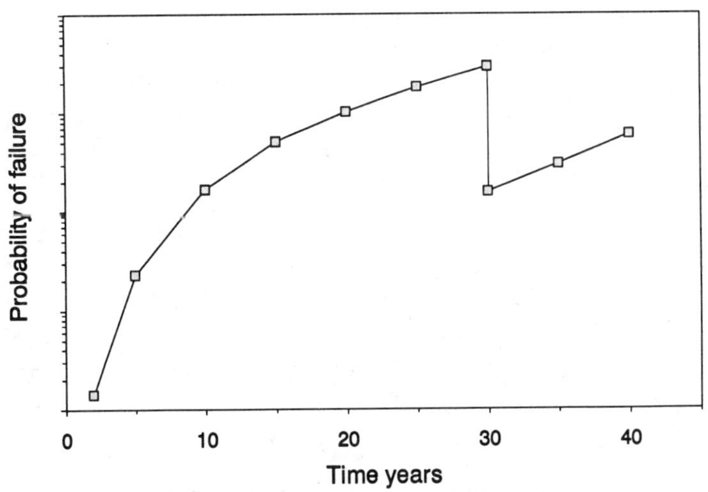

FIG. 6--Probability of failure for an old ammonia storage tank with a 4 mm deep crack detected at inspection in year 30.

4 mm deep crack was found, but the crack was smaller than predicted by the crack growth model. This means that the detection of a crack reduced the probability of failure, as shown in the figure. The model could be revised to reflect the lower crack growth actually observed, but a more conservative approach is to assume continued crack growth after inspection with the higher crack growth rate of the original model. If a crack was found to be larger than predicted by the model, the model could be revised to reflect the higher crack growth rate.

Fig. 7 shows the crack growth that would have to occur in the new storage tank represented in Fig. 5 for the tank to fail in ten years [11]. This is an event with very low probability, only 1×10^{-6}. The curve shows a reduction in crack growth rate during the first years, where the square root of time dependence is the controlling factor. Later, as the crack grows deeper, the square of the stress intensity factor controls. The crack grows exponentially near the end of the vessel life. This implies that the fracture toughness of the steel has very little effect on the lifetime of the tank. For instance, if the fracture toughness was improved and the critical crack depth was increased from 17 to 20 mm, the life of the tank would only be increased by a few months. On the other hand, the initial defect size has a very large influence on the probability of failure. Increasing the initial defect size shifts the entire curve in Fig. 7 to the left, and the crack growth starts at a higher rate, closer to the part of the curve where K^2 is the controlling factor. A 1 mm increase in initial defect size increases the probability of failure in year 10 by almost one order of magnitude [11]. This illustrates that inspection for fabrication defects is an important first step in the prevention of failures.

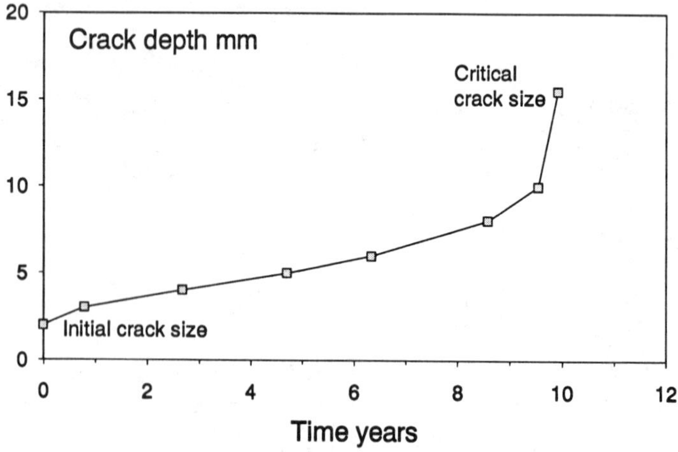

FIG. 7--Crack growth curve resulting in failure of the tank in Fig. 5 in ten years. Probability of occurrence 1×10^{-6}.

CONCLUSIONS

Stress corrosion crack growth studies of carbon steel in liquid ammonia showed that the crack growth rate increases markedly with the stress intensity factor and decreases with time. A model for stress corrosion crack growth of carbon steel in liquid ammonia was developed based on the laboratory results. According to this model, the maximum stress corrosion crack depth is proportional to the square of the stress intensity factor and the square root of the time. Crack growth rates predicted with this model are in good harmony with practical experience.

Initiation of SCC in carbon steel in ammonia is much more difficult at a low temperature of -33 °C than at ambient temperature. Stress corrosion crack growth is also slower at low temperature. The largest crack depths at -33 °C were about one third of the maximum crack depth in comparable specimens tested at ambient temperature. Both laboratory experiments and inspections of ammonia storage tanks have shown that SCC can also occur in low temperature storage tanks, but it is evident that the tendency to cracking is lower in low temperature storage tanks operating at -33 °C and atmospheric pressure than in ambient temperature storage spheres. Several ammonia producers have moved away from spheres to large low temperature ammonia storage tanks.

The model for stress corrosion crack growth in carbon steel in ammonia has been used together with field inspection results and probabilistic methods to obtain a quantitative measure for the probability of tank failure as function of time. Application of this approach for some ammonia storage tanks has produced results consistent with observed service cracks, and has demonstrated how such results can be used for evaluation of inspection results, inspection methods and frequency and continued safe operation of storage tanks. This approach can be used by the ammonia storage tank operators in order to optimize the inspection intervals and procedures while assuring the safety of the tank.

ACKNOWLEDGMENT

The crack growth experiments were carried out in two research projects sponsored by E.I. Du Pont de Nemours and Agricultural Minerals Corporation (USA), BASF AG (Federal Republic of Germany), DSM Research B.V. (The Netherlands), ICI plc and Health & Safety Executive (United Kingdom), Kemira OY (Finland), and Norsk Hydro a.s (Norway). The application of the crack growth model with probabilistic methods has been done by E.I. Du Pont de Nemours and Det Norske Veritas Industry Inc. (USA).

REFERENCES

[1] Lunde, L., Nyborg, R.: "Stress Corrosion Cracking of Different Steels in Liquid and Vaporous Ammonia," Corrosion, Vol. 43, No. 11, November 1987, pp 680-686.

[2] Lunde, L., Nyborg, R.: "Stress Corrosion Cracking of Carbon Steels in Ammonia," Materials Performance, Vol. 28, No. 12, December 1989,pp 29-32.

[3] Lunde, L. and Nyborg, R., "Stress Corrosion Cracking of Carbon Steel Storage Tanks for Anhydrous Ammonia", Proceedings no. 307, The Fertiliser Society, London, 1991.

[4] Byrne, J.R., Moir, F.E., Williams, R.D., "Stress Corrosion in a 12 ktonnes Fully Refrigerated Ammonia Storage Tank," Symposium Safety in Ammonia Plants and Related Facilities, Paper No. 47e, American Institute of Chemical Engineers, New York, NY, 1988.

[5] Appl, M., Fässler, K., Fromm, D., Gebhard, H., Portl, H., "New Cases of Stress Corrosion Cracking in Large Atmospheric Ammonia Storage Tanks", Symposium Safety in Ammonia Plants and Related Facilities, Paper No. 237d, American Institute of Chemical Engineers, New York, NY, 1989.

[6] Selva, R.A., Heuser, A.H., "Structural Integrity of a 12 000-Tonne Refrigerated Ammonia Storage Tank in the Presence of Stress Corrosion Cracks", Symposium Safety in Ammonia Plants and Related Facilities, Paper No. 238a, American Institute of Chemical Engineers, New York, NY, 1989.

[7] Ali, S.B., Smallwood, R.E., "Inspection of an Anhydrous Ammonia Atmospheric Pressure Storage Tank", Symposium Safety in Ammonia Plants and Related Facilities, Paper No. 98g, American Institute of Chemical Engineers, New York, NY, 1990.

[8] Jones, D.A., Kim, C.D., Wilde, B.E., "The Electrochemistry and Mechanism of Stress Corrosion Cracking of Constructional Steels in Liquid Ammonia", Corrosion, Vol. 33, No. 2, February 1977, pp 50-55.

[9] Nyborg, R., Lunde, L., "The Effect of Welding Electrode Composition and Storage Temperature on SCC of Carbon Steels in Liquid Ammonia," CORROSION/91, Paper No. 478, National Association of Corrosion Engineers, Houston, TX, 1991.

[10] Lunde, L., Nyborg, R., "SCC of Carbon Steels in Ammonia - Crack Growth Studies and Means to Prevent Cracking," CORROSION/89, Paper No. 98, National Association of Corrosion Engineers, Houston, TX, 1989.

[11] Conley, M.J., Angelsen, S., Williams, D., "A Modern Approach to Equipment Integrity Solutions," CORROSION/91, Paper No. 169, National Association of Corrosion Engineers, Houston, TX, 1991.

[12] Nyborg, R., Lunde, L., Conley, M.J., "Integrity of Ammonia Storage Vessels - Life Prediction Based on SCC Experience," Materials Performance, Vol. 30, No. 11, November 1991, pp 61-65.

[13] Parkins, R.N., "The Application of Stress Corrosion Crack Growth Kinetics to Predicting Lifetimes of Structures," Corrosion Science, Vol. 29, No. 8, 1989, pp 1019-1038.

[14] Saugerud, O.T., Angelsen, S.O., "Probabilistic Lifetime Analysis of Ammonia Pressure Vessels," Life Prediction of Corrodible Structures, Paper No. 67, National Association of Corrosion Engineers, Houston, TX, 1991.

David C. Silverman[1]

CORROSION PREDICTION FROM ACCELERATED TESTS IN THE CHEMICAL PROCESS INDUSTRIES

REFERENCE: Silverman, D. C., "Corrosion Prediction from Accelerated Tests in the Chemical Process Industries," Application of Accelerated Corrosion Tests to Service Life Prediction of Materials, ASTM STP 1194, Gustavo Cragnolino and Narasi Sridhar, Eds., American Society for Testing and Materials, Philadelphia, 1994.

ABSTRACT: Increasing pressures to move products and processes quickly out of the laboratory coupled with decreased resources for making materials evaluations have forced greater reliance on accelerated tests for making lifetime predictions. This paper discusses how combining both electrochemical and non-electrochemical tests can enable successful lifetime predictions to be made when testing must be limited. Examples are used to show that appropriate use of these tools when combined with experience which may be captured in an artificial neural network, enables the corrosion practitioner to overcome the dilemma of having to make important materials decisions from insufficient data.

KEYWORDS: corrosion, corrosion testing, electrochemical tests, polarization scans, electrochemical impedance spectroscopy, immersion tests, rotating cylinder electrode, artificial neural networks, chemical process industries.

Increasing pressures to move products and processes from concept to market and from the laboratory to the plant in as short a time as possible and with minimum expenditure have

---

[1]Fellow, Monsanto Company, 800 North Lindbergh Blvd., St. Louis, MO 63167

resulted in fewer time and dollar resources being available for materials compatibility studies. However, the goal, prediction of materials performance over the expected life of the product or process, has remained the same. The final materials selections often have to made using only short-term laboratory simulations. Pilot plants are often not constructed. When they are constructed, they are often no more than an enlarged vessel in which several steps are run sequentially. The corrosion tests must often be as simple as possible because of constraints on dollars, test equipment, and personnel. These constraints put much pressure on the corrosion practitioner. Making a lifetime prediction of the performance of materials usually requires the complementary use of experience and of more than one short-term laboratory technique because each provides an imperfect window through which the actual corrosion process is observed.

Two major classes of laboratory techniques, electrochemical and non-electrochemical, are available for predicting corrosion resistance of alloys. Among the electrochemical techniques that have been applied successfully to corrosion prediction in the chemical process industries are potentiodynamic polarization scans[1], polarization resistance methods [2], the rotating cylinder electrode [3], electrochemical impedance spectroscopy [4], corrosion current monitoring[1], and controlled potential tests for cathodic and anodic protection [5]. These techniques can be combined in a number of ways. For example, the rotating cylinder electrode and electrochemical impedance spectroscopy can be combined to provide information on corrosion rates and mechanisms under dynamic conditions [6]. Many electrochemical techniques can be used in the laboratory and pilot plant [2]. Monitoring potential or current fluctuations (electrochemical noise) has been used more for surveillance and monitoring than for prediction of future corrosion but that situation is changing [7,8].

The non-electrochemical technique that has been used most extensively in both the laboratory and the pilot plant in the chemical process industries is direct immersion of alloy samples (coupons) in the process fluids for a certain time period. Corrosion predictions are derived from the final appearance and the mass lost during the exposure. The coupons can be fitted with artificial crevice formers [9] to evaluate if the environment could cause localized corrosion. By including exposure at the vapor/liquid interface and in the vapor, information can often be obtained about the effects of repeated condensation of the liquid or the corrosivity of the vapor phase above the liquid. The shortcoming is that for coupons to provide valid corrosion rates a significant amount of exposure time or a significant number of batch cycles are often required [10]. An increasing need to predict the amount of metal contamination caused by the equipment can greatly increase the required test time. These time requirements might not be consistent with the manpower or time actually available.

Other non-electrochemical techniques are being used. Electrical resistance probes provide an alternative method of estimating general corrosion rates. Such probes require reasonably constant temperatures to provide reliable information. If temperatures change too quickly during the batch cycle, "instantaneous" monitoring might not be possible. Electrical resistance probes are about the only relatively inexpensive, laboratory compatible alternative to coupons when trying to estimate the corrosion in very low conductivity fluids or the vapor phase. Heat transfer is most often examined under static conditions by forcing heat through a coupon acting as a hot wall [11]. Newer devices which evaluate heat transfer under dynamic conditions [12] and which include provision for electrochemical evaluations [13,14] are beginning to be used on a more routine basis. Lastly, if the testing is being done in a pilot plant, the vessels and piping themselves can be inspected for corrosion.

The purpose of this paper is to provide an overview of how limited, "short-term" laboratory testing can be used to provide the information necessary to make "long-term" predictions. Through two examples, the paper examines how combining both electrochemical and non-electrochemical techniques with experience (sometimes as captured in an artificial neural network), the corrosion practitioner can make successful lifetime predictions using the incomplete experimental information provided by limited duration testing.

PREDICTING LIFETIME PERFORMANCE IN A WASTE STREAM WITH MINIMAL TESTING

The combined circumstances of (1) equipment failure causing a process shutdown, (2) required immediate replacement by a similar piece of equipment built of an alternative material of construction, (3) little time to conduct tests before the equipment is placed in service, and (4) a meager budget to determine the best mode of operation are one common combination of constraints within which a lifetime prediction has to be made. Experience becomes a major contributor both to any initial prediction and to the appropriate combination of corrosion tests to generate the data to make a prediction.

Failure of a heat exchanger in a strongly acidic waste stream soon after it was installed provides an example. The exchanger had to be replaced immediately because of its critical relationship to the process. A spare HASTELLOY®B-2 (Haynes International, Inc., Kokomo, Indiana) heat exchanger was located and installed. The alloy is comprised primarily of nominally 69 wt% nickel and 28 wt% molybdenum. The choice was based mainly on the waste stream containing 3 to 10 wt% hydrochloric acid and the extra exchanger existed. However, the stream also would contain 0 to 2 wt% of phosphorous and acetic acids which would vary with plant operation.

Time was critical and funding was extremely limited. The goal of the experimental program was to provide a rapid estimate of expected life with a minimum number of laboratory tests. The approach taken was to use electrochemical impedance spectroscopy to screen the representative conditions rapidly to make sure that imminent failure would not occur and to provide preliminary information as to which condition(s) might cause excessive corrosion. Electrochemical impedance spectroscopy has been shown to provide very rapid estimates of corrosion rates from $2.5x10^{-4}$ to 250 mm/y in relatively short times, about 24 to 48 hours [4,15]. Modeling in terms of one of four simple circuit analogues has been shown to be able to classify the corrosion so that predictions can be made even in the absence of supporting mechanistic corroboration [6]. These tests were followed by several longer term laboratory immersion tests that included examination of corrosion at the vapor/liquid interface and in the vapor while adding further corroboration to the electrochemical results. Finally, coupons were installed in the process to provide some insight into how the exchanger might be performing in actual operation. But, these latter tests were undertaken after the heat exchanger was installed. More sophisticated on-line surveillance was not considered because of a lack of funding to purchase the equipment and lack of personnel to oversee its operation.

Experimental

The solutions tested are shown in Table I. Actual plant waste samples were spiked to the levels shown with reagent grade products. The solutions tested were believed to simulate three classes of environments to which the heat exchanger would be exposed. The temperature was maintained at 55°C which represented the upper temperature limit expected. A nitrogen blanket was used since the waste stream in the plant is maintained under a nitrogen blanket.

TABLE I

Corrosion of HASTELLOY B2 in Waste Water at 55°C

Environment (Spiked with:)	Exposure Time (h)	Corrosion* Rate (mm/y)
HCl at 10 wt%. Phosphite at 1 wt%.	0.5 3-5 22-24	0.014 0.024 0.024
HCl at 10 wt%. Phosphite at 1 wt%. Acetic acid at 1 wt%.	0.5 3-5 22-24	0.069 0.054 0.073
HCl at 1.7 wt%. Phosphite at 0.1 wt%. Acetic acid at 2.2 wt%.	0.5 3.5 22-24	0.040 0.024 0.030

* The electrodes themselves showed a mass loss that ranged between 0.04 mm/y in solution 1 to 0.08 mm/y in solutions 2 and 3. The corrosion rates derived from the mass losses are likely within the expected experimental error because of the short exposure time [10].

The electrochemical impedance equipment and algorithm for generating the spectra have been described in detail previously [4,6]. The tests were run under static conditions in a three electrode cell [4]. The procedure was to immerse the HASTELLOY B-2 electrode (surface sanded with 600 grit silicon carbide paper) and then generate several impedance spectra over 24 hours. The impedance spectra were generated after 30 minutes, 3-5 hours, and 24 hours using a 5 mV amplitude. The first spectrum was generated to 0.01 Hz. The latter spectra were generated to 0.001 Hz. Comparing the spectra as a function of time would give insight into whether the steady state corrosion rates might be higher or lower than those estimated by these "short-term" tests. As discussed in a previous example [6], since the corrosion potential was slowly changing over the first several hours, care had to be exercised to use only those portions of the first two spectra which represented the same corrosion phenomena. Small amplitude d-c polarization curves were generated after the second and third impedance spectra at 3 hours and 24 hours to help to corroborate the impedance results. These curves were generated by the method outlined previously [3] at 0.1 mV/s to -20 mV and +20 mV, each from the corrosion potential. These latter tests were used to provide support for the analysis of the impedance spectra.

A longer term coupon immersion test in which HASTELLOY B-2 was exposed to the liquid phase, vapor phase, and at the vapor/liquid interface was conducted to confirm the impedance results. Finally, coupons were exposed in the plant after the exchanger was installed.

Results and Discussion

Figures 1-3 show the impedance spectra generated after 24 hours of exposure to the three waste stream simulations. Only the Bode format is presented. All of the spectra showed pseudo-inductive behavior. The experimenter must become satisfied that such pseudo-inductance is not caused by experimental artifacts. For example, in this case, this behavior is not believed to be related to non-linearity between voltage input and current output because the amplitude of the excitation was kept small, 5 mV. In a previous study, such a small amplitude was shown to circumvent such non-linearities [16]. The fact that the corrosion potential was constant after several hours strongly suggests that the spectra shown were generated under steady state conditions. The fact that the corrosion potential was the same before and after completion of the second and third spectra is a necessary but not sufficient condition for the system to be classified as stationary. Assuming that the pseudo-inductance is caused by the actual corrosion chemistry, the question is how best to classify these spectra.

The model used for the regression analysis was that originally outlined by Epelboin and Keddam [17] as a model for pseudo-inductive behavior of iron at low pH. It was previously used to successfully model impedance spectra generated for steel in a waste stream showing pseudo-inductance [16].

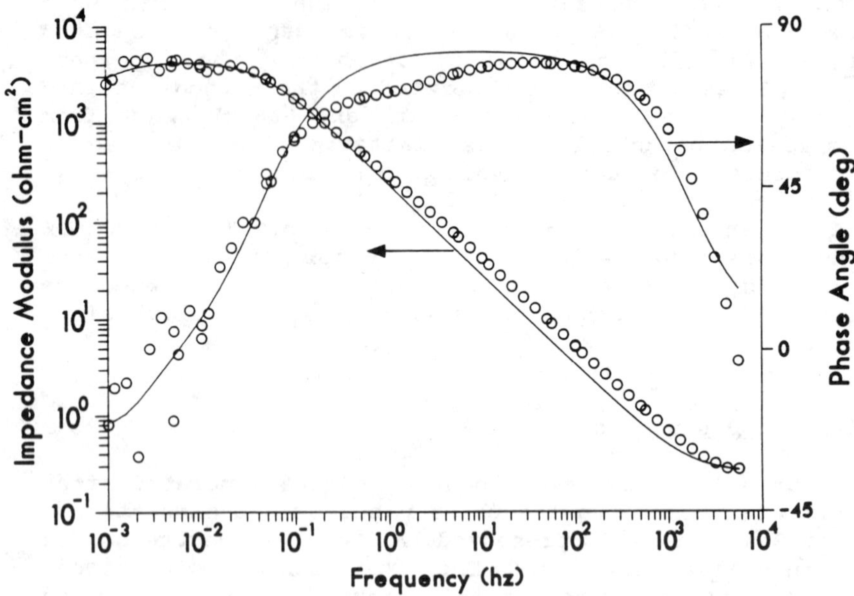

Fig. 1: Impedance spectrum of HASTELLOY B-2 generated after 24 hours of exposure to waste solution spiked to 10 wt% HCl and 1 wt% phosphite at 55°C. (Bode Format) o = Measured, ___ = Calculated.

At that time, the pseudo-inductance was attributed to an electroactive adsorbed intermediate, iminodiacetic acid, present in the waste stream. The equation used is

$$Z = R_s + \left[\frac{1}{R_T} \left[1 + (j\omega\tau_{HF})^{\beta}\right] + \frac{1}{\rho\,(1 + j\omega\tau_{LF})} \right]^{-1} \tag{1}$$

where Z is the impedance (ohm-cm^2), R_s is the solution resistance (ohm-cm^2), R_t is the charge transfer resistance (ohm-cm^2), ω is frequency (rad-s^{-1}), ρ is a resistance (ohm-cm^2), τ_{HF} is a high frequency time constant (s), τ_{LF} is a low frequency time constant (s), β is the constant phase element exponent, and $j = (-1)^{1/2}$.

The hypothesis is that a similar classification of the spectra by using equation (1) is possible in this strongly acidic solution. The parameters obtained from the regression were used to recalculate the spectra shown as the solid line in Figures 1 - 3. Though the empirical model for pseudo-inductance is not expected to be perfect (as seen by the deviation in Figure 1) and may not model the entire electrochemical process, the hope was that it would provide a

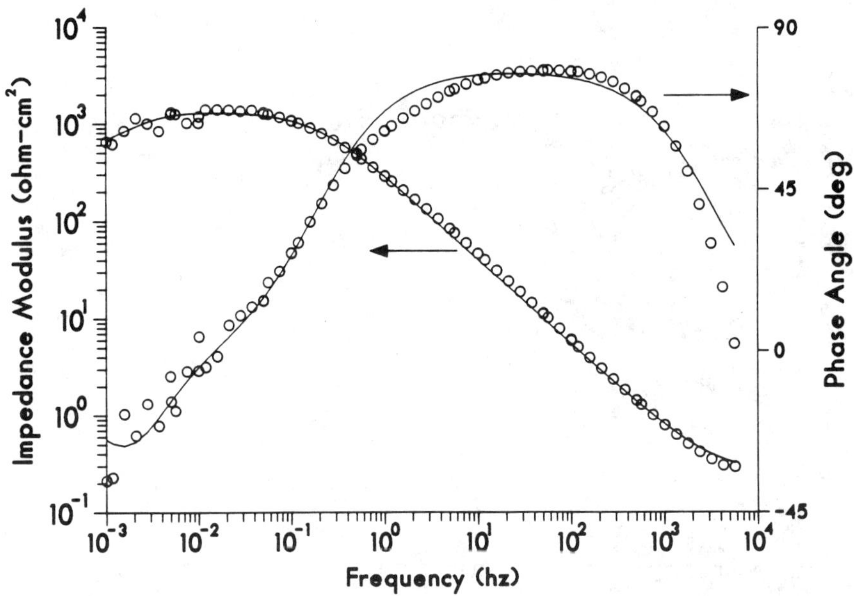

Fig. 2: Impedance spectrum of HASTELLOY B-2 generated after 24 hours of exposure to waste solution spiked to 10 wt% HCl, 1 wt% phosphite, and 1 wt% acetic acid at 55°C. (Bode Format) o = Measured, ___ = Calculated.

framework for obtaining quick estimates of corrosion rates and possibly some insight into the corrosion mechanism which was the goal of the experiments.

Note that the impedance at the lowest applied frequency was not at the d-c limit. There was a non-zero phase angle at 0.001 Hz. This result prevented an assessment of validity from being made by using the Kramers-Kronig relationships by methods as outlined by, for example, Macdonald [18].

Table I shows the corrosion rates for exposure of HASTELLOY B-2 to the three waste stream compositions. The calculation is discussed later. In all cases, steady state was reached within about 3 hours of exposure. The corrosion rate estimated from the impedance spectra were unchanged between 3 and 24 hours. In addition, in all three cases, the corrosion potential changed by no more than 5 to 10 mV

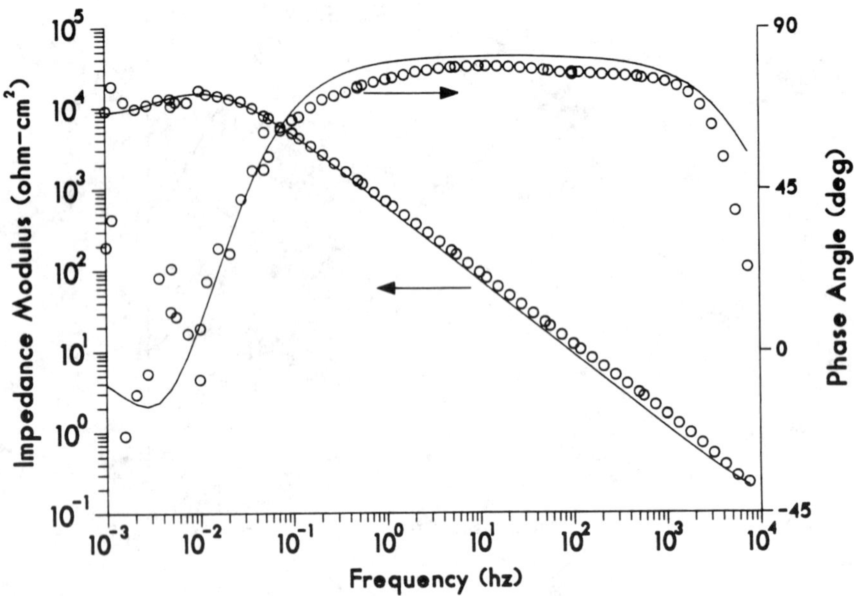

Fig. 3: Impedance spectrum of HASTELLOY B-2 generated after 24 hours of exposure to waste solution spiked to 1.7 wt% HCl, 0.1 wt% phosphite, and 2.2 wt% acetic acid at 55°C. (Bode Format)
o = Measured, ___ = Calculated.

between the times that the second and third spectra were generated. This observation suggests that the corrosion rates estimated after the 24 hours of exposure should provide a reasonable estimate of the long-term corrosion rate.

The resistance used for estimating the corrosion rate (assumed to be the charge transfer resistance as calculated from equation (1)) was obtained by curve-fitting equation (1) to the impedance spectra. Table II compares these resistances to those obtained from using the d-c polarization technique. Figure 4 shows a typical d-c polarization curve generated from the stable corrosion potential in the cathodic and then the anodic direction.

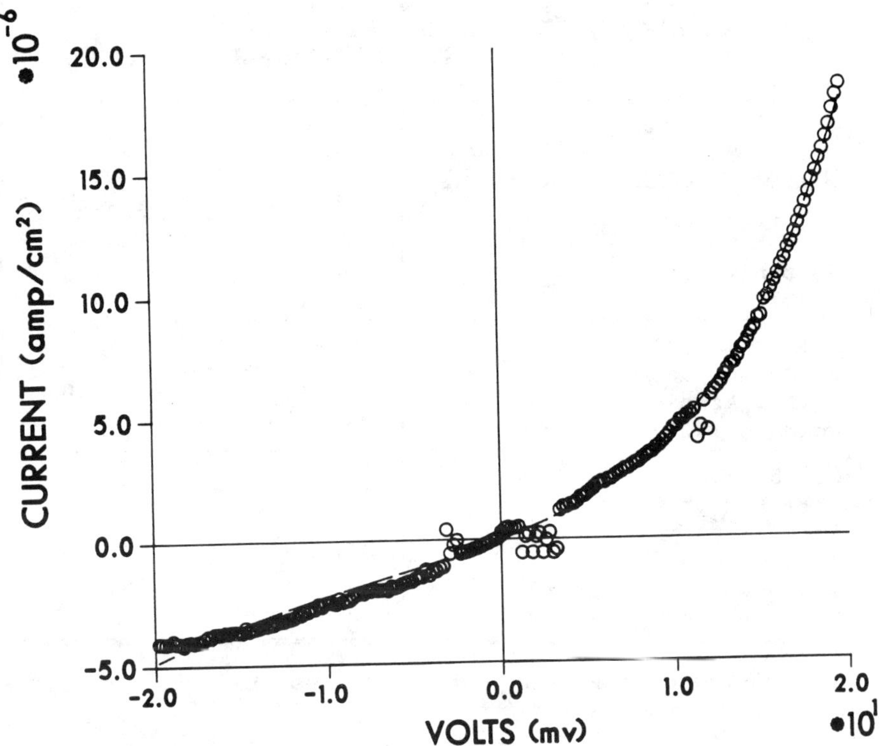

Fig. 4: Typical d-c polarization curve generated near the corrosion potential. Corresponds to conditions in Fig. 1. o = Measured, ___ = Calculated.

The calculated curve in Figure 4 was obtained by assuming that one anodic and one cathodic reaction were controlling the corrosion rate at the sweep rate of 0.1 mV/s. A Butler-Volmer type of equation was fit to the data. The anodic and cathodic Tafel slopes averaged 0.015V and 0.030V respectively for the first two cases in Table II. The Tafel slopes were significantly larger in the third case. The physical significance of the differences between these two Tafel slopes was not explored. The steady state corrosion potentials were about -0.08 to -0.09V (SCE) in all three experiments.

TABLE II

Resistance from EIS And DC Polarization Corrosion of HASTELLOY B2 in Waste Water at 55°C

Environment (Spiked with:)	Exposure Time (h)	Resistance ($ohm-cm^2$) Impedance	DC Polarization
HCl at 10 wt%. Phosphite at 1 wt%.	3-5 22-24	$4.4x10^3$ $4.3x10^3$	$4.1x10^3$ $3.8x10^3$
HCl at 10 wt%. Phosphite at 1 wt%. Acetic acid at 1 wt%.	3-5 22-24	$1.6x10^3$ $1.3x10^3$	$1.6x10^3$ $1.5x10^3$
HCl at 1.7 wt%. Phosphite at 0.1 wt%. Acetic acid at 2.2 wt%.	3-5* 22-24	$1.1x10^4$ $1.6x10^4$	$1.5x10^4$ $1.1x10^4$

* Pseudo-inductance was not fully established for this exposure at this time.

The Tafel slopes were used in conjunction with the charge transfer resistances obtained by the impedance technique to estimate the corrosion rates shown in Table I using

$$I_{corr}=\left(\frac{1}{(2.303)}\right)\left(\frac{1}{R}\right)\left(\frac{b_a b_c}{(b_a+b_c)}\right) \qquad (2)$$

In equation (2) I_{corr} is the corrosion current (Amp/cm^2), R is the charge transfer resistance (ohm-cm^2), b_a is the anodic Tafel slope (V), and b_c is the cathodic Tafel slope (V).

The agreement in Table II between the resistance calculated from the d-c polarization results and the resistance estimated from analysis of the impedance spectra which has been assumed to be a charge transfer resistance serves as a good internal check. However, in this case, the agreement suggests that in some instances, the resistance calculated from the d-c technique may not be the resistance in the limit of zero frequency, the polarization resistance. This point was discussed in detail previously [16]. One reason would be that a surface relaxation process which might cause the pseudo-inductance is occurring at such a slow rate that it cannot respond completely at the frequency corresponding to a scan rate of 0.1 mV/s [19,20], the slowest scan rate available on the PAR 173/276 used here. In this circumstance, the polarization resistance determined experimentally would not be the resistance determined in the limit of zero frequency. Perhaps, the pseudo-inductance is related to an interaction among hydrogen ion, hydrogen, and the alloy surface. However, neither time nor budget allowed for a more thorough examination.

The agreement between measured and calculated impedance spectra strongly suggests that the procedure for classifying the impedance spectra using equation (1) is reasonable. As pointed out elsewhere, this agreement is a necessary but not a sufficient condition for the model used to be reasonable [21]. The agreement in resistances between those obtained from the impedance spectra and d-c polarization results suggests that the electrochemical results are measuring corrosion phenomena. The agreement between the corrosion rates calculated from the electrochemical results and from mass loss of the electrode shown in Table I support the validity of considering the resistance in Table II as a charge transfer resistance and using it in equation (2).

In view of the rather complex spectra and analysis, and the constraints of a limited number of experiments and limited time, agreement with an alternative technique is desirable. The internal consistency shown above suggests that such additional results could be obtained after the initial prediction is made. HASTELLOY B-2 corrosion coupons were immersed in the most corrosive of the solutions (second and third) shown in Table II for about 1500 hours to overcome the errors inherent in short-term mass loss tests [10].

Exposure was to the liquid, vapor, and the vapor/liquid interface. The results again suggested that the general corrosion rate is low upon exposure to the liquid phase in agreement with the impedance results. Some localized corrosion was noted at the vapor-liquid interface.

The hypothesis was made that higher concentrations of acetic acid in the presence of hydrochloric acid might be one cause of localized initiation sites. The presence of acetic acid does seem to increase corrosion as seen in Table I. This hypothesis implies that if exposure to the higher levels of acetic acid could be avoided, the heat exchanger might last longer in service.

Finally, HASTELLOY B-2 corrosion coupons were exposed to the actual waste stream for about 180 days. The process was operated to minimize exposure to the higher levels of acetic acid. The surfaces showed only a slight etch with a slight amount of localized attack under the artificial crevice formers [9] placed on the surface. The heat exchanger was still in service a year later.

CORROSION PREDICTION USING MINIMAL TESTING AND AN ARTIFICIAL NEURAL NETWORK

Information on product performance either in the equipment used for further processing by a customer or in the final application can be included in product bulletins. Such information may be required before any experience has been gained on actual performance. Often, the only way that such information can be obtained is by laboratory simulations in the expected application. Laboratory tests must somehow be meshed with experience to make the prediction because laboratory tests cannot simulate all conditions.

This type of prediction had to made for the short-term performance of 316ss in 50wt% DEQUEST® 2000 (Registered Trademark of Monsanto Company). The product is a 50 wt% solution of aminotrimethylene phosphonic acid in water. A small amount of chloride ion (1 wt%) could be present in this acidic product. This composition is that in which this product is normally supplied to customers [22]. After steel, alloy 316ss is one of the most common materials of construction found in the process industries. The original recommendation was that this product should not be handled in 316ss because of the risk of localized corrosion.

However, localized corrosion in the form of pitting and crevice corrosion often develops over time. Therefore, though long term storage in 316ss is to be avoided, the question arose if short term exposures of the order of several days would be possible. This type of exposure would occur in transfer lines and short term (e. g. 1 day) storage. If 316ss could be used for such service, the amount and availability of equipment suitable for handling this product would be increased. Testing had to be limited to fit within the allowed budget.

The question was addressed experimentally by using E vs. log(i) polarization scans to obtain the needed short-term information. Though longer-term immersion tests were to follow, they really would only provide imperfect corroboration because they would show longer term (2 to 4 week) behavior, not short term (1 to 4 day) behavior. The desire was to use polarization scans generated after short (1 day) and longer (4 day) exposures to forecast differences in behavior. One of the problems with using polarization scans is that interpretation of the results can be difficult. Important subtleties might be missed. Therefore, certain parameters extracted from the polarization scans were passed through a specially constructed artificial neural network [23] to aid in the prediction. This approach would allow the interpretation to take advantage of the cumulative experiences of interpreting a number of polarization scans.

Experimental

E vs log(i) polarization scans were generated after one and four days of exposure to this product under ambient conditions (assumed to be 49°C). The scans were generated using a Princeton Applied Research 273 potentiostat that was controlled by a Hewlett-Packard 9816S microcomputer with in-house developed software. The scan rate was 0.5 mV/s which should have provided reasonable estimates of the repassivation potential and pitting potentials [24] used for the analysis. The scan direction was reversed at 100 $\mu A/cm^2$.

Coupon immersion tests were run in this product for 840 hours. The 316ss specimens were exposed to the liquid, vapor/liquid interface, and the vapor. The reason for the three exposures is that in most storage and transport, the containment vessel is exposed to an interface and a vapor phase at least part of the time. Corrosion in these regions can be very different from liquid exposures. The specimens were fitted with artificial crevice formers [9]. Some of the coupons were removed at the midpoint of the exposure period.

Artificial Neural Network

The construction and training of the artificial neural network and how the numerical results are used by an Expert System built around the network are explained in detail elsewhere [23] but are summarized here. Artificial neural networks are computer simulations which can learn and classify the patterns between a group of input and output observations in the absence of a specific model. They can use that learning to predict the appropriate outputs for input data not used in the training [25,26]. The inputs and outputs used for this application are shown in Table III. The input variables are certain features extracted from the polarization scans. These features are a "repassivation potential" (E_{prot}) (also referred to as a "protection potential"), "pitting potential" (E_{pit}), "current density estimated at the corrosion potential", presence or absence of an "anodic nose", type of "hysteresis" between the forward and reverse portion of the polarization scan, and the potential at which the current on the reverse scan changes from "anodic" or "oxidizing" to "cathodic" or "reducing" ($E_{a \rightarrow c}$). All potentials are referenced to the corrosion potential (E_{corr}). The output variables are the prediction of crevice corrosion, prediction of pitting, and if general corrosion would be measurable or large. See reference 23 for details on these variables.

The artificial neural network was trained using 87 polarization scans. Whenever possible, the actual coupon exposure results or actual plant experience was used as the outputs. In this way, the artificial neural network would be set up to function like the corrosion practitioner, to classify parameters measured in the laboratory with expectations in the field. The algorithm used for the supervised learning was based on the back-propagation method [26,27]. A commercial, PC-based software package NeuralWare Professional IIPlus™ (NeuralWare, Inc., Pittsburgh, Pennsylvania) was used for the training on an IBM Personal System/2 Model 70 386 computer [28]. The network was checked by using 17 examples not used in the original training.

TABLE III

Data Inputs for Neural Network

Input Parameter	Value of Feature
Repassivation Potential	$E_{prot} - E_{corr}$
Pitting Potential	$E_{pit} - E_{corr}$
Hysteresis	+1 = Positive 0 = None -1 = Negative
Anodic "Nose"	+1 = Yes 0 = No
Passive Current Density	$\mu amp/cm^2$
Potential at Anodic-to-Cathodic Transition	$E_{a \to c} - E_{corr}$

Data Outputs for Neural Network*

Output Parameter	Value of Feature
Crevice Corrosion Predicted	+1 = Yes 0 = No
Pitting Predicted	+1 = Yes 0 = No
Should General Corrosion Be Considered?	+1 = Yes 0 = No

* For the purposes of interpretation, an actual output value of less than 0.2 is assumed to be the same as 0 (No) and an actual output value of greater than 0.8 is assumed to be 1 (Yes). Numerical values falling in the region between 0.2 and 0.8 is assumed to lie in the borderline region.

The network was placed within an expert system to provide an easy way to input the data, to provide simple consistency checks between electrode and scan appearances, and to interpret the numerical output of the neural network to make the final prediction[23]. This expert system used rules to translate the numerical output from the neural network into a description of the type of attack that might be observed in the field. Such description was not derived from the training. It was developed to capture the expertise used in interpreting what the output from the artificial neural network means in a practical sense in field applications.

Results and Discussion

Figure 5 shows the polarization scan generated after one day and Figure 6 shows the polarization scan generated after 4 days of exposure. Though at first glance the polarization scans seem to be similar, there are differences to which the artificial neural network was sensitive to make somewhat different predictions. The scans have slight differences in the hysteresis between the forward and reverse portions of the scan. The point at which the reverse scan changes sign (anodic to cathodic current) moves closer to the corrosion potential the longer the exposure. These differences were reflected in different values used for the "repassivation potential", the potential at the "anodic-to-cathodic transition" in the reverse portion of the scan, and the estimated current at the corrosion potential.

The expert system which interpreted the numerical results from the artificial neural network predicted that after one day, the risk of localized corrosion in the form of crevice corrosion only is likely. The output from the expert system also mentioned the possibility of longer term pitting especially in splash zones, at vapor/liquid interfaces, and under deposits. The polarization scan generated after four days was predicted to indicate more severe localized corrosion. The risk of localized corrosion in the form of both crevice corrosion and pitting were predicted. This change in prediction was derived from the fact that the output for pitting from the neural network changed from the defined numerical border-line region (0.2 to 0.8) for the first scan to the defined numerical high region (>0.8) for the second. The risk of localized corrosion increases with

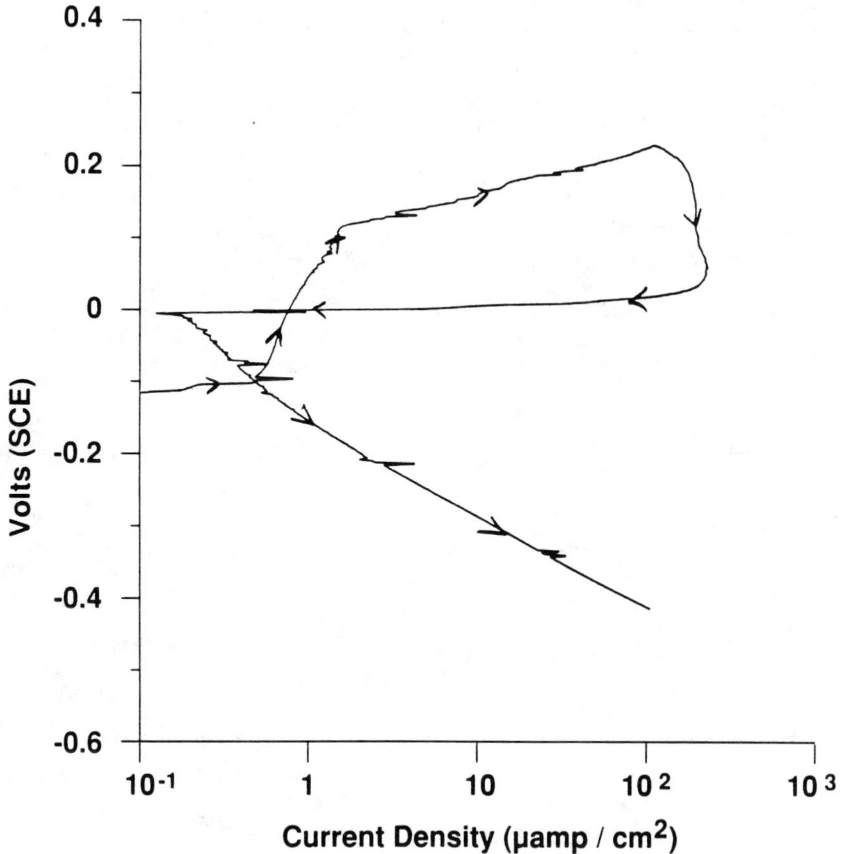

Fig. 5: Polarization scan for 316ss generated after 1 day of exposure to DEQUEST 2000 product at 49°C.

time of exposure of 316ss to this environment. Since such localized corrosion often takes time to develop, the practical prediction is that the risk of localized corrosion of 316ss to this product increases with time but short, several day exposures could be acceptable.

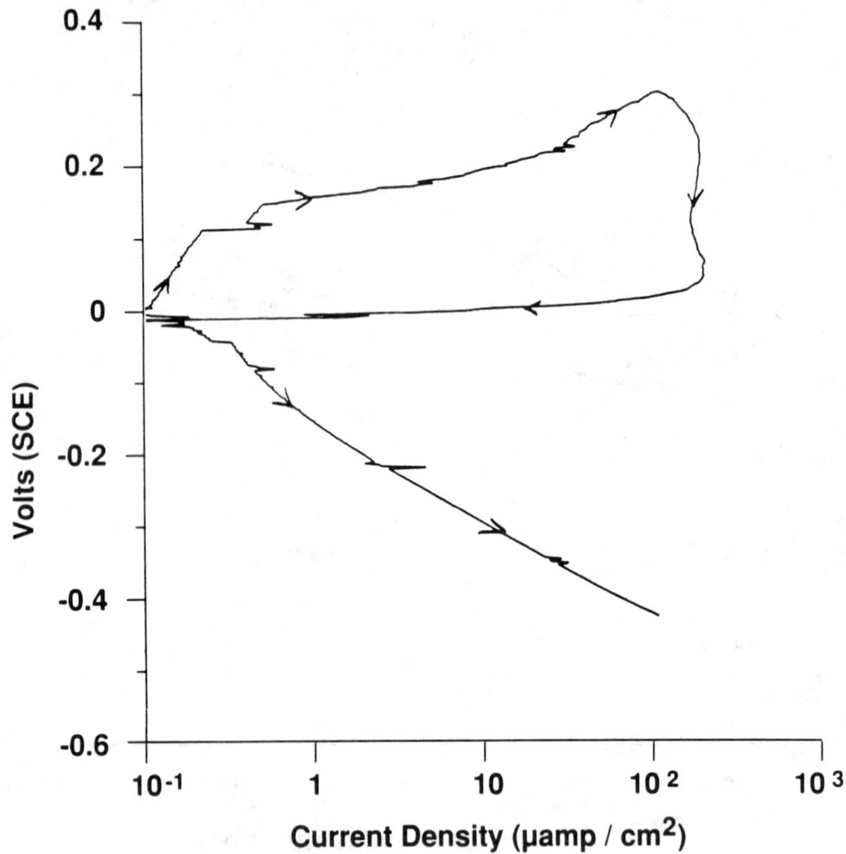

Fig. 6: Polarization scan for 316ss generated after 4 days of exposure to DEQUEST 2000 product at 49°C.

Coupon immersion tests confirmed the longer term predictions. Slight attack was found under the special artificial crevice formers in the complete liquid exposure. However, pits were found on the coupons mounted at the vapor/liquid interface especially under deposits. Some pits were found on the coupon mounted in the vapor region. Thus, localized corrosion is expected after long term exposure to this environment as predicted by the artificial neural network.

The goal of this study was the prediction of exposure after short times of 1 to 4 days. The similarity in polarization scans would have made them somewhat difficult to use as stand-alone results to make this prediction. The expertise captured in the artificial neural network was able to differentiate between them. The artificial neural network when coupled with the rule based expert system was able to differentiate between the scans and was able to make the proper prediction like the expert. This result strongly suggests that artificial neural networks can provide a practical way to recognize patterns that relate experimental results to field observations in the absence of a specific model. They thus provide an additional link between the laboratory measurement and the lifetime prediction.

For this example, the prediction would be that crevice corrosion might initiate after short exposure times but should not be severe. However, long term exposure should be avoided because both pitting and crevice corrosion would be expected. At present, experience is that 316ss can be used for short (less than 3 to 4 day) exposures to this product as long as the equipment is cleaned after usage so as not to leave residual material behind. Long term exposure of 316ss to this product is still to be avoided.

CONCLUSIONS

1. Time and budget constraints often dictate the amount of testing that can be done to make lifetime corrosion predictions. Usually these constraints result in a very limited number of tests being run before the predictions are made.

2. Lifetime predictions in the chemical process industries can be made using laboratory techniques as long as the limitations of those techniques are fully understood. Electrochemical techniques can play an important role. However, two independent measures of corrosion should be made even if in the same experimental apparatus.

3. Experience becomes an important asset when making life-time predictions from laboratory measurements. Artificial neural networks which can classify patterns between laboratory data and experience should enable more confident predictions to be made from laboratory data.

ACKNOWLEDGMENT

I want to thank John E. Carrico who ran the laboratory tests.

REFERENCES

[1] Liening, E. L., "Trouble-Shooting Industrial Corrosion Problems with Electrochemical Testing Techniques", in Corrosion Monitoring in Industrial Plants Using Nondestructive Testing and Electrochemical Methods (G. C. Moran and P. Labine, ed.), ASTM STP 908, American Society for Testing and Materials, Philadelphia, PA, 1986, p. 289.

[2] Liening, E. L., "Field Applications of Electrochemistry", in Process Industries Corrosion-The Theory and Practice (B. J. Moniz and W. I. Pollock, ed.), p. 85, National Association of Corrosion Engineers, Houston, TX, 1986.

[3] Silverman, D. C., "Rotating Cylinder Electrode - An Approach for Predicting Velocity Sensitive Corrosion", in Flow-Induced Corrosion: Fundamental Studies and Industry Experience (K. J. Kennelley, R. H. Hausler, and D. C. Silverman, ed.), p. 20-1, National Association of Corrosion Engineers, Houston, TX, 1991.

[4] Silverman, D. C. and Carrico, J. E., Corrosion, Vol. 44, No. 5, 1980, p. 280.

[5] Riggs, O. L. and Locke, C. E., Anodic Protection - Theory and Practice in the Prevention of Corrosion, Plenum Press, NY, 1981.

[6] Silverman, D. C., Corrosion, Vol. 46, No. 7, 1990, p. 589.

[7] Smith, S. and Francis, R. British Corrosion Journal, Vol. 25, No. 4, 1990, p. 285.

[8] Williams, D. E., Fleischmann, M., Stewart, J., and Brooks, T., "Some Characteristics of the Initiation Phase of Pitting Corrosion of Stainless Steels", in Electrochemical Methods in Corrosion Research (M. Duprat, ed.),p. 151, Materials Science Forum, Vol. 8, 1986.

[9] Streicher, M. A., Materials Performance, Vol. 22, No. 5, 1983, p. 37.

[10] Freeman, R. A. and Silverman, D. C., Corrosion, Vol. 48, No. 6, 1992, p. 463.

[11] Fisher, A. O., Corrosion, Vol. 17, No. 5, 1961, p. 93.

[12] Ross, T. K., British Corrosion Journal, Vol. 2, 1967, p. 131.

[13] Alwash, S. H., Ashworth, V., Shirkhanzadeh, M., and Thompson, G. E., Corrosion Science, Vol. 27, No. 4, 1987, p. 383.

[14] Francis, P. E. and Mercer, A. D., "Corrosion of a Mild Steel in Distilled Water and Choride Solutions: Development of a Test Method", in Laboratory Corrosion Tests and Standards (G. S. Haynes and R. Baboian, ed.), ASTM STP 866, American Society for Testing and Materials, Philadelphia, PA, 1985, p. 184.

[15] Silverman, D. C., "Simple Models/Practical Answers Using the Electrochemical Impedance Technique", in Corrosion Testing and Evaluation: Silver Anniversary Volume (R. Baboian and S. W. Dean, ed.), ASTM STP 1000, American Society for Testing and Materials, Philadelphia, PA, 1990, p. 379.

[16] Silverman, D. C., Corrosion, Vol. 45, No. 10, 1989, p. 824.

[17] Epelboin, I. and Keddam, M., Journal of the Electrochemical Society, Vol. 117, 1970, p. 1052.

[18] Macdonald, D. D. and Macdonald, M., Journal of the Electrochemical Society, Vol. 132, Vol. 10, 1985, p. 2316.

[19] Gabrielli, C., Keddam, M., and Takenouti, H., "The Use of AC Techniques in the Study of Corrosion and Passivity", in Treatise of Materials Science and Technology (J. C. Scully, ed.), Vol. 23, San Diego, CA, Academic Press, 1983, p. 395.

[20] Mansfeld, F. and Kendig, M., Corrosion, Vol. 37, No. 9, 1981, p. 545.

[21] Silverman, D. C., Corrosion,Vol. 47, No. 2, 1991, p. 87.

[22] "DEQUEST® 2000 and 2006 Phosphonates For Scale and Corrosion Control, Chelation, Dispersion", Monsanto Company Technical Bulletin No. 9023.

[23] Rosen, E. M. and Silverman, D. C., Corrosion, Vol. 48, No. 9, 1992, p. 734.

[24] Atrens, A., "Application of Electrochemical Pitting Measurements to Pitting and Crevice Corrosion of X20CrMoV121 in Deaerated Chloride Solutions", Proceedings of the 8th International Congress on Metallic Corrosion", p. 206, DECHEMA, Frankfurt, 1981.

[25] Lippmann, R. P., "An Introduction to Computing with Neural Nets", IEEE ASSP Magazine, 1987, p. 4.

[26] Dayhoff, J. E., Neural Network Architectures An Introduction, Van Nostrand Reinhold, New York, 1990.

[27] Rumelhart, D. E. Hinton, G. E., and Williams, R. J., Nature, Vol. 323, 1986, p. 523.

[28] NeuralWorks Professional IIPlus™, NeuralWare, Inc., Pittsburgh, PA, Release 4.05, 1990-1991.

Richard H. McCuen[1] and Pedro Albrecht[1]

Composite Modeling of Atmospheric Corrosion Penetration Data

REFERENCE: McCuen, R. H. and Albrecht, P., "**Composite Modeling of Atmospheric Corrosion Penetration Data,**" Application of Accelerated Corrosion Tests to Service Life Prediction of Materials, ASTM STP 1194, Gustavo Cragnolino and Narasi Sridhar, Eds., American Society for Testing and Materials, Philadelphia, 1994.

ABSTRACT: Corrosion penetration of steel depends on the chemical composition of the material, the environment at the exposure site, and the extent to which debris accumulates on the surface of the specimen. Corrosion penetration data have traditionally been modeled with a power function over the full length of exposure time. The power model lacks the flexibility to describe a variety of behaviors and often underpredicts corrosion penetration at the end of the service life of a structure. Composite models are shown herein to better represent corrosion penetration data. Thirty-eight sets were fitted with power and composite models. In all cases, the composite model fitted the data at least as well as the power model, and in most cases better judging by goodness-of-fit statistics and errors between the predicted and measured penetrations. Errors in penetration predicted with the power model often show large local biases, suggesting the model structure is incorrect. Local biases also serve as a warning that projections of penetration to the end of the service life may be highly inaccurate. Conversely, composite models have unbiased coefficients and better represent the factors that govern corrosion penetration, thus giving greater confidence in service life projections.

KEY WORDS: atmospheric, corrosion, penetration, composite modeling, power model, linear model, weathering steel.

INTRODUCTION

Any model for fitting corrosion penetration data should capture the actual behavior of the process. As an example, Figs. 1 through 3 show 38 sets of data for

---

[1]Professor, Department of Civil Engineering, University of Maryland, College Park, Maryland 20742-3021

weathering steel supplied to *ASTM Standard Specification for High-Strength Low-Alloy Structural Steel* (A242). Six sets are for exposures in rural (Fig. 1), 20 in industrial (Fig. 2), and 12 in marine environments (Fig. 3). Table 1 lists the identification number, the maximum exposure time, the exposure site, and the source from which the data were taken for each curve. The data are summarized in detail in Ref. 1, and the original sources are listed in the Appendix. The curves in each set exhibit the following three different behaviors:

- Negative curvature during the full range of exposure time, with a steadily increasing radius of curvature. A negative curvature means that the center of the circle that best fits the curve at a point lies below the curve. An increasing radius of curvature means that the slope of the curve decreases. The growing rust layer increasingly protects the underlying steel base and continuously slows the corrosion process albeit at a declining rate. Examples of this behavior are curves No. 4 for rural South Bend, Pennsylvania; No. 1 for industrial Bayonne, New Jersey; and No. 7 for moderate marine Kure Beach, North Carolina.

- Negative curvature for short exposure times of, say, up to 3 to 5 years, followed by a quasi-straight line for longer exposure times. A straight line has an infinite radius of curvature. This behavior reflects a change in the corrosion mechanism from film building to steady-state kinetics within the duration of the test. Natural processes such as wind and rain erode the rust layer on the surface at the same rate as new rust forms at the interface with the base metal. Examples are curves No. 2 for rural Potter County, Pennsylvania; No. 17 for industrial Kearny, New Jersey; and No. 4 for marine Block Island, Rhode Island.

- Negative curvature for short exposure times of, say, up to 3 to 5 years, followed by positive curvature for longer exposure times. A positive curvature means that the center of the circle that best fits the curve at a point lies above the curve. Reasons for an increase in corrosion rate are changes in micro-environment, accumulation of debris on the surface of the specimen, and contamination with airborne salt or industrial pollutants. Examples are curves No. 1 for rural South Bend, Pennsylvania; No. 20 for industrial Kearny, New Jersey; and No. 11 for moderate marine Kure Beach, North Carolina.

Clearly, a single model cannot describe best all three behaviors. Each behavior calls for a different model.

Corrosion penetration data that exhibit the three behaviors described above have traditionally been fitted with a power model using the logarithmic transformation of both exposure time and corrosion penetration. This approach, hereafter called the *logarithmic power model*, has the advantages that it linearizes the least-squares fit, yields closed-form equations for the regression coefficients, and simplifies the

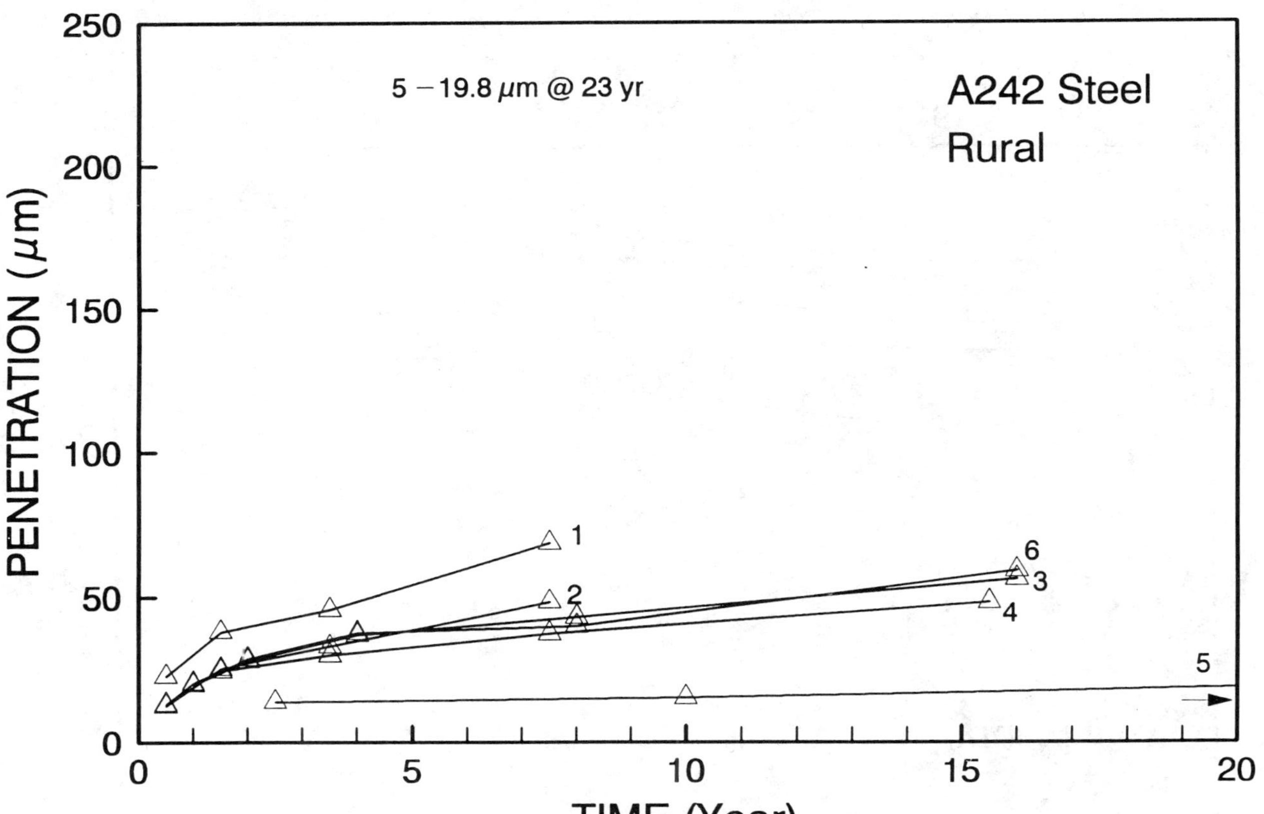

FIG. 1 -- *Corrosion penetration data for A242 steel exposed in rural environments (see Table 1 for identification of curves).*

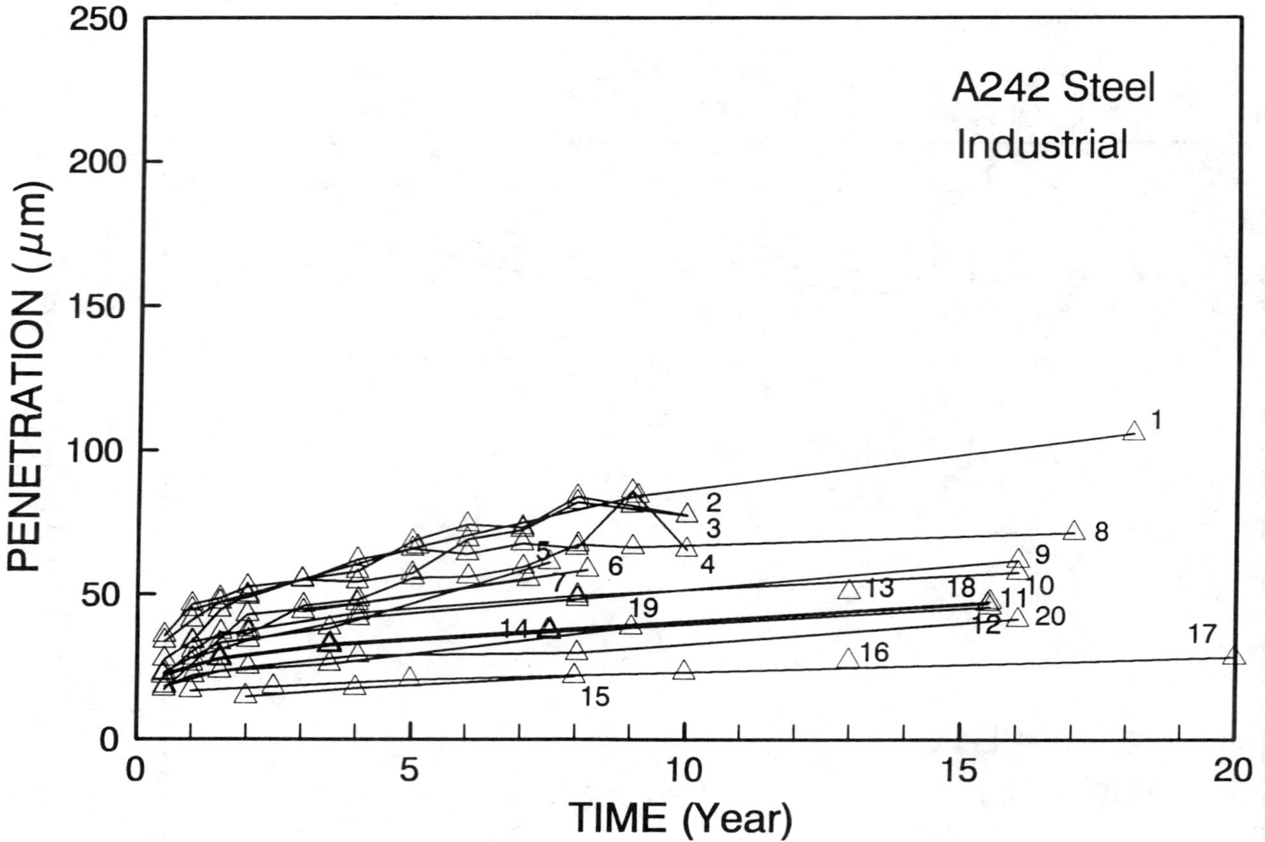

FIG. 2 -- *Corrosion penetration data for A242 steel exposed in industrial environments (see Table 1 for identification of curves).*

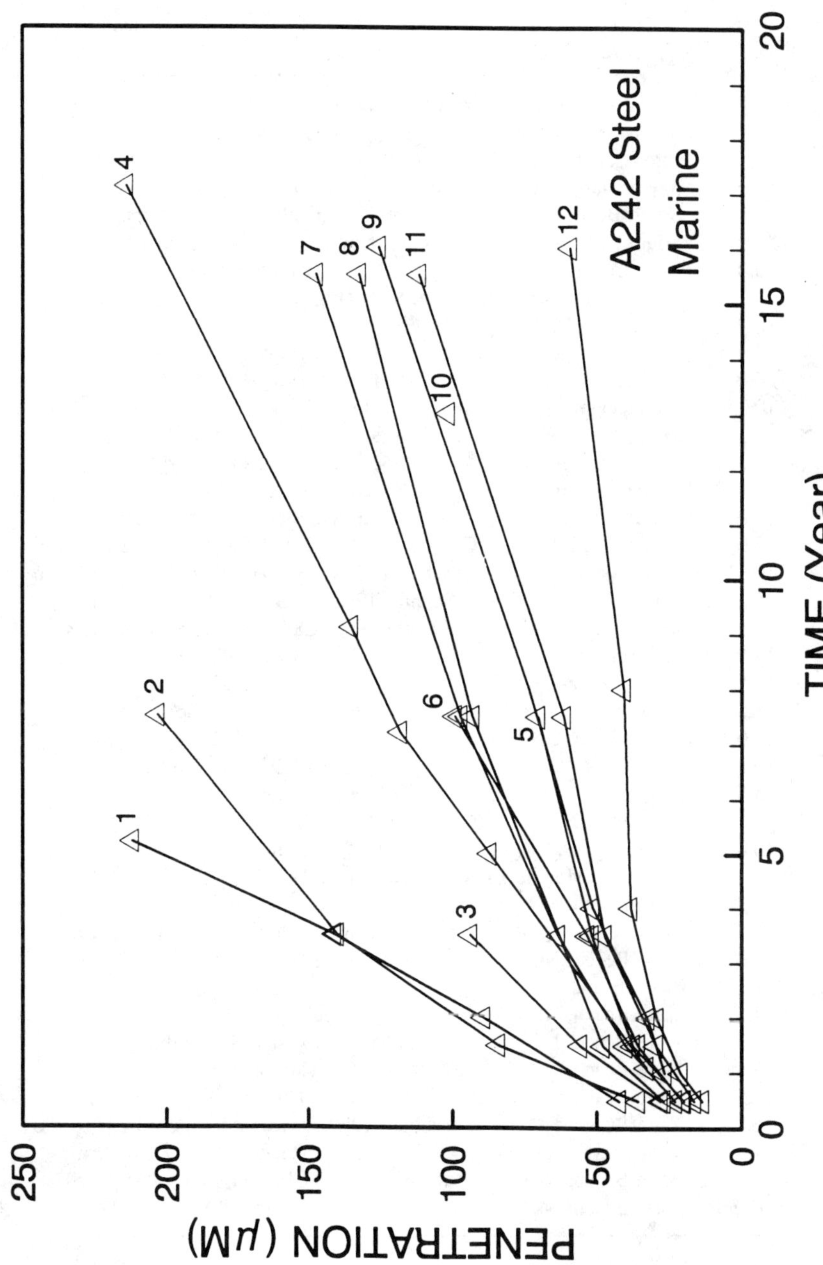

FIG. 3 -- *Corrosion penetration data for A242 steel exposed in marine environments (see Table 1 for identification of curves).*

TABLE 1 -- *Description of corrosion penetration curves for A242 steel.*

Curve No.	Exposure Time (yr)	Exposure Site	Source of Data (see Appendix)
		Rural Environments (Fig. 1)	
1	7.5	South Bend, PA	Gallagher 1978, 1982; Vrable 1985
2	7.5	Potter County, PA	Gallagher 1978, 1982; Vrable 1985
3	16.0	Saylorsburg, PA	Cosaboom 1979, Townsend 1982 , Shastry 1986
4	15.5	South Bend, PA	Schmitt 1965, Gallagher 1976, Komp 1987
5	23.0	South Bend, PA	Larrabee 1959, 1962; Cruz 1962
6	16.0	Potter County, PA	Komp 1988
		Industrial Environments (Fig. 2)	
1	18.1	Bayonne, NJ	Copson 1960
2	10.0	Rankin, PA	Horton 1964, Horton 1965
3	10.0	Columbus, PA	Horton 1964, Horton 1965
4	10.0	Bethlehem, PA	Horton1 964, Horton 1965
5	7.5	Monroeville, PA	Gallagher 1978, 1982; Vrable 1985
6	8.2	Detroit, MI	Zoccola 1976, Tinklenberg 1986
7	7.1	Detroit, MI	Zoccola 1976, Tinklenberg 1986
8	17.0	Rankin, PA	Horton 1964, Horton 1965
9	16.0	Newark, NJ	Cosaboom 1979, Townsend 1982, Shastry 1986
10	16.0	Bethlehem, PA	Cosaboom 1979, Townsend 1982, Shastry 1986
11	15.5	Newark, NJ	Schmitt 1965, Gallagher 1976, Komp 1987
12	15.5	Monroeville, PA	Schmitt 1965, Gallagher 1976, Komp 1987
13	13.0	Los Angeles, CA	Reed 1982
14	7.5	Kearny, NJ	Larrabee 1953
15	8.0	Detroit, MI	Zoccola 1976, Tinklenberg 1986
16	13.0	Sacramento, CA	Reed 1982
17	20.0	Kearny , NJ	Larrabee 1959, 1962;Cruz 1962
18	15.5	Kearny, NJ	Gallagher 1976
19	9.0	Kearny, NJ	Vrable 1985, Komp 1991
20	16.0	Kearny, NJ	Komp 1988
		Marine Environments (Fig. 3)	
1	5.2[a]	Kure Beach, NC	Cruz 1962
2	7.5[a]	KureBeach, NC	Schmitt 1965, Gallagher 1976, Komp 1987
3	3.5[a]	Kure Beach, NC	Gallagher 1978, 1982; Vrable 1985
4	17.1	Block Island, R.I.	Copson 1960
5	7.5[b]	Kure Beach, NC	Larrabee 1953
6	7.5[b]	Kure Beach, NC	Gallagher 1978, 1982; Vrable 1985
7	15.5[b]	Kure Beach, NC	Copson 1960
8	15.5[b]	Kure Beach, NC	Cruz 1962
9	16.0[b]	Kure Beach, NC	Cosaboom 1979, Townsend 1982, Shastry 1986
10	13.0	Point Reyes, CA	Reed 1982
11	15.5[b]	Kure Beach, NC	Schmitt 1965, Gallagher 1976, Komp 1987
12	16.0[b]	Kure Beach, NC	Komp 1988

[a]Specimen exposed at 25-m lot.
[b]Specimen exposed at 250-m lot.

calculations. McCuen et al. [2] showed that logarithmic transformation has several disadvantages: the sum of the squares of the errors in penetration is not minimized even though the sum of the squares of the logarithmic errors is minimized, the resulting model gives biased estimates of penetration, and the logarithmic transformation gives too much weight to the penetration data for shorter exposure times.

Instead of fitting the coefficients of the power model logarithmically, McCuen et al. [2] recommended fitting them numerically to the actual values of exposure time and corrosion penetration. Although this approach, hereafter called the *numerical power model*, requires nonlinear least-squares fitting and numerical optimization of the regression coefficients, it reduces the standard error of estimate, eliminates the overall bias, and more accurately predicts penetration for longer exposure times.

Yet, whether fitted logarithmically or numerically, the power model still exhibits local biases. It often underpredicts penetration at short and long exposure times and overpredicts penetration at intermediate exposure times. This was apparent from an analysis of 32 sets of data [2].

Models are commonly used to described the effect of the physical processes of corrosion on penetration and to extrapolate penetration to the end of service life. Hence, the presence of local biases suggests that the power model may lead to erroneous conclusions and predictions. A more rational and flexible model structure is, therefore, needed.

Building on the previous work, the authors demonstrate in the present study how fitting composite models to corrosion penetration data further improves accuracy.

CORROSION MODELS

Logarithmic Power Model

The power model has the form

$$\hat{p} = b_o t^{b_1} \tag{1}$$

where

$\hat{p}$ = predicted value of corrosion penetration p,
t = exposure time, and
b_o, b_1 = fitting coefficients.

The coefficients of Eq 1 are calculated by taking the logarithms of the two variables p and t, fitting a straight line to the transformed data by the linear least-squares method, and transforming the coefficients from the log-log space back to the linear space. Eq 1 is called hereafter the *logarithmic power model*.

Numerical Power Model

The numerical power model has the same functional form as the logarithmic power model of Eq 1. In this case, however, the coefficients are fitted numerically with the nonlinear least-squares method directly to the actual values of the variables p and t, not to the logarithms of the variables. The solution is optimized until the overall bias becomes zero. The procedure, described in Ref. 3 and in the appendix of Ref. 2, is called the *numerical power model.*

Constant-Intercept Power-Linear Model

A composite model is derived below for the second behavior, that is, a curve with negative curvature for short exposure times followed by a straight line for longer exposure times. It is given by the following two functions shown in Fig. 4:

$$\hat{p}_1 = b_o t^{b_1} \qquad for\ t \le t_c \tag{2a}$$

$$\hat{p}_2 = b_2 + b_3 t \qquad for\ t > t_c \tag{2b}$$

where

b_i = fitting coefficients (i = 0,1,2 and 3), and
t_c = exposure time at transition from power function to linear function; referred to as *intersection time.*

Equations 2a and 2b are called the *constant-intercept (C.I.) power-linear model* because the intercept coefficient b_0 of the power function is constant. They are spliced at $t = t_c$ so that both give equal corrosion penetrations (ordinate):

$$\hat{p}_1 = \hat{p}_2 \quad at \quad t = t_c \tag{3a}$$

and corrosion rates (slope):

$$\frac{d\hat{p}_1}{dt} = \frac{d\hat{p}_2}{dt} \quad at \quad t = t_c \tag{3b}$$

The constraints of Eqs 3a and 3b provide continuity of corrosion penetration and rate, respectively. Substituting Eqs 2a and 2b for the corrosion penetrations in Eq 3a, and the derivatives of Eqs 2a and 2b for the corrosion rates in Eq 3b, gives

$$b_o t_c^{b_1} = b_2 + b_3 t_c \tag{4}$$

and

$$b_3 = b_o b_1 t_c^{b_1 - 1} \tag{5}$$

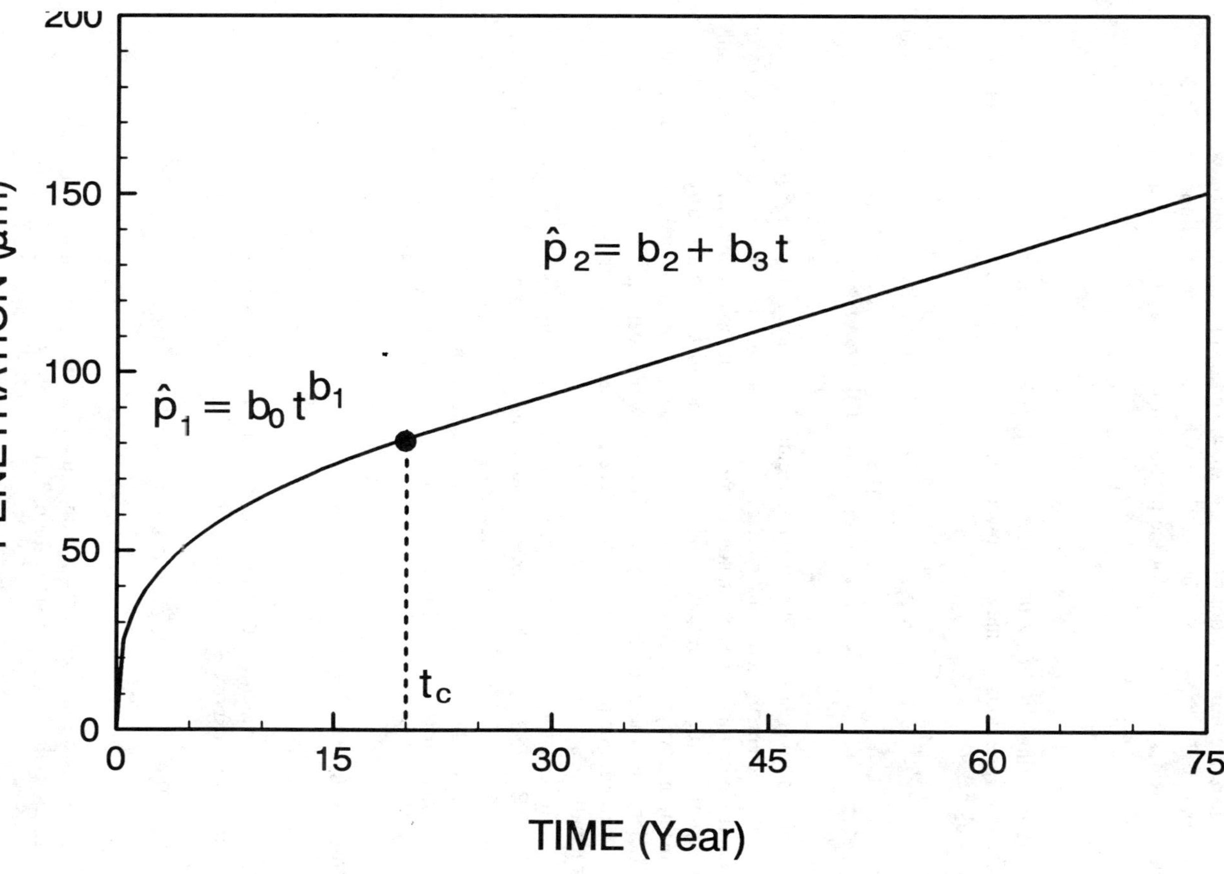

FIG. 4 -- *Composite power-linear model with constant intercept.*

There are four equations (Eqs 2a, 2b, 4, and 5) in the four unknown coefficients b_0, b_1, b_2, and b_3. The coefficients b_2 and b_3 are now eliminated. Eq 5 already provides an expression for the coefficient b_3, so substituting Eq 5 for b_3 in Eq 4 and solving for b_2 gives:

$$b_2 = b_0 t_c^{b_1}(1 - b_1) \tag{6}$$

Finally, inserting into Eqs 2 the values of b_2 from Eq 6 and b_3 from Eq 5 gives the following expressions for the C.I. power-linear model in terms of the coefficients b_0 and b_1 alone:

$$\hat{p}_1 = b_o t^{b_1} \qquad for\ t \leq t_c \tag{7a}$$

$$\hat{p}_2 = b_o t_c^{b_1}\left[1 + \frac{1}{t_c} b_1(t - t_c)\right] \qquad for\ t > t_c \tag{7b}$$

For a given a set of corrosion penetration data, the nonlinear Eqs 7 are solved for the coefficients b_0 and b_1 by numerical optimization. The intent is to find values of the coefficients for which the standard error of estimate is a minimum and the overall bias is zero. Typically, these two conditions are not met simultaneously. Numerical optimization always searches for the minimum standard error. Once this is found, the coefficients are modified to achieve a zero bias while keeping the standard error near the minimum. For the corrosion data analyzed in the present study, the standard error was usually near minimum for the zero-bias solution.

The constraints imposed by Eqs 3 limit the flexibility of the C.I. power-linear model of Eqs 7. Because the intercept coefficient b_o is constant, the model sometimes does not describe accurately the behavior of the data at short exposure times of, say, up to 5 years--where the slope of the curve changes rapidly. Thus, the C.I. power-linear model can result in local biases at short exposure times.

Variable-Intercept Power-Linear Model

Local bias can be reduced with a so-called *variable-intercept (V.I.) power-linear model* in which the coefficient b_0 is multiplied by an exponential-decay term that varies with exposure time. One possible V.I. power-linear model consists of:

$$\hat{p}_1 = b_o e^{-b_2 t} t^{b_1} \qquad for\ t \leq t_c \tag{8a}$$

$$\hat{p}_2 = b_3 + b_4 t \qquad for\ t > t_c \tag{8b}$$

Splicing the two functions at $t = t_c$ with equal corrosion penetrations and rates (Eqs 3) leads to the following expressions for b_3 and b_4:

$$b_3 = b_o e^{-b_2 t_c} t_c^{b_1}(1 - b_1 + b_2 t_c) \tag{9}$$

$$b_4 = b_o e^{-b_2 t_c} t_c^{b_1 - 1}(b_1 - b_2 t_c) \tag{10}$$

Equations 9 and 10 are substituted for b_3 and b_4 in Eqs 8, yielding the V.I. power-linear model

$$\hat{p}_1 = b_o e^{-b_2 t} t^{b_1} \qquad for\ t \le t_c \qquad (11a)$$

$$\hat{p}_2 = b_o e^{-b_2 t_c} t_c^{b_1} \left[1 + \frac{1}{t_c}(b_1 - b_2 t_c)(t - t_c)\right] \qquad for\ t > t_c \qquad (11b)$$

The intersection time t_c and the coefficients b_o, b_1, and b_2 are obtained by numerically fitting Eqs 11 to the corrosion penetration data so that the standard error of estimate is near minimum for the zero-bias solution.

INDICES FOR ASSESSING ACCURACY

The accuracy of fitting a model is assessed with the standard error of estimate, the standard error ratio, and the bias. These measures of accuracy are calculated for the four models described above.

Standard Error of Estimate

The standard error of estimate is computed from the sum of the squares of the errors, i.e., the differences between the predicted and measured penetrations, $\hat{p}_i$ and p_i:

$$S_e = \left(\frac{1}{n - q} \sum_{i=1}^{n} (\hat{p}_i - p_i)^2\right)^{1/2} \qquad (12)$$

where

n = sample size, i.e., number of penetration measurements, and
q = number of unknowns fitted with least squares method.

If the predicted values $\hat{p}_i$ are equal to the measured values p_i, then $S_e = 0$ and the fit is perfect.

Standard Error Ratio

The relative accuracy can be assessed with the ratio of the standard error of estimate, S_e, to the standard deviation, S_y, of the measured values of penetration. The latter is calculated as:

$$S_y = \left(\frac{1}{n-1} \sum_{i=1}^{n} (p_i - \bar{p})^2\right)^{1/2} \qquad (13)$$

where $\overline{p}$ is the mean of the measured penetrations.

Bias

In addition to the accuracy statistics S_e and S_e/S_y, the model bias is important. The overall bias is defined as the mean error:

$$\bar{e} = \frac{1}{n} \sum_{i=1}^{n} (\hat{p}_i - p_i) \tag{14}$$

Ideally, a model should be neither overall biased nor locally biased. The former implies a mean error $\overline{e} = 0$, the latter a frequent and random change in sign of the residuals $e = \hat{p}_i - p_i$. A model can have very significant local biases even though it has no overall bias. Local biases usually indicate an inadequate model structure and, therefore, are possibly more important than the overall bias. The residuals can be tested for randomness with a *nonparametric run* test or the Dubbin-Watson test.

SENSITIVITY ANALYSIS

Sensitivity analysis helps to understand the assumptions underlying a corrosion model. Expressions are derived next for the relative sensitivities of the coefficients b_0 and b_1 of the power model (Eq 1); b_0 and b_1 of the first of the two equations for the C.I. power-linear model (Eq 7a); and b_0, b_1, and b_2 of the first of the two equations for the V.I. power-linear model (Eq 11a).

Power Model

The absolute sensitivities of the intercept and slope coefficients of the power model (Eq 1) are:

$$\frac{dp}{db_o} = t^{b_1} \tag{15}$$

$$\frac{dp}{db_1} = b_o t^{b_1} \ln t \tag{16}$$

Because the absolute sensitivities depend on the units and are therefore not comparable, the relative sensitivities are more useful for assessing importance of the coefficients. Relative sensitivity, R_s, is the rate of change of penetration, dp/p, with respect to a relative change in a coefficient, db/b:

$$R_s = \frac{dp/p}{db/b} = \frac{dp}{db}\frac{b}{p} \tag{17}$$

As Eq 17 shows, the relative sensitivity is computed from the absolute sensitivity, dp/db. Since the relative sensitivity is invariant to the dimensions of b_o and b_1, it is one way of assessing the importance of the coefficients. For the power model, the relative sensitivities of b_o and b_1 are:

$$R_s(b_o) = t^{b_1}\left(\frac{b_o}{b_o t^{b_1}}\right) = 1 \tag{18}$$

$$R_s(b_1) = b_o t^{b_1} \ln t \left(\frac{b_1}{b_o t^{b_1}}\right) = b_1 \ln t \tag{19}$$

Clearly, the relative sensitivity of the intercept coefficient b_0 is invariant to exposure time, while the relative sensitivity of b_1 increases with time. For small exposure times, b_1 is much less important than b_o. For typical values of b_1 and an exposure time of 75 years, b_1 may be twice as important as b_o.

C.I. Power-Linear Model

The equations for the relative sensitivities of the coefficients b_0 and b_1 of the C.I. power-linear model are identical to those of the power model. They are given by Eqs 18 and 19.

V.I. Power-Linear Model

The equations for the relative sensitivities of the coefficients b_0 and b_1 of the V.I. power-linear model are also identical to those of the power model (Eqs 18 and 19). However, the overall sensitivity is different because the V.I. power-linear model includes the decay coefficient b_2, with absolute and relative sensitivities of:

$$\frac{dp}{db_2} = -\, b_o e^{-b_2 t} t^{b_1+1} \tag{20}$$

$$R_s(b_2) = \frac{dp}{db_2}\frac{b_2}{p} = -\, b_o e^{-b_2 t} t^{b_1+1}\left(\frac{b_2}{b_o e^{-b_2 t} t^{b_1+1}}\right) = -\, b_2 t \tag{21}$$

The negative sign of R_s indicates the direction of change, and the value quantifies the relative importance.

The sensitivity of the variable intercept, $b_o \exp(-b_2 t)$, is a function of exposure time because the exponential-decay part is a function of time. A time-dependent intercept reflects the time-dependent formation of the rust coating. Thus, the V.I. power-linear model with a linear portion for $t > t_c$ has a rational model structure over the entire range of exposure times.

COMPARISON OF CORROSION MODELS

The authors have analyzed atmospheric corrosion data reported by Horton [*4,5*], and Komp and Coburn [*6*]. They have chosen the first data because it contains measurements of penetration at yearly removals, and the second because it contains duplicate measurements at the standard removal times.

Horton's Data [4,5]

Corrosion Model -- The authors fitted Horton's data with the logarithmic power model, the numerical power model, and the C.I. power-linear model.

Atmospheric Exposure Tests -- The data came from atmospheric exposure tests of carbon steel, copper-bearing steel, and A242 weathering steel exposed in industrial Rankin, Pennsylvania. The carbon and copper steels had 0.02 and 0.22% Cu respectively. The weathering steel is supplied to the *ASTM Standard Specification for High-Strength Low-Alloy Structural Steel* (A242). The type used in Horton's study had a chemical composition richer than is common today for A242 steel.

The specimens consisted of 300 mm long strips of 22 gauge sheet. They were mounted on racks at an angle of 30 deg, facing south, as is recommended in the *ASTM Standard Practice for Conducting Atmospheric Corrosion Tests on Metals* (G50). Multiple specimens were removed from the racks at predetermined exposure intervals. The removal schedules were: 0.5, 1, 1.5, 2, 3, 4, 5, 6, 7, 8, 9, and 10 or 17 years.

The specimens were chemically stripped of corrosion products. The corrosion penetration was calculated from the measured mass loss as is recommended in the *ASTM Standard Practice for Preparing, Cleaning, and Evaluating Corrosion Test Specimens* (G1). All values of corrosion penetration and corrosion rate are per exposed surface.

Horton averaged the corrosion penetrations of the replicate specimens that were exposed for the same length of time, and reported the mean value at each exposure interval. This practice, commonly followed in studies by steel industry researchers, is inadmissible from the viewpoint of statistical analysis. All values must be reported, and a curve of corrosion penetration versus exposure time must be fitted through all data points, not just through the points for average penetration in replicate tests. Fig. 5 shows the single data points for all three steels exposed in Rankin.

Carbon Steel; Rankin, PA -- Fitting the carbon steel data with the three models yielded the following results (Fig. 5):

Logarithmic power model: .... $\hat{p} = 72.03\ t^{0.6156}$ (22)

Numerical power model: .... $\hat{p} = 59.73\ t^{0.7353}$ (23)

C.I. power-linear model:

For $t \leq 1.25$ yrs: .... $\hat{p}_1 = 70.96\ t^{0.4570}$ (24a)

For $t > 1.25$ yrs: .... $\hat{p}_2 = 42.668 + 28.728\ t$ (24b)

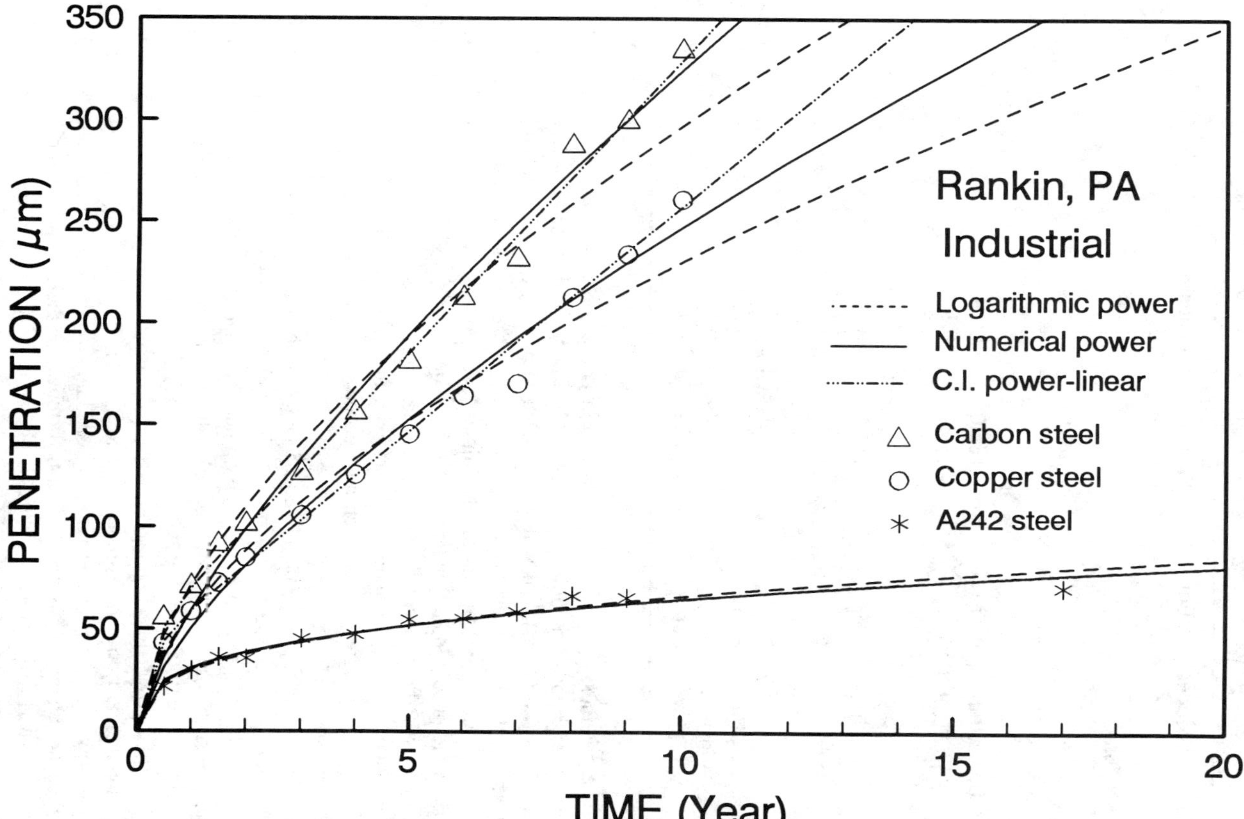

FIG. 5 -- *Corrosion penetration data for carbon, copper, and A242 steels exposed in Rankin, PA, industrial atmosphere.*

TABLE 2 -- *Accuracy of models for steel exposed in Rankin, PA.*

Model	Equation No.	Standard Error of Estimate S_e (μm)	Standard Error Ratio S_e/S_y	Mean Error (Bias) $\bar{e}$ (μm)
		Carbon Steel		
Logarithmic power	22	18.36	0.194	-2.20
Numerical power	23	12.24	0.119	0
C.I. power-linear	24	6.65	0.064	0
		Copper Steel		
Logarithmic power	25	13.52	0.191	-1.24
Numerical power	26	10.21	0.131	0
C.I. power-linear	27	6.25	0.080	0
		A242 Steel		
Logarithmic power	28	3.73	0.238	0.01
Numerical power	29	3.28	0.162	0
C.I. power-linear	30	3.28	0.162	0

The goodness-of-fit statistics -- Se, S_e/S_y, and $\bar{e}$ -- are summarized in Table 2. The logarithmic power model (Eq 22 and dashed curve in Fig. 5), is locally biased. As Table 3 shows, the residual is negative at the shortest exposure time of one-half year. A string of positive residuals follow that first increase and then decrease in magnitude. Finally, the residuals for the longest exposure times are negative. Clearly, the logarithmic power model underpredicts the penetration for short and long exposure times and overpredicts for intermediate exposure times. The logarithmic power model cannot provide the sharp curvature at short exposure times and the nearly constant slope at longer exposure times, which are needed to accurately fit the carbon steel data.

The numerical power model (Eq 23 and solid curve in Fig. 5) is unbiased and has a one-third smaller standard error of estimate than the logarithmic power model (Table 2). While it is more accurate, the numerical power model still has long positive and negative strings of residuals (Table 3), which again suggest the model structure is not sufficiently flexible to fit the data.

The C.I. power-linear model (Eqs 24 and dashed-dotted curve in Fig. 5) greatly reduces the standard error of estimate, S_e and S_e/S_y, below those of the logarithmic and numerical power models. Furthermore, the residuals for the C.I. power-linear model are smaller and change signs more often than those for the two power models (Table 3). The residuals for the power models change signs only two and four times in sequence, while those for the composite model change signs six times. Since the residuals are small and apparently random, the C.I. power-linear model is acceptable for prediction. The power models violate the least-squares assumption of independence, and thus are unacceptable.

Independence of residuals implies a lack of correlation between adjacent values. In modeling penetration, correlation between residuals indicates an incorrect model structure, one that does not follow the trend of the data and, therefore, does not reflect the underlying physical processes.

The significance of the trend in each set of residuals listed in Table 3 was checked with the one-sample run test for randomness. According to the results given in Table 4, the null hypothesis of independence can be safely rejected for the logarithmic power model. On the other hand, the null hypothesis is accepted for the numerical power and C.I. power-linear models, even at 17 and 34 % levels of significance respectively. This statistically verifies that the logarithmic power model does not satisfy the assumption of independence for this data set.

Copper Steel; Rankin, PA -- The following models were fitted to the copper steel data (Fig. 5):

Logarithmic power model: . . . . $\hat{p} = 59.00\ t^{0.5917}$ (25)

Numerical power model: . . . . $\hat{p} = 50.86\ t^{0.6870}$ (26)

C.I. power-linear model:

For $t \leq 1.25$ yrs: . . . . $\hat{p}_1 = 59.70\ t^{0.4188}$ (27a)

For $t \geq 1.25$ yrs: . . . . $\hat{p}_2 = 38.1 + 21.96\ t$ (27b)

The goodness-of-fit statistics are summarized in Table 2. As was the case for the carbon steel data, the logarithmic power model of Eq 25 is biased and has a high standard error ratio, $S_e/S_y = 0.191$. The numerical power and C.I. power-linear models of Eqs 26 and 27 are unbiased and have much smaller standard error ratios, as low as $S_e/S_y = 0.080$ for the latter.

The residuals for the three models are listed in Table 3. Again, they are smallest for the C.I. power-linear model. This is apparent from comparing the medians of the absolute values of the residuals, which are 6.8 for the logarithmic and numerical power models, and 2.3 for the C.I. power-linear model. Also, the largest absolute values of the residuals are 38.5, 22.6, and 18.0 for the logarithmic power, numerical power, and C.I. power-linear models respectively.

The residuals for the logarithmic power model again show significant local biases -- a negative value followed by a long string of positive values followed by a string of negative values. The numerical power model shows similar local biases, but with smaller residuals. Finally, the C.I. power-linear model yields greater randomness of residuals than is the case for the two power models.

According to the results of the run test (Table 4), the hypothesis of independence should be rejected at the 5% level for the two power models. The residuals of the C.I. power-linear model pass the test for independence. Both the residuals and the goodness-of-fit statistics indicate that the C.I. power-linear model better fits the copper-steel data than either of the power models.

TABLE 3 -- *Prediction error of models for steel exposed in Rankin, PA.*

Exposure Time	Measured Penetration	Residuals		
		Logarithmic Power e	Numerical Power e	C.I. Power-Linear e
(yr)	(μm)	(μm)	(μm)	(μm)
		Carbon Steel (Eqs 22 to 24)		
0.5	55.1	-8.1	-19.2	-3.4
1	70.6	1.4	-10.9	0.4
1.5	91.0	1.5	-10.5	-5.2
2	101	9.4	-1.6	-0.9
3	126	15.7	8.0	2.9
4	156	13.1	9.5	1.6
5	181	13.0	14.0	5.3
6	212	5.1	11.0	3.0
7	231	7.7	18.8	12.8
8	287	-27.9	-11.4	-14.5
9	299	-20.4	1.5	2.2
10	334	-36.7	-9.3	-4.1
		Copper Steel (Eqs 25 to 27)		
0.5	43.4	-4.2	-11.8	1.3
1	59.0	0.0	-8.1	0.7
1.5	73.1	1.9	-5.9	-2.2
2	85.3	3.6	-3.4	-3.6
3	106	7.0	2.2	-2.9
4	126	8.0	5.8	-1.4
5	146	6.7	7.7	0.0
6	165	5.3	9.2	2.5
7	171	15.6	22.6	18.0
8	213	-11.1	-0.8	-2.6
9	234	-17.4	-3.9	-2.1
10	261	-38.5	-13.6	-7.6
		A242 Steel (Eqs 28 to 30)		
0.5	21.9	1.4	3.1	3.1
1	29.9	-0.2	1.3	1.3
1.5	36.3	-2.0	-0.8	-0.8
2	35.7	2.2	3.2	3.2
3	45.6	-1.9	-1.3	-1.3
4	47.6	0.7	1.0	1.0
5	55.1	-2.8	-2.9	-2.9
6	55.5	0.2	-0.2	-0.2
7	58.9	-0.1	-0.8	-0.8
8	66.7	-5.1	-6.1	-6.1
9	65.8	-1.6	-2.9	-2.9
17	70.9	9.4	6.2	6.2

TABLE 4 -- *Testing for independence with run test; steel exposed in Rankin, PA.*

Model	No.of Positive Residuals n_1	No.of Negative Residuals n_2	No.of Runs U	Rejection Probability[a] P_u
	Carbon Steel			
Logarithmic power	8	4	3	0.024
Numerical power	6	6	5	0.175
C.I. power-linear	7	5	7	0.348
	Copper Steel			
Logarithmic power	8	4	3	0.024
Numerical power	5	7	3	0.015
C.I. power-linear	5	7	4	0.076
	A242 Steel			
Logarithmic power	5	7	8	0.146
Numerical power	5	7	7	0.348
C.I. power-linear	5	7	7	0.348

[a]Probability of rejecting a true hypothesis of independence.

A242 Steel; Rankin, PA -- The exposure tests of the A242 steel specimens differed from those of the carbon and copper steel specimens in that the last removal was at 17 years instead of 10 years. The results of fitting the A242 data are (Fig. 5):

Logarithmic power model: . . . . $\hat{p}_1 = 29.72\ t^{0.3506}$ (28)

Numerical power model: . . . . $\hat{p}_1 = 31.22\ t^{0.3191}$ (29)

C.I. power-linear model:

For $t \leq 20$ yrs: . . . . $\hat{p}_1 = 31.22\ t^{0.3191}$ (30a)

For $t > 20$ yrs: . . . . $\hat{p}_2 = 55.293 + 1.2957\ t$ (30b)

The goodness-of-fit statistics are given in Table 2. The logarithmic power model has a negligible bias (0.01). Its standard error of estimate is about 14 % greater than that for the other two models, an insignificant amount as indicated by the difference 3.73 - 3.28 = 0.45 μm. The numerical power and C.I. power-linear models are unbiased ($\bar{e} = 0$).

The residuals for all three models are small and change sign often (Table 3), which suggests random variation rather than variation due to an incorrect model structure. The residuals are no locally biased. The results of the run test (Table 4) indicate that the hypothesis of independence should be accepted for all three models, even at a 14 % level of significance. The models thus give equally valid representations of the physical processes.

The C.I. power-linear model does not increase the prediction accuracy beyond that provided by the power model. For this data set, the intersection point of the C.I. power-linear model is greater than the largest exposure time in the data set, so it is essentially the same model as the numerical power model, but only within the range of the measured data. The models differ beyond the range of the data as Fig. 5 shows.

Komp and Coburn's Data [6]

Corrosion Model -- The authors fitted Komp and Coburn's data with the three models cited above plus the V.I. power-linear model.

Atmospheric Exposure Tests -- Komp and Coburn reported atmospheric corrosion data for carbon steel, copper steel, and A588 and A242 weathering steels exposed in rural Potter County, Pennsylvania; industrial Kearny, New Jersey; and moderate marine Kure Beach, North Carolina. The carbon and copper steels had 0.014 and 0.26% Cu, respectively. A588 steel is supplied to the *ASTM Standard Specification for High-Strength Low-Alloy Structural Steel with 345 MPa Minimum Yield Point to 100 mm Thick.*

The specimens consisted of 150 mm long by 100 mm wide by 0.75 mm thick coupons. They were mounted on racks at an angle of 30 deg, facing south. Duplicate specimens were removed from the racks after exposure times of 1, 2, 4, 8, and 16 years. The corrosion penetration was determined by the mass loss method. Fig. 6 shows, as an example, the duplicate data points for copper steel exposed in Kure Beach.

Copper Steel; Kure Beach, NC -- The results of the authors' analysis of Komp and Coburn's data for copper steel exposed in Kure Beach are as follows:

Logarithmic power model: . . . . $\hat{p} = 38.66\ t^{0.7184}$ (31)

Numerical power model: . . . . $\hat{p} = 34.60\ t^{0.7804}$ (32)

C.I. power-linear model:

For $t \leq 5$ yrs: . . . . $\hat{p} = 38.26\ t^{0.7002}$ (33a)

For $t > 5$ yrs: . . . . $\hat{p} = 35.40 + 16.54\ t$ (33b)

V.I. power-linear model:

For $t \leq 4$ yrs . . . . $\hat{p} = 39.12\ e^{-0.01167\ t}\ t^{0.7121}$ (34a)

For $t > 4$ yrs . . . . $\hat{p} = 33.53 + 16.67\ t$ (34b)

The data points and the predicted curves are plotted in Fig. 6.

The goodness-of-fit statistics for Eqs 31 to 34 are given in Table 5. The numerical power model has a standard error of estimate about 75% of that of the logarithmic power model. The two composite models have the same standard error, which is about 40% of that of the logarithmic power model. The standard error is

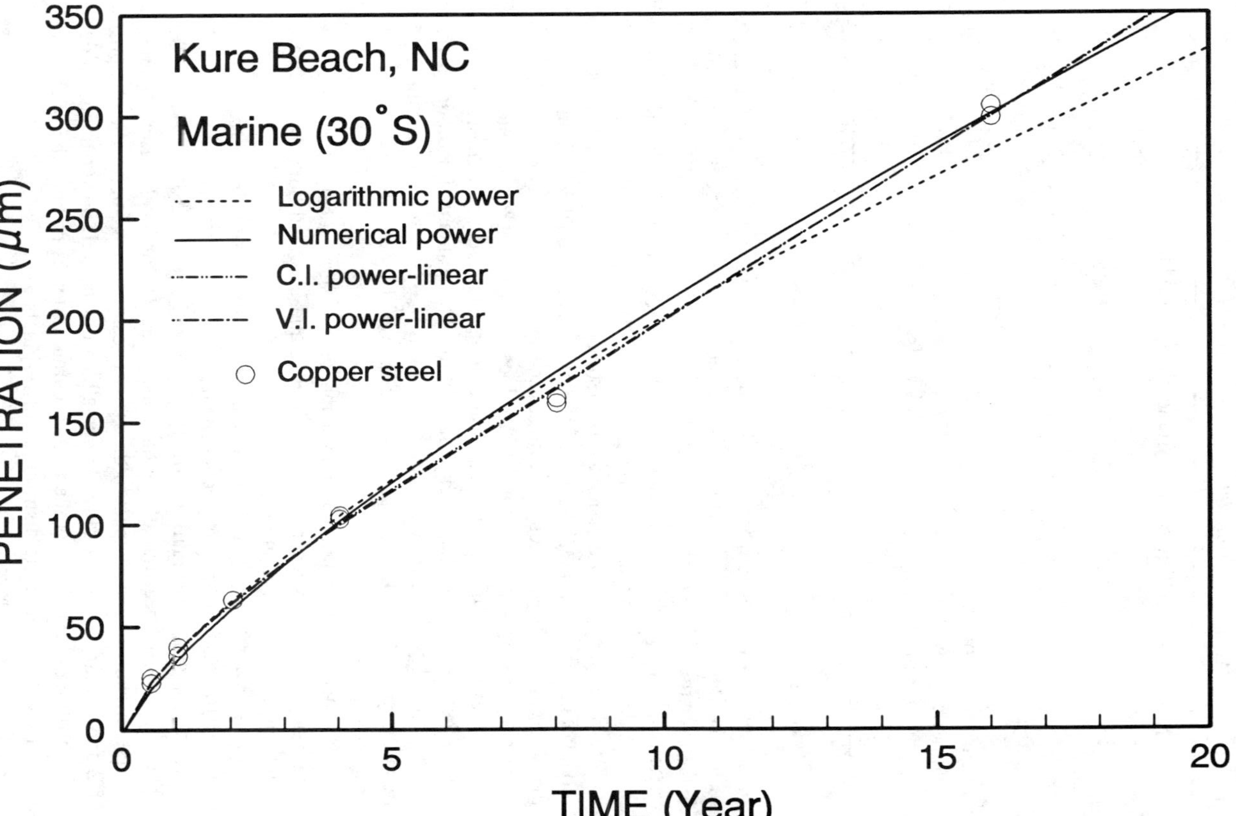

FIG. 6 -- Corrosion penetration data for copper steel exposed in Kure Beach, NC, moderate marine atmosphere.

TABLE 5 -- *Accuracy of models for steel exposed in Kure Beach, NC.*

Model	Equation No.	Standard Error of Estimate S_e (μm)	Standard Error Ratio S_e/S_y	Mean Error (Bias) $\bar{e}$ (μm)
		Copper Steel		
Logarithmic power	31	10.08	0.102	-1.11
Numerical power	32	7.41	0.075	0
C.I. power-linear	33	3.93	0.040	0
V.I. power-linear	34	3.94	0.040	0

based on 10 degrees of freedom for the C.I. power-linear model, but only 9 degrees of freedom for the V.I. power-linear model. Thus, the error sum of squares is smaller for the V.I. than for the C.I. power-linear model.

The residuals for the four models are listed in Table 6. The logarithmic power model has the largest residuals, with four exceeding 10 μm in absolute value. The numerical power model has two residuals exceeding 10 μm, as does the C.I. power-linear model. The largest residual for the V.I. power-linear model is 7.3 μm. On this basis, the V.I. power-linear model is the most accurate.

The residuals must also be checked for local biases. Since duplicate specimens were exposed, residuals with opposite signs at any one exposure time indicate that the model passes between the points, a desirable feature. The residuals for the logarithmic power model are all positive for exposure times from 2 to 8 years, while both residuals are large and negative for an exposure time of 16 years. So the model has significant local biases.

The residuals of the numerical power model also show local biases. All residuals are negative for exposure times of 4 years and less, and both residuals are positive and large for an exposure time of 8 years.

The C.I. power-linear model has four large positive residuals for exposure times of 8 and 16 years. It underpredicts penetration for the intermediate times of 2 and 4 years, although the residuals are small.

The V.I. power-linear model has the least tendency for local biases, as none are evident at either short or long exposure times. The residuals for intermediate exposure times of 2 and 4 years are all negative but small. Thus, the V.I. power-linear model with its three fitting coefficients is the most accurate in terms of both the overall goodness-of-fit statistics and the distribution of residuals.

Carbon, Copper, A588, and A242 Steels; Potter County, PA; Kearny, NJ; and Kure Beach, NC -- Table 7 summarizes the goodness-of-fit statistics for all 12 sets of data that Komp and Coburn [*6*] reported for specimens exposed at an angle of

TABLE 6 -- *Prediction error of models for steel exposed in Kure Beach, NC.*

Exposure Time (yr)	Measured Penetration p (µm)	Residuals: Logarithmic Power e (µm)	Numerical Power e (µm)	C.I. Power-Linear e (µm)	V.I. Power-Linear e (µm)
		Copper Steel (Eqs 31 to 34)			
0.5	25.5	-2.0	-5.4	-1.9	-1.8
0.5	23.0	0.5	-2.9	0.6	0.7
1.0	40.3	-1.6	-5.7	-2.0	-1.6
1.0	36.2	2.5	-1.6	2.1	2.5
2.0	63.3	0.3	-3.9	-1.1	-0.7
2.0	63.3	0.3	-3.9	-1.1	-0.7
4.0	102.8	1.9	-0.7	-1.8	-2.6
4.0	104.5	0.2	-2.4	-3.5	-4.3
8.0	159.6	12.6	15.7	12.1	7.3
8.0	162.1	10.1	13.2	9.6	4.8
16.0	305.2	-21.9	-4.1	8.0	-5.0
16.0	299.5	-16.2	1.6	13.7	0.7

30 deg facing south. Overall the V.I. power-linear model fitted the data best, followed by the C.I. power-linear model, the numerical power model, and last the logarithmic power model. The notable exceptions are:

- A242 steel exposed in rural Potter County, with about equal standard error ratios and random variation of residuals for all four models.

- A588 steel exposed in industrial Kearny, also with about equal standard error ratios and random variation of residuals for all four models.

- Carbon steel exposed in marine Kure Beach, with about equal standard error ratios but random variation of residuals only for the composite models.

The various model structures reflect differences in environmental factors that influence corrosion penetration. Since the cumulative effects of the environmental factors are difficult to predict, the data is allowed to dictate which model structure best reflects the factors causing corrosion at the exposure site. Fitting all sets of data with the same model -- such as the logarithmic power model in the *ASTM Standard Guide for Estimating the Atmospheric Corrosion Resistance of Low-Alloy Steels* (G101) --

incorrectly suggests that environmental factors do not vary with type of environment and location.

INTERSECTION TIME

The intersection time t_c of the composite models must be fitted to the data along with the regression coefficients (Eqs 7 and 11). However, t_c differs from b_0 and b_1 in that it is a location parameter rather than a shape or scale parameter. The parameter t_c influences the goodness-of-fit statistics as well as the values of the regression coefficients, so it is part of the fitting process. The easiest solution method is to assume a value of t_c, determine numerically the regression coefficients b_0 and b_1, and then calculate the standard error ratio. This procedure is repeated until the lowest value of the standard error ratio is obtained. While this may seem cumbersome, it usually requires only a few extra runs.

To demonstrate that t_c affects the goodness-of-fit statistics and the residuals, the V.I. power-linear model was fitted to the data for carbon steel exposed in rural Potter County [*6*]. The standard error ratio, S_e/S_y, is shown in Fig. 7 as a function of t_c varying from 1 to 16 years. It is smallest at $t_c = 7$ years. The error increases rapidly for $t_c < 7$ years. For this example, the residuals showed very little local bias for $t_c \approx 7$ years, but the local biases in the residuals became more pronounced as t_c deviated from the optimum value of 7 years. The parameter t_c thus affects the distribution of the residuals, a fitting criterion that should be considered along with S_e/S_y in selecting the optimum value.

The intersection point separates the p versus t space into two regions: one for the power function and the other for the linear function. However, the data points from the region of the power function influence the fitted coefficients of the linear function, and the data points from the region of the linear function influence the fitted coefficients of the power function. All coefficients depend on all data points. Table 7 lists the intersection time for the Komp and Coburn data. Five sets have a larger and 19 sets a smaller value of t_c than the maximum exposure time of 16 years in the test.

PROJECTION TO END OF SERVICE LIFE

For engineers, the main purpose of fitting corrosion penetration data is to predict the thickness loss of a structural member at the end of its service life. Having estimated the thickness loss, the engineer can then add a corrosion allowance to the member thickness arrived at by stress calculation. This is a risky proposition, because corrosion penetration data for a specific application are usually collected over 2 to 3 years at most, a length of time much shorter than the 75-year service life, say, of a structure. Engineers must therefore ensure they have an accurate model.

The effect of type of model on corrosion penetration and rate at the end of an assumed 75-year service life is shown next for two ideal cases, Horton's 10- and 17-year data and Komp and Coburn's 16-year data. The effect of length of exposure time on 75-year predictions was shown in Ref. 2.

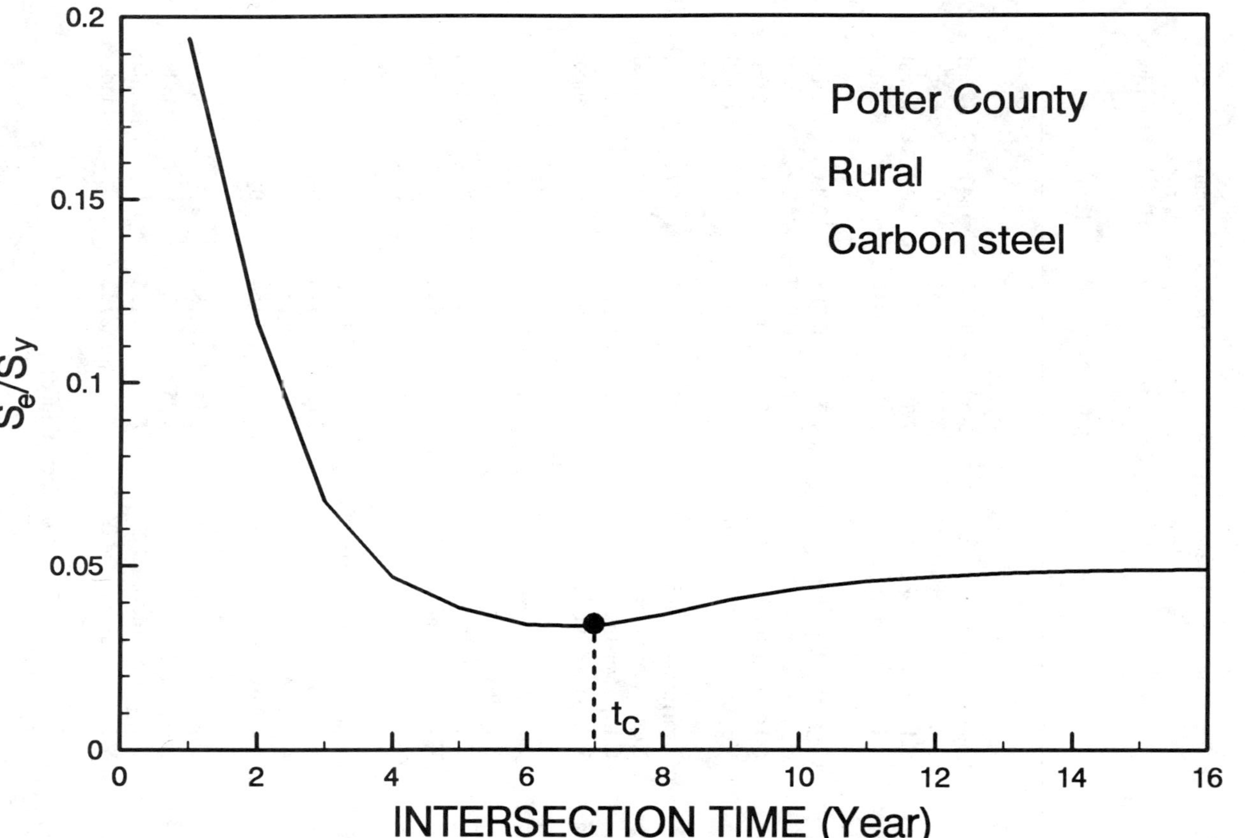

FIG. 7 -- *Standard error ratio for power-linear model fitted to copper steel data; Kure Beach, NC, moderate marine atmosphere.*

TABLE 7 -- *Accuracy of models for steel exposed in Potter County, PA; Kearny, NJ; and Kure Beach, NC.*

Type of Steel	Model	Intersection Time t_c (yr)	Standard Error of Estimate S_e (μm)	Standard Error Ratio S_e/S_y	Mean Error (Bias) $\bar{e}$ (μm)	Variation of Residuals
		Potter County, PA				
Carbon	Logarithmic power	...	3.60	0.079	0.43	Local
	Numerical power	...	2.32	0.051	0	Local
	C.I. power-linear	20	2.32	0.051	0	Local
	V.I. power-linear	7	1.53	0.034	0	Random
Copper	Logarithmic power	...	3.71	0.088	0.43	Local
	Numerical power	...	2.92	0.069	0	Local
	C.I. power-linear	9	2.71	0.064	0	Local
	V.I. power-linear	6	1.85	0.044	0	Local
A588	Logarithmic power	...	4.29	0.163	0.27	Local
	Numerical power	...	3.31	0.125	0	Local
	C.I. power-linear	17	3.31	0.125	0	Local
	V.I. power-linear	6	1.40	0.053	0	Random
A242	Logarithmic power	...	7.47	0.451	0.27	Random
	Numerical power	...	7.36	0.444	0	Random
	C.I. power-linear	5	6.76	0.408	0	Random
	V.I. power-linear	2.5	6.99	0.422	0	Random
		Kearny, NJ				
Carbon	Logarithmic power	...	3.63	0.175	-0.10	Local
	Numerical power	...	3.62	0.175	0	Local
	C.I. power-linear	6	3.18	0.153	0	Local
	V.I. power-linear	6	3.19	0.154	0	Random
Copper	Logarithmic power	...	3.31	0.186	-0.13	Random
	Numerical power	...	3.25	0.183	0	Random
	C.I. power-linear	5	2.40	0.135	0	Random
	V.I. power-linear	4.25	2.36	0.132	0	Random

(to be continued)

TABLE 7 (continued) -- Accuracy of models for steel exposed in Potter County, PA; Kearny, NJ; and Kure Beach, NC.

Type of Steel	Model	Intersection Time t_c (yr)	Standard Error of Estimate S_e (µm)	Standard Error Ratio S_e/S_y	Mean Error (Bias) $\bar{e}$ (µm)	Variation of Residuals
	Kearny, NJ (continued)					
A588	Logarithmic power	...	2.59	0.229	-0.07	Random
	Numerical power	...	2.59	0.229	0	Random
	C.I. power-linear	17	2.32	0.205	0	Random
	V.I. power-linear	3	2.15	0.191	0	Random
A242	Logarithmic power	...	2.53	0.344	-0.11	Random
	Numerical power	...	2.46	0.335	0	Random
	C.I. power-linear	5	1.72	0.233	0	Random
	V.I. power-linear	2.5	1.81	0.246	0	Random
	Kure Beach, NC					
Carbon	Logarithmic power	...	71.25	0.315	-6.58	Local
	Numerical power	...	69.47	0.306	0	Local
	C.I. power-linear	15	69.47	0.306	0	Random
	V.I. power-linear	15	73.03	0.322	0	Random
Copper	Logarithmic power	...	10.08	0.102	-1.11	Local
	Numerical power	...	7.41	0.075	0	Local
	C.I. power-linear	5	3.93	0.040	0	Local
	V.I. power-linear	4	3.94	0.040	0	Random
A588	Logarithmic power	...	9.00	0.150	-1.05	Local
	Numerical power	...	6.14	0.102	0	Local
	C.I. power-linear	3	1.83	0.030	0	Random
	V.I. power-linear	3	1.84	0.031	0	Random
A242	Logarithmic power	...	5.32	0.145	-0.50	Local
	Numerical power	...	4.10	0.112	0	Local
	C.I. power-linear	3.75	1.81	0.049	0	Random
	V.I. power-linear	3.75	1.87	0.051	0	Random

Horton's Data [4,5]

The corrosion penetration and corrosion rate at 75 years were calculated with Eqs 22 to 24 for carbon steel, Eqs 25 to 27 for copper steel, and Eqs 28 to 30 for A242 steel. The steels were exposed in Rankin, PA. The rates were obtained from the time derivatives of the equations for penetration. The results are given in Fig. 8 and Table 8.

The C.I. power-linear model predicts the deepest penetration of the three models, and more so for the carbon and copper steels than for A242 steel. The type of model has an even greater effect on the prediction of corrosion rate than on corrosion penetration. Because no measurements exist for exposure times of 75 years, it is not possible to say precisely which estimates are the most accurate. However, the goodness-of-fit statistics strongly suggest that the C.I. power-linear model more accurately projects both penetration and rate. A basic assumption in statistical analysis is that the goodness-of-fit statistics reflect the accuracy of values predicted with the fitted equation. Of course, interpolation is always more accurate than extrapolation.

The power models predict a lower corrosion rate at 75 years than does the C.I. power-linear model, because their structure (especially that of the logarithmic power model) is dominated by a rapid initial rise, thus pulling down the tail at longer exposure times. Recognizing that the power models produce local biases (again especially the logarithmic power model), they may not accurately describe the corrosion processes.

When the power models are fitted to carbon and copper steel data, the residuals at the longest exposure time are always among the largest. For example, the -38.5 μm residual for the copper steel (Table 3) is twice as large in absolute value as any of the other residuals for copper steel. Large residuals at long exposure times suggest that the model diverges from the data and becomes increasingly inaccurate. Estimates made with the power models appear to be less reliable than those made with the C.I. power-linear model.

Komp and Coburn's Data [6]

The 75-year corrosion penetration and the corrosion rate for the Komp and Coburn data were predicted with all four models. The results are given in Table 9. Again, the composite models predict much higher penetrations for the four steels (carbon, copper, A588 and A242) in the three environments (Potter County, Kearny, and Kure Beach) than do the power models. Carbon steel exposed in Kure Beach is the only exception. Fig. 9 shows, as an example, the extrapolation of corrosion penetration of copper steel exposed in Kure Beach.

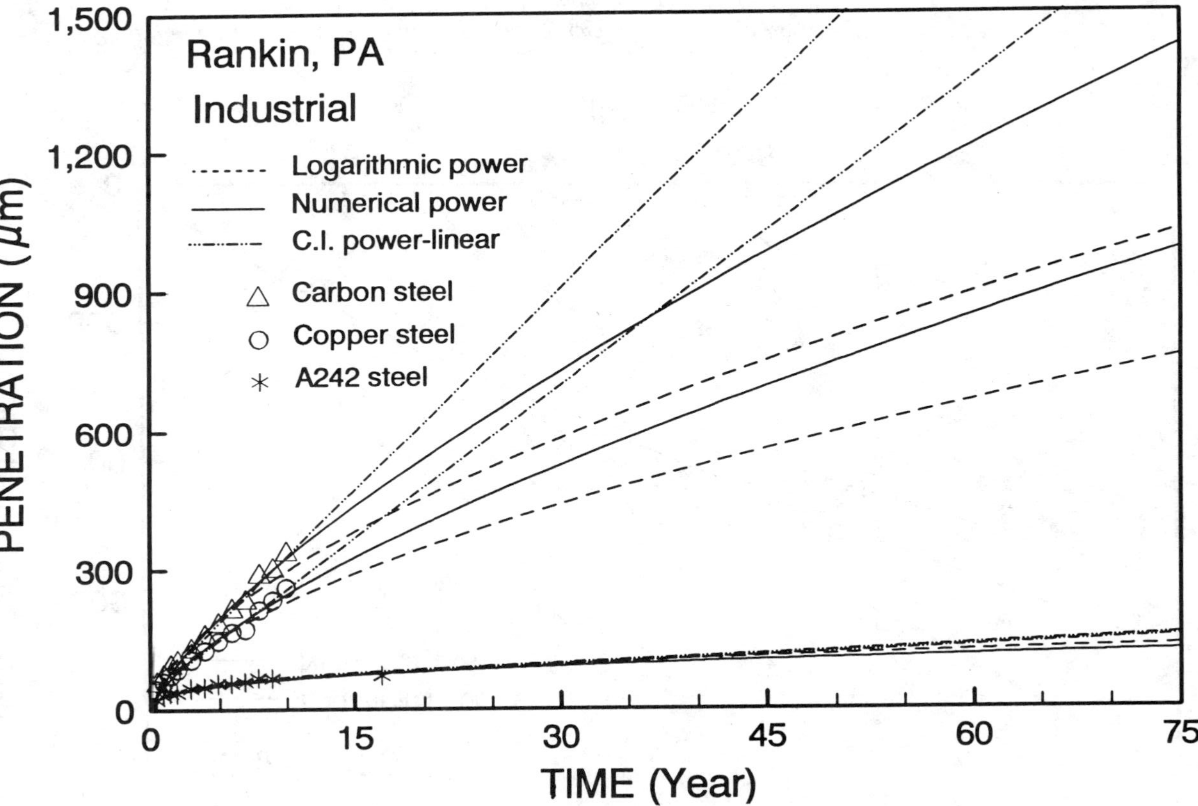

FIG. 8 -- *Predicted corrosion penetration at end of 75-year service life; carbon, copper, and A242 steels exposed in Rankin, PA, industrial atmosphere.*

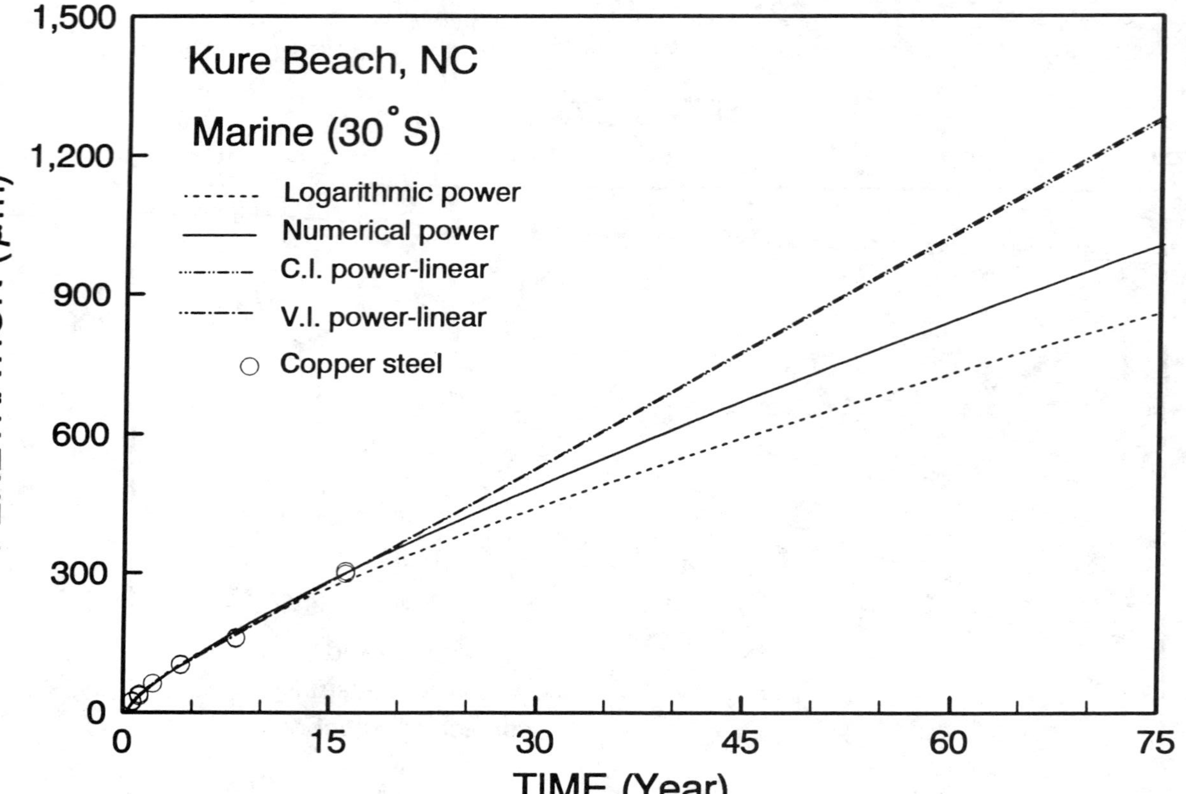

FIG. 9 -- *Predicted corrosion penetration at end of 75-year service life of copper steel exposed in Kure Beach, NC, moderate marine atmosphere.*

TABLE 8 -- *Predicted corrosion penetration and corrosion rate at end of 75-year service life; for steel exposed in Rankin, PA.*

Type of Steel	Model	Corrosion Penetration $\hat{p}$ (µm)	Corrosion Rate $\hat{dp/dt}$ (µm/yr)
	Rankin, PA		
Carbon	Logarithmic power	1028	8.43
	Numerical power	1429	14.01
	C.I. power-linear	2197	28.73
Copper	Logarithmic power	759	5.99
	Numerical power	988	9.05
	C.I. power-linear	1685	21.96
A242	Logarithmic power	135	0.63
	Numerical power	124	0.53
	C.I. power-linear	152	1.30

SUMMARY AND CONCLUSIONS

Comparison of Models

Composite models have a more flexible structure than the power models, regardless of the method of fitting the latter. The greater flexibility improves prediction accuracy and provides coefficients that more rationally represent the underlying processes. The disadvantage of a more flexible model is the initial effort needed to derive expressions for the coefficients, such as Eqs 7 for the C.I. power-linear model and Eqs 11 for the V.I. power-linear model. Once the equations are derived, however, the coefficients are easily fitted to corrosion penetration data using numerical optimization.

The composite models better fitted most sets of data analyzed in this study than did the power models. Accuracy was improved, and residuals were not locally biased. The improvement in accuracy was greatest for the carbon and copper steel data, and less for the A588 and A242 steel data. At the very least, the accuracy of a composite model will equal that of the power model within the range of the data. The power model offers no benefit over the composite model.

TABLE 9 -- *Predicted corrosion penetration and corrosion rate at end of 75-year service life for steel exposed in Potter County, PA; Kearny, NJ; and Kure Beach, NC.*

Type of Steel	Model	Corrosion Penetration $\hat{p}$ (μm)	Corrosion Rate $d\hat{p}/dt$ (μm/yr)	Corrosion Penetration $\hat{p}$ (μm)	Corrosion Rate $d\hat{p}/dt$ (μm/yr)	Corrosion Penetration $\hat{p}$ (μm)	Corrosion Rate $d\hat{p}/dt$ (μm/yr)
		Potter County, PA		**Kearny, NJ**		**Kure Beach, NC**	
Carbon	Logarithmic power	397	3.34	138	0.65	2379	28.85
	Numerical power	360	2.86	140	0.66	2662	33.57
	C.I. power-linear	432	4.88	251	2.84	2781	36.66
	V.I. power-linear	502	6.08	239	2.66	2615	33.84
Copper	Logarithmic power	378	3.31	117	0.56	860	8.23
	Numerical power	342	2.84	122	0.61	1005	10.46
	C.I. power-linear	489	6.05	224	2.61	1276	16.54
	V.I. power-linear	489	6.05	219	2.53	1284	16.67
A588	Logarithmic power	211	1.47	77	0.28	434	3.40
	Numerical power	184	1.16	77	0.28	536	4.76
	C.I. power-linear	238	2.53	134	1.39	792	10.16
	V.I. power-linear	273	3.13	139	1.47	789	10.11
A242	Logarithmic power	97	0.50	49	0.14	253	1.75
	Numerical power	107	0.60	52	0.16	300	2.30
	C.I. power-linear	193	2.29	102	1.07	477	6.00
	V.I. power-linear	195	2.32	102	1.08	474	5.94

Numerical Fitting of Models

The numerical power model and the composite models are easily fitted numerically with any least-squares program. Such programs are referred to by any number of terms, including steepest descent, pattern search, nonlinear least squares, and conjugate gradient algorithms. They should yield essentially identical results as long as the program minimizes the sum of the squares of the errors. A few guidelines on fitting the models are given below, with more detail given in Ref. 3.

Numerical Power Model (Eq 1):

1. Estimate the coefficients b_0 and b_1 with linear least-squares regression analysis using the logarithms of the measured penetrations and exposure times; that is, by minimizing the sum of the squares Σe^2 in the log y-space.
2. Calculate the optimum (minimum Σe^2 in the y-space) values of b_0 and b_1 with, for example, the NUMOPT program. Several executions may be needed. A copy of this program can be obtained by mailing a diskette and a self-addressed mailer to the first author.
3. Using the response surface option, unbias the model such that the Σe^2 is a minimum for any unbiased solution. For an example of response surfaces see Fig. 18.2 of Ref. 3.

Constant-Intercept Power-Linear Model (Eqs 7):

1. Select an initial value for the intersection time t_c. From the results listed in Table 7, a value of 5 years seems reasonable.
2. Estimate the coefficients for b_0 and b_1. Values obtained with the numerical power model are reasonable estimates.
3. Calculate the optimum values of b_0 and b_1 with a numerical algorithm.
4. Repeat steps 1 through 3 until the response surface of S_e/S_y versus t_c is delineated (see Fig. 7 for an example).
5. Retain the solution for t_c for which S_e/S_y is a minimum.

Variable-Intercept Power-Linear Model (Eqs 11):

The procedure for fitting the V.I. power-linear model is the same as that for fitting the C.I. power-linear model, except that an initial estimate is also needed for the coefficient b_2. So only step 2 has to be modified as follows:

2. Estimate the coefficients for b_0 and b_1. Values obtained with the numerical power model are reasonable estimates. Estimate the coefficient b_2. The median value of $b_2 = 0.03$ for all 12 sets of data fitted in Table 7 with the V.I. power-linear model seems to be a reasonable estimate.

POSTSCRIPT

Power-Power Model

The statistical analysis for this paper, completed in early 1992, is based on empirical rather than physio-chemical representation of corrosion kinetics. During the peer review of the paper, the authors were sent in early 1993 a copy of a just-published paper in which Morcillo et al. [7] also provide empirical evidence of the power model's inability to represent corrosion kinetics, in their case for mild steel exposed in El Escorial and Madrid, Spain. They reported deviations from the behavior predicted by the power model and proposed instead a power-power model consisting of

$$\hat{p}_1 = b_0 t^{b_1} \qquad for\ t \le t_c \tag{35a}$$

$$\hat{p}_2 = b_2 t^{b_3} \qquad for\ t > t_c \tag{35b}$$

Without the restriction of $t \le t_c$, Eq 35a is the same as Eq 1. For ease of comparison, the authors have changed the notation in Morcillo et al. to that used in the present study.

Constant-Intercept Power-Power Model

Imposing only the constraint of equal penetrations at $t = t_c$, the constraint of Eq 3a leads to the power-power model in Morcillo et al.

$$\hat{p}_1 = b_0 t^{b_1} \qquad for\ t \le t_c \tag{36a}$$

$$\hat{p}_2 = b_0 t_c^{b_1 - b_3} t^{b_3} \qquad for\ t > t_c \tag{36b}$$

Because Morcillo et al. do not impose the second constraint of equal corrosion rates at $t = t_c$ (Eq 3b), one is left with a discontinuity in corrosion rate at $t = t_c$ for which there is no rational physio-chemical explanation. If the constraint of equal corrosion rate at $t = t_c$ is imposed, then the exponents become equal, $b_3 = b_1$, and Eqs 36 reduce to the simple power model given by Eq 1. It follows that the power-power model with two constraints is not possible in the form of either Eqs 35 or Eqs 36.

Morcillo et al. fitted their data in the log space and found that the exponent of the logarithmic power-power model changed from a higher value b_1 for $t \le t_c$ to a lower value b_3 for $t > t_c$. Having also observed in scanning electron micrographs that the rust layer was denser at 13 years than at 2 years, they concluded that the rust layer was more compact during the second period, resulting in a reduced diffusion of reactants and progressively lower corrosion rate. They attributed the "anomalous" behavior to environmental conditions favorable to the formation of a protective rust coating; this includes a mildly corrosive atmosphere, long annual sunlight periods, and high air temperatures, all contributing to shorter drying times. The change between Eqs 1 and 35 occurred at about 4 years for steel exposed in Madrid and about 6 years in El Escorial, values comparable to many intersection times listed in Table 7 for 12 data sets analyzed by the authors.

It appears that Morcillo et al. plotted their data points on log-log scales, separated the data visually into two groups, fitted the power functions -- Eqs 36a and 36b -- to each group separately, and determined the intersection time at $\hat{p}_1 = \hat{p}_2$. This approach does not provide a model that is either unbiased or has a minimum mean square error. Better accuracy can be achieved by numerically fitting the coefficients b_0, b_1 and b_3 as well as the intersection time t_c. Simultaneous numerical fitting of all four parameters can also provide an unbiased model. Regardless of whether the model of Eqs 36 is fitted in the manner of Morcillo et al. or numerically, it still yields a discontinuous corrosion rate at the intersection time $t = t_c$.

Variable-Intercept Power-Power Model

The slope discontinuity in the power-power model with a single constraint of equal displacements at $t = t_c$ (Eqs 36) is easily eliminated by adding one more unknown coefficient to the prediction equation (Eq 35a). Also, as the data analyzed in this study shows, prediction accuracy is improved by choosing a variable intercept for the power function in the region $t \le t_c$ and fitting the data numerically. Using the same function as in Eq 8a, the authors propose the following V.I. power-power model:

$$\hat{p}_1 = b_o e^{-b_2 t} t^{b_1} \qquad \textit{for } t \le t_c \tag{37a}$$

$$\hat{p}_2 = b_3 t^{b_4} \qquad \textit{for } t > t_c \tag{37b}$$

Splicing Eqs 37 with both equal corrosion penetrations and equal corrosion rates at $t = t_c$ yields the following equation for the V.I. power-power model in terms of the fitting coefficients b_0, b_1 and b_2 as well as the intersection time t_c:

$$\hat{p}_1 = b_o e^{-b_2 t} t^{b_1} \qquad \textit{for } t \le t_c \tag{38a}$$

$$\hat{p}_2 = b_o e^{-b_2 t_c} t_c^{b_2 t_c} t^{b_1 - b_2 t_c} \qquad \textit{for } t > t_c \tag{38b}$$

Comparison of Power-Power Models

The work reported by Morcillo et al. has led to three new models. The power-power model of Eqs 35 has no constraints imposed at the intersection time t_c. If the corrosion penetrations are equated (Eq 3a), then Eqs 35 yield the C.I. power-power model described by Eqs 36. If both the corrosion penetrations (Eq 3a) and the corrosion rates are equated (Eq 3b), the power-power model reduces to the simple power model of Eq 1. The V.I. power-power model enables both constraints of Eqs 3 to be met.

In addition to the model form, the fitting method affects the values of the coefficients as well as the goodness-of-fit statistics of Eqs 12 and 14. Logarithmic fitting of separate power functions, after separating the data visually into two groups, may result in less than optimal accuracy. It also creates an uncertainty when the data points are so distributed that a point near the apparent transition time could be assigned to one or the other group. See for example the data point at four years of exposure in Madrid, Fig. 3 of Morcillo et al. Numerical fitting is much preferable because it optimizes accuracy and yields the transition time as part of the fitting process. It does not require the user to choose a transition time or arbitrarily assign points to one of two groups.

Value of Statistical Modeling

Empirical analysis of data is not sufficient by itself. Studies of the physio-chemical processes are needed to validate empirical findings. However, statistical modeling can provide direction for studies of the physio-chemical processes and the means for representing observed phenomena. Both statistical modeling and studies of the physio-chemical processes are important to advancements in knowledge of corrosion penetration.

REFERENCES

[*1*] Albrecht, P. and Lee, H.Y., "Corrosion Performance of Boldly Exposed Weathering Steel." Civil Engineering Report, University of Maryland, College Park, Maryland, 1989, 129 p.

[*2*] McCuen, R.H., Albrecht, P., and Cheng, J.G., "A New Approach to Power-Model Regression of Corrosion Penetration Data." *Corrosion Forms and Control for Infrastructure*, ASTM STP 1137, Victor Chaker, Ed., American Society for Testing and Materials, Philadelphia, 1992, pp. 46-76.

[*3*] McCuen, R.H., "*Microcomputer Applications in Statistical Hydrology*," Prentice-Hall, Englewood Cliffs, New Jersey, 1993.

[*4*] Horton, J.B., "The Composition, Structure and Growth of Atmospheric Rust on Various Steels," Ph.D. Dissertation, Lehigh University, Bethlehem, Pennsylvania, 1964.

[*5*] Horton, J.B., "The Rusting of Low Alloy Steels in the Atmosphere," Presented at San Francisco Regional Technical Meeting, American Iron and Steel Institute, 1965.

[*6*] Komp, M.E., and Coburn, S.K., "Private communication," 1988.

[*7*] Morcillo, M., Feliu, S., and Simancas, J., "Deviation from Bilogarithmic Law for Atmospheric Corrosion of Steel," *British Corrosion Journal*, Vol. 28, No. 1, 1993, pp. 50-52.

APPENDIX

Copson, H.R. (1960). "Long-Time Atmospheric Corrosion Tests on Low-Alloy Steels." American Society for Testing and Materials, 60, 650-667.

Cosaboom, B., Mehalchick, G. and Zoccola, J.C. (1979). "Bridge Construction with Unpainted High-Strength Low-Alloy Steel, Eight-Year Progress Report." New

Jersey Department of Transportation, Division of Research and Development, Trenton, New Jersey.

Cruz, I.S. and Mullen, C.X. (1962). "The Architectural Application of Bare USS COR-TEN High-Strength Low-Alloy Steel." Technical Report, Applied Research Laboratory, United States Steel, Monroeville, Pennsylvania.

Gallagher, W.P. (1976). "Long-Term Corrosion Performance of USS COR-TEN B Steel in Various Atmospheres." Technical Report, Research Laboratory, United States Steel, Monroeville, Pennsylvania.

Gallagher, W.P. (1978). "One and One-half Year Atmospheric Corrosion Performance of Modified COR-TEN B Steel." Research Laboratory, United States Steel Corporation.

Gallagher, W.P. (1982). "Performance of USS COR-TEN Steels in Chloride-Contaminated Atmospheres." Technical Report, Research Laboratory, United States Steel Corporation, Monroeville, Pennsylvania.

Horton, J.B. (1964). "The Composition, Structure and Growth of Atmospheric Rust on Various Steels." Ph.D. Dissertation, Lehigh University, Bethlehem, Pennsylvania.

Horton, J.B. (1965). "The Rusting of Low Alloy Steels in the Atmosphere." Presented at San Francisco Regional Technical Meeting, American Iron and Steel Institute.

Komp, M.E. (1987). "Atmospheric Corrosion Ratings of Weathering Steels - Calculation and Significance." Materials Performance, National Association of Corrosion Engineers. CORROSION/87, Paper No. 423, NACE, San Francisco, California, 1987, 42-44.

Larrabee, C.P. (1953). "Corrosion Resistance of High-Strength Low-Alloy Steels As Influenced by Composition and Environment." CORROSION, 9(8), 259-271.

Larrabee, C.P. (1959). "Twenty-Year Results of Atmospheric Corrosion Tests on USS Cor-Ten Steel, Structural Copper Steel, and Structural Carbon Steel." Applied Research Laboratory, United States Steel, Monroeville, Pennsylvania.

Larrabee, C.P., and Coburn, S.K. (1962). "The Atmospheric Corrosion of Steels as Influenced by Changes in Chemical Composition." First International Congress on Metallic Corrosion, London, 276-285.

Reed, F.O. and Kenderick, C.B. (1982). "Evaluation of Weathering Effects on Structural Steel." Research Report No. FHWA/CA/TL-82/04, California Department of Transportation, Division of Construction, Office of Transportation Laboratory, Sacramento, California.

Schmitt, R.J. and Mullen, C.X. (1965). "Corrosion Performance of Mn-Cr-Cu-V Type USS COR-TEN High-Strength Low-Alloy Steel in Various Atmosphere." Technical Report, Applied Research Laboratory, United States Steel, Monroeville, Pennsylvania.

Shastry, C.R., Friel, J.J. and Townsend, H.E. (1986). "Sixteen-Year Atmospheric Corrosion Performance of Weathering Steels in Marine, Rural and Industrial Environments." *Degradation of Metals in the Atmosphere*, ASTM STP 965, S.W. Dean and T.S. Lee, Eds., American Society of Testing and Materials, Philadelphia, 1988, pp. 5-15.

Tinklenberg, G.L. (1986). "Evaluation of Weathering Steel in a Detroit Freeway Environment, --Second Eight-Year Study." Research Report No. R-1277, Research Laboratory Section, Materials and Technology Division, Lansing.

Townsend, H.E. and Zoccola, J.C. (1982). "Eight-Year Atmospheric Corrosion Performance of Weathering Steel in Industrial, Rural, and Marine Environments." *Atmospheric Corrosion of Metals*, ASTM STP 767, S.W. Dean, Jr., and E.C. Rhea, Eds., American Society for Testing and Materials, 45-59.

Vrable, J.B. (1985). "Seven and One-Half Years Atmospheric Corrosion Performance of Improved USS COR-TEN B Steel." Technical Report 33-C-207 (005-1), Research Laboratory, United States Steel Corporation.

Zoccola, J.C. (1976). "Eight Year Corrosion Test Report - Eight Mile Road Interchange." Bethlehem Steel Corporation, Bethlehem, Pennsylvania.

Darrel R. Duncan,[1] Dewey A. Burbank, Jr.,[1] Burdell C. Anderson,[2] and James A. Demiter[1]

APPLICATION OF SERVICE EXAMINATIONS TO TRANSURANIC WASTE CONTAINER INTEGRITY AT THE HANFORD SITE

REFERENCE: Duncan, D. R., Burbank, D. A., Jr., Anderson, B. C., and Demiter, J. A., "Application of Service Examinations to Transuranic Waste Container Integrity at the Hanford Site," Application of Accelerated Corrosion Tests to Service Life Prediction of Materials, ASTM STP 1194, Gustavo Cragnolino and Narasi Sridhar, Eds., American Society for Testing and Materials, Philadelphia, 1994.

ABSTRACT: Transuranic waste containers in retrievable storage trenches at the Hanford Site and their storage environment are described. The containers are of various types, predominantly steel 0.21-m^3 (55-gal) drums and boxes of many different sizes and materials. The storage environment is direct soil burial and aboveground storage under plastic tarps with earth on top of the tarps. Available data from several transuranic waste storage sites are summarized and degradation rates are projected for containers in storage at the Hanford Site.

KEYWORDS: corrosion, drum, soil, soil exposure, steel

INTRODUCTION

The Hanford Site is one of 14 U.S. Department of Energy (DOE) sites located throughout the United States that generate and/or store radioactive transuranic (TRU) waste from national defense programs. This waste is stored in trenches of several different designs or configurations and in aboveground interim-storage areas.

The disposal at the Waste Isolation Pilot Plant (WIPP) of TRU waste stored at the Hanford Site is required by DOE Order 5820.2A [1], and the Hanford Site Defense Waste Final Environmental Impact Statement Record of Decision [2]. Disposal will consist of retrieving the waste and repackaging and treating as necessary, in the Waste Receiving and Processing (WRAP) Facility to certify for shipment to the WIPP.

---

[1]Principal Engineer, Principal Engineer, and Principal Scientist, respectively, Westinghouse Hanford Company, P. O. Box 1970, Richland, WA 99352.

[2]Retired Principal Engineer, Westinghouse Hanford Company, P. O. Box 1970, Richland, WA 99352, and Westinghouse Waste Isolation Division, Carlsbad, NM.

The integrity of containers is of interest from two standpoints: (1) assessment of current or future contamination leakage with potential spread to the surrounding environment, and (2) the impact on environmentally safe handling during retrieval of containers with breaches or lost structural integrity. This paper will describe the types of TRU waste containers in retrievable storage at the Hanford Site and their expected environment, the available data relating to corrosion rates of the containers, and expected degradation of the containers.

CONTAINER ENVIRONMENT

Transuranic-bearing wastes were packaged in sealed containers and segregated from low-level waste (LLW) in retrievable storage trenches from May 1, 1970 until the end of 1988; after that time wastes have been stored in aboveground buildings. Before 1970, the TRU wastes were commingled and buried with the LLW. About 15,421 m^3 (544,500 ft^3) of wastes designated as TRU waste have been retrievably stored at the Hanford Site. These wastes consist of dry waste (e.g., soiled clothing; laboratory supplies; and tools packed in cardboard, wood, or metal containers) and industrial waste (primarily items of failed process equipment packaged in plastic shrouds, wood, fiberglass-reinforced polyester [FRP], metal, or concrete boxes). The containers include 37,461 drums of 0.21 m^3 (55 gal) capacity, 329 metal boxes, 202 FRP boxes, 58 concrete boxes, 37 plywood boxes, and 462 other miscellaneous containers. The drums are painted carbon steel and galvanized carbon steel, procured according to the U.S. Department of Transportation (DOT) 17-C or 17-H specifications. The painted carbon steel drums were used before 1981, approximately, and galvanized drums were used thereafter.

The containers have been stored in four different configurations. The first configuration used had containers stacked in a horizontal position in a gravel bottom "v" trench covered with soil. Later a concrete "v" trench with a metal cover over drum containers was used for a short period of time. This mode of storage was abandoned because of the cost. The third configuration consisted of wide-bottom and "v" trenches with plywood bottoms and plywood sheeting between vertical stacks of drums. A plastic tarp was used to cover the top layer of containers and approximately 1.3 m of earth was backfilled on top of the tarp. The fourth configuration was wide-bottom trenches similar to the third configuration with asphalt bottoms rather than plywood. Over 70% of the waste containers are in trenches of the fourth configuration. A depiction of the fourth configuration used for storage is given in Figure 1. In most cases, the waste drums were stacked in modules 12 drums long by 12 drums wide by 4 to 5 drums high, for an approximately 8-m-long by 8-m-wide by 5-m-high (26.2-ft by 26.2-ft by 16.4-ft) geometry. Boxes were stacked in single layers in the same modules as drums. Modules were separated by several feet in the trench. Each trench varied in length but could be nearly 200 m (656 ft) long. The bottom of the trench was 4 m to 6 m (13.1 ft to 19.7 ft) below grade.

AVAILABLE DATA

A large and widely varying body of data exists on soil and atmospheric corrosion of steel. The most applicable data would be those of similar containers in similar environments. Other DOE sites have examined waste drums stored underground [3,4,5,6]. At the Idaho National Engineering Laboratory (INEL) in Idaho Falls, Idaho, 65% of the drums stored for 18 to 21 years were breached, with several containers having lost structural integrity. This corresponds to a median corrosion rate in the breached containers of approximately 0.102 mm/year

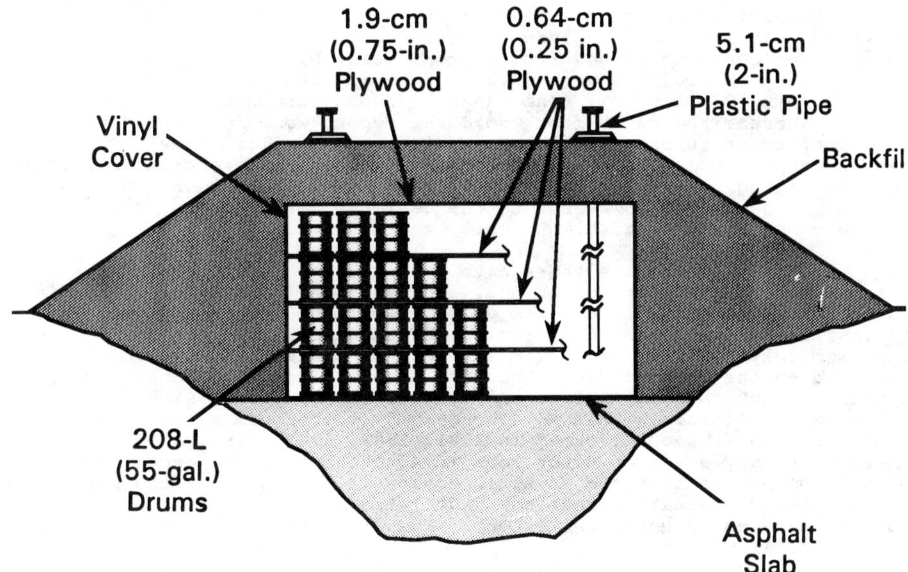

FIG. 1--Typical drum storage with atmospheric environment.

(4 mil/year), assuming a DOT 17-C nominal 1.52-mm (60-mil) wall thickness. This number was estimated from the assumption that the drum breaches represent a median corrosion rate, a rate at which roughly half of the drums would be breached. The assessment that 65% were breached with several of the containers having lost integrity implies two different failure modes with respect to physical effects of corrosion on the drums. Breach means wall penetration in a small area, while loss of structural integrity means large areas have been penetrated or nearly penetrated such that normal drum-handling stresses would collapse the drum. This is consistent with the usual progress of corrosion of steels in soil – some localized areas corrode much more rapidly than others and appear as pits while the rest of the surface corrodes more slowly and more evenly. As corrosion products accumulate with time, the pitting rate slows down. In contrast, drum retrieval at INEL after 6 to 9 years of burial under plastic covering yielded 2 of 102 drums with serious corrosion but no breaches. This would be consistent with the corrosion rates previously described for INEL.

At the Savannah River Site in South Carolina, plastic covering of drums in aboveground soil mounds (with soil contacting the drums under the plastic) resulted in a maximum of approximately 52% wall thickness penetration after 14 years of exposure. This behavior is not comparable to the Hanford Site, however, because of the much higher soil moisture and the use of galvanized drums at the Savannah River Site. Only painted carbon steel drums, which should be less protective than galvanized drums, were exposed to soil at the Hanford Site.

There have been several assessments and examinations of corrosion of buried steel at the Hanford Site, in the two environments of direct soil burial and under plastic tarp covering. The oldest and perhaps the most comprehensive survey of direct soil effects was performed by Jaske [7], for a variety of buried piping materials. The observed corrosion rates averaged 0.127 mm/year (5 mil/year) for general corrosion and 0.229 mm/year (9 mil/year) for pitting corrosion. A more recent investigation of corrosion of 16 large underground steel storage tanks [8] resulted in much less observed corrosion. The maximum penetration observed ranged from approximately 0.025 mm/year (1 mil/year) to 0.089 mm/year (3.5 mil/year), with an average of nearly 0.05 mm/year (2 mil/year). The tanks had been coated with coal tar enamel, whereas the rates from the Jaske report were for bare metal. Soil properties reported for both studies were similar. Resistivity values ranged from 5,000 to 50,000 ohm-cm and chloride content was very low, averaging 0.01 mg-eq per 100 g. The difference between the two sets of data obviously was due to the coal tar coating; the data should not be considered applicable to bare metal corrosion.

At the Hanford Site there have been several investigations of corrosion under the tarp coverings. In 1982, an inspection was carried out on drums stored over 8 years using visual and ultrasonic techniques [9]. The maximum measured corrosion corresponded to a rate of approximately 0.025 mm/year (1 mil/year). The maximum corrosion occurred on the drum surfaces touching the tarp probably because of a wicking effect that maintained moisture on the drum. In situ photography of drums within a storage module, through a vertical inspection port, was performed in 1981, 1982, 1983, 1988, and 1992. There were many areas of paint loss (most paint loss had occurred from weathering before tarp and backfill coverage) on drum surfaces but there was no visual indication that the underlying metal was penetrated by corrosion to any significant extent. The drums have been in the storage module for 11 years.

Expected corrosion from a theoretical assessment of the environment will be discussed next. The Hanford Site soil moisture content varies with depth and surface precipitation but at the depth of the buried containers could range from near 0% to as much as 10%. Typical Hanford Site soils are slightly alkaline with a pH near 8.0, resistivity approximately 5-50,000 ohm-cm, and low chloride content. It is difficult to predict the corrositivity of soils from chemical and physical analyses, particularly a definite corrosion rate, but the stated characteristics of Hanford Site soils would place them in a relatively nonaggressive category [10,11]. Higher moisture contents, lower pH, lower resistivity, and greater chloride content would tend to cause greater relative corrosivity. The most extensive survey of soil corrosion was performed by the National Bureau of Standards (now the National Institute of Standards and Technology [NIST]). Jaske reported that soils from the NIST study similar to Hanford Site soils had rates of maximum general corrosion penetration from approximately 0.076 mm/year (3 mil/year) to 0.229 mm/year (9 mil/year) [7]. These rates were determined from linear calculations of penetration rate divided by the time of exposure. A linear regression analysis by NIST of a larger body of data (with several exposure times) from the same soils, however, led to a predicted average maximum penetration rate of only 0.05 mm/year (2 mil/year). The difference between the two analyses probably arises from the fact that Jaske effectively extrapolated rates linearly from a single point whereas the linear regression analysis incorporated several exposure times.

In comparison to the Hanford Site soils, the soils at the INEL site have a higher moisture content and ionic content and are less alkaline. These characteristics would be expected to lead to greater corrosion rates. This is in contrast to the Jaske data which yielded higher rates than those at the INEL site. Comparisons with NIST soil corrosion data have led to contrasting corrosion estimates, as done by Jaske and the NIST. This illustrates the problems arising when comparing corrosion data from different locations.

A markedly different environment in the storage modules from soil exposure is found in areas which have tarp covers with plywood sheeting between each drum layer. The tarps were placed between the drums and the soil backfill to avoid the accelerated corrosion of soil exposure. This leads to a slower atmospheric type of corrosion. Atmospheric corrosion rates for a time span of 10 years in an urban-industrial environment have been reported as averaging 0.013 mm/year (0.5 mil/year) [12]. Measurements taken on a small number of drums underneath tarp coverage after 8 years of exposure yielded an estimate of 0.025 mm/year (1 mil/year) in the worst case [9]. A still different environment, for which no observations exist, would be provided by the plywood sheeting that touches the tops and bottoms of each drum. Exposure to wood has been known to accelerate corrosion [13].

HYPOTHETICAL CONTAINER INTEGRITY SCENARIO

In light of the differing environments within the storage modules, it is of interest to forecast their effects on the waste drum population from the meager and qualitative data available. While the data do not support a statistical approach, a probabilistic approach can provide estimates of the probability of failure given assumptions of corrosion rates and numbers of drums.

A model has been developed to project the probability of corrosion failure in retrievably stored drums at the Hanford Site. The model applies the Poisson probability distribution to data on storage container type and age, and uses estimated corrosion rates for the various storage configurations and failure modes to calculate the projected number of failures that can be expected.

The stored drums are divided into four populations. Group 1 drums are the oldest, emplaced between 1970 and 1972, using random placement and direct soil-contact burial. Group 2 drums were emplaced between 1973 and 1975, stacked on asphalt pads, under plastic tarp covers. Group 3 drums were stored using the same method as group 2, but the model reflects the use of thicker-walled type 17-C drums during the emplacement period from 1976 to 1980. The group 4 drums were emplaced between 1981 and 1988, using galvanized drums instead of the painted steel drums used previously. The rates for Group 1, soil contact, were based on the data from Jaske [7]. The general corrosion rates for Groups 2 and 3, trench burial with tarp coverage to avoid direct soil contact with the drums, were based on the Morton study [9] results. An acceleration factor of two, for pitting corrosion, was chosen as a best estimate to represent the more rapid and more local corrosion expected to occur from impingement of wood upon the stored drums. The less rapid corrosion expected for galvanized drums, Group 4, compared to painted carbon steel drums was represented by halving the carbon steel corrosion rate as a best estimate.

The Poisson probability distribution used for the model [14], is a probabilistic description of the frequency of "arrivals" in time. In this model, an arrival is considered to be the loss of one mil (0.025 mm) of drum wall thickness, either at a particular location (where pitting is occurring) or across the entire drum surface (for general corrosion). The equation is a function of the starting drum wall thickness, the rate of corrosion, and the age of the drum. The probability of observing X arrivals in a period of time T is given by

$$P(x) = \frac{\lambda^x}{x!} e^{-\lambda} \qquad x=0,1,2,... \qquad (1)$$

and the probability of equalling or exceeding the wall thickness in time T is given by Equation 2.

$$P(x \geq L) = \sum_{x=L}^{\infty} P(x) = 1 - \sum_{x=0}^{L} P(x) \quad (2)$$

where

L = Starting total drum wall thickness, mil,
α = Corrosion rate, mil/yr,
T = Drum age, years, and
$\lambda = \alpha T$.

Thus, the model can be used to calculate the probability of a drum failure when give the original drum wall thickness, drum age, and a representative corrosion rate. The calculated probabilities are multiplied by the total number of drums in the population of interest to get the projected number of drum failures for that population. Conversely, by using a known failure rate, wall thickness, and drum age, the inferred average corrosion rate can be backcalculated.

RESULTS AND CONCLUSIONS

The model was used to calculate the total projected number of drum failures as a function of the date of retrieval. The corrosion rates listed in Table 1 were used to calculate the probability of the corresponding failure mode occurring.

TABLE 1--Drum corrosion model parameters.

Group	Emplacement time span	Drum wall thickness	Corrosion rate mm/yr (mil/yr)		Comment
			Pitting/ crevice	General	
1	1970-1972	1.27 mm (50 mil)	0.229 (9)	0.127 (5)	Painted 17-H drum, direct soil contact
2	1973-1975	1.27 mm (50 mil)	0.051 (2)	0.025 (1)	Painted 17-H drum, stacked under tarp
3	1976-1980	1.52 mm (60 mil)	0.051 (2)	0.025 (1)	Painted 17-C drum, stacked under tarp
4	1981-1988	1.52 mm (60 mil)	0.025 (1)	0.013 (0.5)	Galvanized 17-C drum, stacked under tarp

Figure 2 shows that all group 1 drums are expected to currently have pitting failures, and that groups 2 and 3 could begin to show a significant number of pitting (or possibly crevice-type corrosion caused by wood contact) failures during the next few years. Galvanized drums are not expected to be breached by corrosion for many decades.

Figure 3 shows that the group 1 drums are projected to have already lost their structural integrity from large scale corrosion effects. Loss of structural integrity is defined as occurring when the average drum wall thickness has been reduced to 0.254 mm (10 mils). Figure 3 also shows that group 2 drums could begin losing structural integrity at a significant rate soon after the year 2000. The drums in group 3 are expected to retain their structural integrity until at least the year 2015.

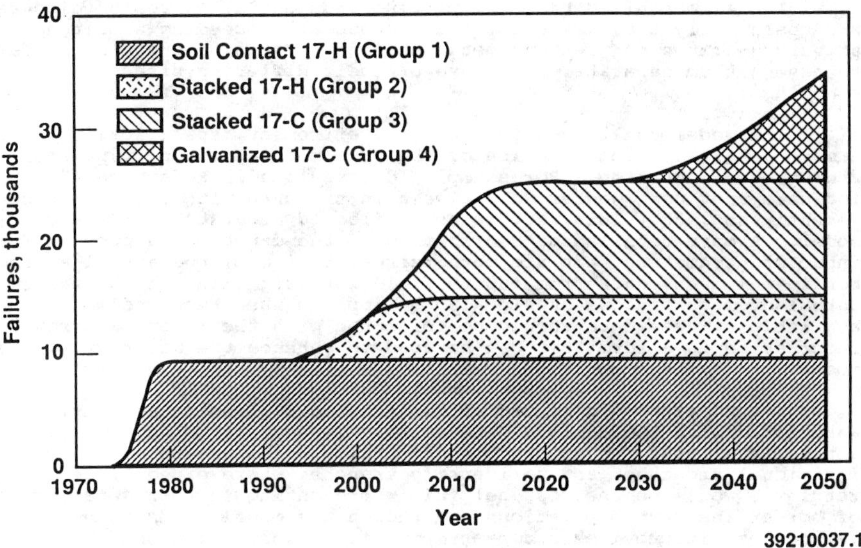

FIG. 2--Projected drum failures (pitting corrosion).

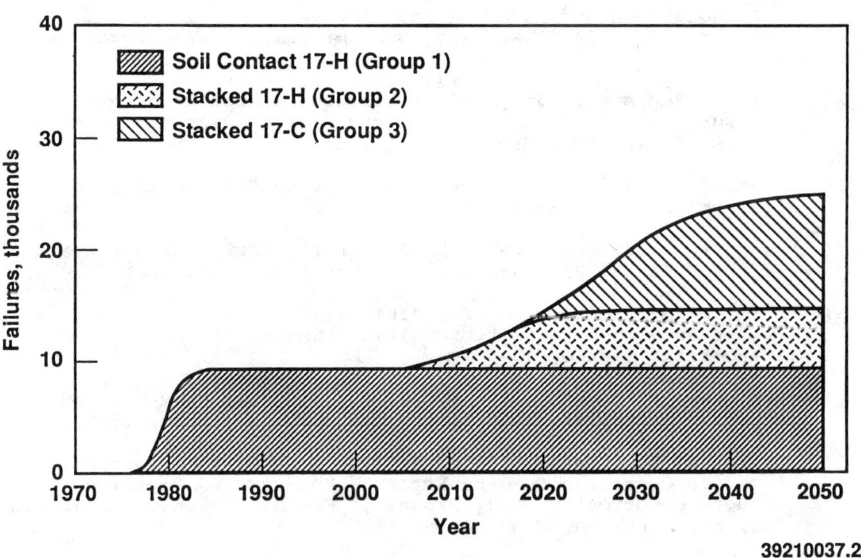

FIG. 3--Projected drum failures (loss of structural integrity).

It is recognized that the pitting and general corrosion discussed are essentially the same corrosion processes, proceeding simultaneously at varying rates on the same metal surface. It is of practical interest to assess them separately, because of their differing effects on drum handling.

The model could use some further refinement, specifically through a more detailed categorization of the drum populations by type and storage configuration. For example, discussions in Reference 15 indicate that some wastes were placed in trenches with gravel bottoms as recently as 1988. This fact is not reflected accurately in the current model. A more detailed categorization of the drums would provide for input of parameters that are more representative of the actual storage conditions. Any data from future field investigations which measure actual drum conditions should also be applied when they become available. These data should be collected with the intent of providing information in a form that can be used to enhance the accuracy of the model.

The model should be useful for strategic planning activities associated with the Hanford Site retrieval program. With the appropriate input, it can be used to estimate failure rates in individual trenches, and to identify trenches where expeditious retrieval would be the most helpful in preventing further deterioration of borderline drum populations. Although the model needs some refinement, it provides a conservative first approximation for assessing the expected condition of drums in the storage trenches. A pilot scale retrieval investigation of conditions and contents of over 200 of the drums is in the planning stage at the Hanford Site. The results of this investigation will be a check of the validity of the corrosion rate assumptions.

REFERENCES

[1] U.S. Department of Energy, "Radioactive Waste Management," DOE Order 5820.2A, U.S. Department of Energy, Washington, D.C., 1988.

[2] Federal Register, "Disposal of Hanford Defense High-Level, Transuranic, and Tank Wastes, Hanford Site, Richland, Washington, Record of Decision," Federal Register, Vol. 53, p. 12449, 1988.

[3] Bishoff, J. R., "INEL Transuranic Storage Cell Penetration and Inspection," TREE-1311, EG&G Idaho, Idaho Falls, Idaho, 1979.

[4] Bishoff, J. R. and Hudson, R. J., "Early Waste Retrieval Final Report," TREE-1321, EG&G Idaho, Idaho Falls, Idaho, 1979.

[5] Card, D. H. and Wang, D. K., "Initial Drum Retrieval Interim Report July 1974 to September 1976," TREE-1079, EG&G Idaho, Idaho Falls, Idaho, 1976.

[6] McKinley, K. B. and McKinney, J. D., "Initial Drum Retrieval Final Report," TREE-1286, EG&G Idaho, Idaho Falls, Idaho, 1978.

[7] Jaske, R. T., "Evaluation of Soil Corrosion at Hanford Atomic Products Operation Summary Report - Underground Pipeline and Structure Corrosion Study Program," HW-33911, General Electric Company, Richland, Washington, 1955.

[8] Carlos, W. C., "Underground Fuel Storage Tank Corrosion Study," WHC-EP-0507, Westinghouse Hanford Company, Richland, Washington, 1992.

[9] Morton, R. L., "Inspection of Retrievably-Stored Transuranic Waste Containers," SD-WM-TRP-002, Westinghouse Hanford Company, Richland, Washington, 1991.

[10] Romanoff, M., "Underground Corrosion," National Bureau of Standards Circular 579, National Bureau of Standards, Washington, D.C., 1957.

[11] Miller, F. P., Foss, J. E., and Wolf, D. C., "Soil Surveys: Their Synthesis, Confidence Limits, and Utilization for Corrosion Assessment of Soil," in Underground Corrosion, Special Technical Publication 741, American Society for Testing and Materials, 1981.

[12] American Society for Metals, "Metals Handbook," 9th ed., Vol. 13, American Society for Metals, 1987.

[13] Evans, U. R., "Metallic Corrosion, Passivity, and Protection," Edward Arnold & Company, London, England, 1945.

[14] Perry, et al., "Chemical Engineer's Handbook'" 6th ed. McGraw-Hill, New York, New York, 1984.

[15] Anderson, B. C., Anderson, J. D., Demiter, J. A., Duncan, D. R., and McCann, D. C., "Contact-Handled Transuranic Waste Characterization Based on Existing Records," WHC-EP-0225, Westinghouse Hanford Company, Richland, Washington, 1990.

Life Prediction Techniques in Various Applications

Ellis D. Verink, Jr.[1]

HIGH LEVEL NUCLEAR WASTE MANAGEMENT IN U.S.A.

REFERENCE: Verink, E. D., Jr., **"High Level Nuclear Waste Management in the U.S.A.,"** Application of Accelerated Corrosion Tests to Service Life Prediction of Materials, ASTM STP 1194, Gustavo Cragnolino and Narasi Sridhar, Eds., American Society for Testing and Materials, Philadelphia, 1994.

ABSTRACT: Designing a system for the safe disposal of spent fuel and high-level defense waste will require being able to predict repository performance over time spans exceeding the total known time of human existence on this planet.

Historically, primary reliance has been placed on natural (geological) barriers to prevent or retard release of radionuclides to the accessible environment. These have included crystalline rock (Sweden, Finland, Switzerland and Canada), salt domes (Germany), clay (Belgium), and tuff (U.S.A.). In all countries except the United States, the potential disposal environments are saturated with moisture. By contrast, the site at Yucca Mountain, Nevada, currently being characterized by the United States, is located in an arid climate, and the zone proposed for the repository is an unsaturated zone well above the water table. As a result, one can say that the site under evaluation in the U.S.A. is unique. It provides special opportunities and challenges with respect to radioactive waste disposal.

Recently, there has been some "erosion" of confidence in international circles about the ability of geological barriers to provide the degree of containment originally assumed. This has led to an increased interest in a number of countries in the development of engineered barriers of various kinds that work in concert with the natural barriers. Designs under consideration cover a wide range of concepts, for example, special materials for more robust waste containers; new concepts for the emplacement of waste containers--such as room or drift emplacement; chemical control of the container environment; use of thermal energy as a design parameter, and growing interest in a more "systems oriented approach" to the storage, transportation and emplacement of spent fuel and high-level waste through the use of multi-purpose containers.

Because of increased interest in multi-purpose containers and because they may allow us to improve our ability to predict radionuclide containment over long periods of time, corrosion behavior of the various materials remains a major consideration. As experiments to predict and measure corrosion behavior are designed, we must also develop statistical methods to evaluate the data collected during these experiments.

KEYWORDS: high level waste, HLW, engineered barrier system, EBS, natural barriers, Yucca Mountain, thermal-loading, hydrology

[1] Professor, Department of Materials Science and Engineering, University of Florida, Gainesville, FL 32611

INTRODUCTION

In the 1987 amendment to the Nuclear Waste Policy Act, Congress designated a site at Yucca Mountain, Nevada, for characterization. In the same legislation, Congress created the Nuclear Waste Technical Review Board to advise it and the Secretary of Energy on site characterization and on issues related to the handling and transport of spent fuel. The Nuclear Waste Technical Review Board has been functioning since March of 1989. The subject matter presented herein reflects the deliberations and observations of the Board and particularly of the panels concerned with the Engineered Barrier System (EBS) and the Transportation and Systems (T & S). The Board is mandated to report to Congress and the Secretary of Energy at least twice a year. To date, five such reports have been published and are available to interested parties.

In the early 1980's, the Nuclear Regulatory Commission established geologic repository criteria (10CFR60) that include radionuclide containment standards for engineered barriers (here reference is primarily to the waste package and any backfill immediately surrounding the waste package). The requirement was for containment for 300 to 1000 years. The Environmental Protection Agency (EPA) set environmental radiation protection standards for spent fuel disposal (40CFR191) for the total repository (i.e., engineered and natural barriers within the repository). It is the Board's view that these should be considered as minimum standards and that the waste management program should strive to exceed them to enhance safety and increase public confidence.

As presently contemplated, the repository would be located approximately 300 meters below the eastern flank of Yucca Mountain, would occupy approximately 1520 acres, and would contain approximately 100 miles of underground tunnels. Congress has mandated a limit of 70 000 metric tons of spent fuel and high-level waste for this proposed site. The baseline plan calls for vertical boreholes in the floors of drifts. The repository would remain open to permit retrieval of emplaced waste packages for 50 years following initial emplacement (according to 10CFR60). The areal power density (APD), a measure of thermal-loading at the time of emplacement, would be approximately 57 kilowatts per acre. A requirement of EPA standard 40CFR191 is that the repository system must isolate radionuclides from the accessible environment effectively for 10 000 years.

BASELINE THERMAL-LOADING STRATEGY

With these parameters in mind, the temperature of the host rock adjacent to the waste package would exceed the boiling point of water shortly after emplacement and remain above the boiling point of water for a period of 300 to 1000 years, at which time the temperature would fall below the boiling point of water. According to this concept, the above-boiling temperatures would drive away any moisture that might otherwise reach the containers and presumably should prevent or at least greatly retard aqueous corrosion for at least 300 years. This has not yet been tested

The geology of portions of the Yucca Mountain being characterized for suitability for locating a repository consist of tufaceous rock (consolidated volcanic ash) located in an unsaturated zone that is as much as 760 meters thick. In the event that waste containers in a repository are breached, the most likely pathway for non-gaseous radionuclides to the accessible environment is thought to be via groundwater. Of course, gaseous releases do not require the presence of liquid water. Locating the repository in an unsaturated zone will result in increased confidence that liquid water will be kept away from waste containers for a period of time, thus preventing or at least greatly retarding aqueous corrosion of containers for that period. In addition, it is believed that maintaining temperatures above the boiling point of water around the containers will keep the containers dry for the duration of the time above boiling, and for a considerable time thereafter, and thereby retard aqueous corrosion.

(1) According to the base line plan, there will be two thermal regimes along with their associated uncertainties: above-boiling for 300 to 1000 years and below-boiling for the period thereafter.

(2) The baseline strategy incorporating relatively short-lived containers relies primarily on geologic barriers to isolate the radionuclides beyond the 300 to 1000 year period.

(3) Both the technical and non-technical communities might perceive the current above-boiling strategy as entailing greater uncertainties with regard to geotechnical, hydrologic, and geochemical aspects than below-boiling strategies because many conceptual aspects associated with the above-boiling conditions have yet to be tested and validated.

(4) The baseline strategy involves young, spent fuel and requires a substantial hot cell for fuel handling to obtain the appropriate mixtures of burn-up and age. The attendant uncertainties and human factors, seismic and other risks, as well as system costs must be recognized.

ALTERNATIVES TO THE CURRENT U.S. BASELINE STRATEGY

Below-Boiling Strategy

The below-boiling strategy does not rely on a thermal shield to protect the waste package from liquid water. In return, however, the below-boiling strategy gives a temperature regime where rock properties and behaviors may be more predictable; it also leads to less thermal expansion of the rock and, concomitantly, to lower levels of stress within the host rock. The below-boiling strategy can be achieved by a combination of spent fuel aging, placing less waste in each package, and/or wider spacing of waste packages in the repository. A potential disadvantage could be the reduction in the ultimate capacity of the repository or the need for more repository area outside the currently envisioned 1520 acres. Because temperatures are below the boiling point of water, materials such as bentonite, a natural clay, could be used to surround the waste containers. Properly placed bentonite is virtually impervious to water, however, bentonite might lose its sorptive properties and impermeability if long-term temperatures remain above boiling.

Long-Term Above-Boiling Strategy

The long-term above-boiling strategy[2] postulates that if it proves beneficial to keep moisture away from the waste package for 300 to 1000 years, why not keep the waste at above-boiling temperatures for 10 000 years? This, of course, is based on the same unproven theory as the U.S. baseline strategy....that keeping moisture away from the waste package will prevent aqueous corrosion and transport of non-gaseous radionuclides.

Initial very high rate of heat generation of young fuel decreases significantly during the first 60 or so years of aging, although the waste still remains highly radioactive. If the waste were aged for 60 or more years, with correct repository loading configuration, rock temperatures near the waste packages would remain above-boiling for periods in excess of 10 000 years. According to this strategy, moisture would be kept away from the waste for over 10 000 years thereby extending protection to cover the 10 000-year period required in 40CFR191. There are other implications for this strategy with regard to the system as a whole.

(1) The 1520-acre repository planned for Yucca Mountain could accommodate spent fuel in excess of 150 000 metric tons (rather than being limited to 70 000 metric tons), thereby possibly eliminating the need for a second repository site or at least postponing it for several decades.

(2) To reduce the high initial temperatures and maintain above-boiling conditions, the spent fuel would have to be aged for approximately 60 years before emplacement and packed more densely in the repository.

(3) Like the current U.S. baseline strategy, the long-term above-boiling strategy might be perceived by the technical and non-technical communities as entailing more uncertainties than the below-boiling strategy.

WASTE PACKAGE ISSUES AND UNCERTAINTIES

The present reference design calls for radioactive waste going into the repository in two forms: (1) waste packages containing spent-fuel assemblies encased in a thin-walled metal container, and (2) high-level waste from reprocessing incorporated in borosilicate glass, encased in a thin-walled metal canister with a thin-walled metal container overpack. The effects of thermal-loading strategies on the waste package can be evaluated by investigating how the different strategies will affect the performance of the container material, the zircaloy cladding (the outer shell of the individual fuel rods that make up the fuel assemblies), the fuel pellets in the fuel rods (in the case of spent fuel), and the borosilicate glass within the canister (in the case of high-level waste). Two generic thermal-loading strategies may be considered: above-boiling, which would provide a hotter/dry environment, and below-boiling, which would provide a cooler/wet environment.

1. Below-Boiling Strategies

(a) In the below-boiling regime, depending on the alloy selection, generalized corrosion, localized corrosion such as pitting and crevice corrosion, stress corrosion cracking, hydrogen embrittlement, and even microbiological corrosion conceivably could take place, leading to an early release of radionuclides. Thin-walled metal containers provide little shielding to contain the radiation emanating from the waste itself. Radiolysis could occur, the products of which may exacerbate corrosion damage. Although some radiolysis could occur at above-boiling temperatures, the potential adverse effects would be more severe because of the assumed presence of liquid water in the below-boiling regime.

(b) The potential concerns for zircaloy (fuel) cladding in the below-boiling regime, assuming that moisture can reach the fuel assemblies, are similar to those above given for the container material.

(c) In both the below- and above-boiling regimes, if oxygen is present, oxidation of fuel pellet fragments of UO_2 to U_3O_8 may take place. The consequent decrease in spent-fuel particle size and increase in spent-fuel surface area can, in the presence of water, lead to an increase of colloid formation and dissolution rate. This could be the case in the below-boiling environment if both the containers and the cladding were breached. However, oxidation of the fuel pellet fragments may proceed more slowly in this environment than in the above-boiling one.

(d) If both the containers and canisters have been breached, the possibility of water in the below-boiling regime creates the potential for dissolution and hydrolysis of borosilicate glass, which in turn could result in the mobilization of radionuclides contained in the glass (the rate of hydrolysis may be slower at these low temperatures).

2. Above-Boiling Strategies

(a) At elevated temperatures, there could be microstructural changes caused by long-term aging, and possibly grain-boundary migration which could possibly weaken the container and increase its susceptibility to corrosion. In addition, if liquid water somehow came in contact with the hot container and evaporated, mineral deposition could exacerbate corrosion. On the other hand, above-boiling temperatures might encourage the growth of surface oxide layers, which could provide greater protection against container corrosion once temperatures eventually fall below-boiling. The above-boiling temperatures also could help relieve residual fabrication stresses for the containers. Because of the absence of liquid water in the above-boiling regime, existing and well-known general oxidation models probably could be used with some confidence to predict container performance over long periods of time, even more than 10 000 years.

(b) The potential concerns for zircaloy cladding in the above-boiling regime are the same as those for the container material. Hydride precipitation in the cladding is an additional concern since it could lead to premature failure. On the other hand, at higher temperatures associated with the above-boiling regime, cladding stresses (usually the result of radiation hardening) also could be relieved. As with the container material, growth of a zircaloy oxide layer probably could be modeled with some confidence.

(c) If both the containers and cladding have been breached and oxygen is present, oxidation of fuel pellet fragments may occur faster at higher temperatures than in below-boiling environments. A period of concern could occur after the above-boiling temperatures have been lowered, if liquid water were to come in contact with oxidized fuel. The increased spent fuel surface area resulting from oxidation may enhance radionuclide transport by dissolution and colloid formation. This period could occur after 300 to 1000 years in the current baseline thermal-loading strategy and after 10 000 years in the proposed extended above-boiling strategy. Using thick-walled robust containers would isolate waste for long periods of time and could mitigate these concerns.

HYDROLOGIC ISSUES AND UNCERTAINTIES

A key scientific issue associated with thermal-loading is to determine the extent to which the different thermal-loading strategies can increase or decrease the likelihood that groundwater will come in contact with radioactive waste containers and subsequently serve as a vehicle to transport radionuclides to the accessible environment.

1. Below Boiling Strategies

For any strategy that assumes below-boiling temperatures during any part of, the 10 000-year regulatory lifetime, the fundamental problem becomes one of understanding and characterizing the essentially ambient hydrologic regime during that period. This presents four important questions which must be addressed:

(a) What is an appropriate conceptual model for describing ground-water flow in an unsaturated zone such as the one beneath Yucca Mountain?

Although modeling saturated flow in porous media is considered straightforward, the modeling of unsaturated flow in fractured media is much more difficult. Firstly, flow of water between the rock matrix and the fractures that separate that matrix is complex and not easily measured in the field. Secondly, in saturated porous media, the flow may attain steady-state conditions, however the flow in unsaturated fractured media is transient and episodic.

(b) Assuming the development of a satisfactory conceptual model for ground-water flow, the next question is can this model be expressed as a computational model that will allow meaningful prediction of ground-water flow over the next 10 000 years? (To date, most ground-water modelling has relied on simple models).

To date, most modeling of flow in fractured media has made use of the equilibrium continuum model. This model assumes that flow properties of both fractures and matrix can be approximated by using a porous medium with gross properties chosen to reflect the presence of fractures. A somewhat more sophisticated model is the dual porosity model, which assumes two average porosities, one reflecting the rock matrix and the other representing the fractures. The approach which appears to be most direct is use of the discrete fracture model, which assumes and makes use of knowledge of individual fracture locations and characteristics. Although this latter model may be more accurate, it represents a major computational effort, even in two dimensions, let alone in three dimensions. To date these models have been used primarily for theoretical studies.

(c) Assuming that a computational model is feasible, can we obtain all the porosity, permeability, fracture, saturation, etc., data needed to run the model with confidence over the entire space affected by the repository?

Exploration at depth by means of an exploratory studies facility (ESF) will greatly increase our knowledge of the existing fracture network, however, the question remains whether it will be sufficient to provide a robust and unchallengeable view of the flow regime. Presumably extrapolation of the data away from the exploratory drifts always will be necessary.

(d) A related problem is with determining the upper and lower bounds of future climate in the Yucca Mountain region.

This will be necessary to establish the precipitation (and the resultant infiltration) of water into the repository block. In the geologic past, wetter climates have existed in the region during glacial periods.

2. Above-Boiling Strategies

According to proponents of the above-boiling strategy, the main hydrologic advantage is the creation of a dry-out zone within the repository such that no water can reach the waste containers[3]. Presumably, this will avoid most of the problems and uncertainties associated with characterizing the ambient hydrologic regime. An additional level of assurance would be provided by the creation of a hydrothermal "umbrella," or umbrellas, above the repository. This umbrella effect assumes that when the water vapor driven away from the repository eventually reaches cool enough rock and condenses, it will flow down the edges of the repository as if there were a giant umbrella protecting the repository. A modification of this umbrella effect would occur if the thermal effects were more localized. In this case, the condensed water vapor would flow through the repository between, and therefore avoiding, the heated regions around individual waste packages. This raises several questions.

(a) How valid are predictions of an extensive, long-lasting dry-out region when they are, as is currently the case, based on relatively simple models?

(b) How well can *in-situ* testing validate predictions of extensive, long-lasting dry-out?

(c) If the current baseline thermal-loading strategy is adopted, both above-boiling and below-boiling environments will have to be accounted for. This raises at least two questions: (i) will the need to characterize two different hydrologic regimes, both of which change in time and space, result in a large increase in uncertainty?, and (ii) will the below-boiling hydrologic regime, occurring after extensive heating of the rock, be significantly different from the current ambient regime?

(d) The existing site suitability (10CFR960) and licensing (10CFR60) regulations include groundwater travel time criteria that call for a pre-emplacement travel time of greater than 1000 years from the disturbed zone to the accessible environment. The "disturbed zone" is defined as that portion of the surrounding rock whose physical or chemical properties have changed as a result of construction or "as a result of heat generated by the emplaced radioactive waste such that the resultant change in properties may have significant effect on the performance of the geological repository" (10CFR60). Scientists need to evaluate how a thermally induced extension of the disturbed zone will affect the ability of the Yucca Mountain site to satisfy this criterion.

THE ENGINEERED BARRIER SYSTEM

The "Engineered Barrier System" (EBS) includes all barriers designed or engineered by humans. As such, the engineered barrier system refers to all components of the waste disposal system that have been designed, engineered, or constructed to prevent the release of radionuclides into the accessible environment. Components of the EBS include the waste form (spent-fuel-rod assemblies or borosilicate-glass which contains, high-level waste from reprocessing); the waste package, which includes the waste form and any other containers; any material placed over and around the waste package; and materials that could be used to backfill the openings in a repository. This EBS working together with the natural, geologic setting, is expected to prevent, or greatly retard, the migration of radionuclides to the accessible environment. Given this system definition, the repository design and the choice of thermal-loading strategy play critical roles in the design of the EBS.

One approach to achieving such a program might be to design a robust long-lived waste package that could be shown to have a reasonable assurance of containing radioactive wastes of all forms (solid, liquid and gaseous) for thousands of years. This would provide a repository system of redundant natural and engineered barriers to radionuclide release for a much longer period of time. The use of such a multi-barrier, defense-in-depth system would appear to be an effective way to improve scientists' ability to confidently predict repository performance over long periods of times of the order of thousands of years.

The choice of a thermal-loading strategy for the repository is of great significance to the design and fabrication of radioactive waste containers. In fact, it is potentially advantageous to consider the repository thermal-loading as a constructive attribute of the engineered barrier system. For example, it may be desirable to keep the entire repository waste inventory dry for over 10 000 years by "engineering" a relatively high thermal load. Or perhaps maintaining consistent conditions at either high or low temperatures will prove useful for reducing uncertainties about containers lasting thousands of years. The type of container chosen for emplacement in a repository also will affect, to varying degrees, many other parts of the waste management system. Research aimed at verifying and optimizing the roles of various components of the engineered barrier system (including thermal loading) has not yet been carried out and is sorely needed.

A thorough understanding of the hydrogeology is critical to early determination of site suitability. It is also critical to decisions pertaining to thermal-loading strategy, repository design, waste package design, and other elements of the waste management system. Early assumptions have been that the natural (geologic) barriers would have the capability to isolate radionuclides for very long periods of time. However, drilling and subsurface studies might show that the long-term capabilities of the natural geologic barriers to isolate radioactive waste may be more uncertain and harder to predict than some may have supposed. If this were to be the case, it is crucial that data be available on potential contributions of alternative waste package designs and other engineered barriers.

The baseline waste package is designed to meet minimum regulatory requirements (300 to 1000 years of "substantially complete containment" according to 10CFR60). The Board has seen nothing to indicate that a waste package lifetime of 10 000 years, and perhaps more, would be unattainable in the U.S. program. However, achieving such a goal with an appropriately high confidence level requires new experimental data, particularly from multi-year experiments. Such experiments must begin soon for results to be ready in time to meet the current schedule for repository application in 2001. Although engineered barriers may have shorter lifetimes than do natural geologic barriers, it may be possible to predict the performance of engineered barriers with greater confidence than is possible for natural geologic barriers. The current lack of emphasis on developing and testing long-lived waste packages and other elements of the engineered barrier system may prove to be counterproductive to the goal of building public trust and confidence.

Although robust long-lived containers may prove to have a higher first cost than thin-walled containers such as used for the baseline design, they could provide overall system cost reductions. For example, costs could be reduced through use of alternative emplacement modes.

A redundant multi-barrier system could offer more flexibility to the overall waste management system as it evolves. For example, initial studies of repository configuration by Swedish scientists (KBS II studies) originally contemplated emplacement of waste packages in vertical boreholes in the floors of mined tunnels (similar to DOE current strategy). However, for a number of reasons, such as lower excavation costs, the possible need to be able to retrieve the waste, and increased flexibility to modify the environment, there has been a notable shift towards considering "drift", or "room" emplacement rather than borehole emplacement. There also would be advantages for developing robust long-lived packages which are self-shielding and would facilitate the use of drift or room emplacement, thereby providing direct and simple solutions to other potential safety problems.

Recognizing that overall near- and long-term safety is a primary goal, and that the repository should be based on multibarrier, defense-in-depth principles, I conclude with the following observations:

CONCLUDING OBSERVATIONS

1) Substantial underground excavation is critical to early determination of site suitability. It also is critical for the collection and analysis of the isolation potential of the natural (geologic) barrier.

2) Consideration of both below-boiling and above-boiling conditions provides huge conceptual and experimental challenges for materials scientists faced with modelling corrosion behavior of waste-package materials over periods of thousands of years.

3) The research and development program aimed at the development of the engineered barrier system should be enhanced and receive higher priority, giving emphasis to the development of robust, long-lived containers.

4) The immediate surroundings of the emplaced waste containers represent an opportunity to reduce uncertainties about waste package stability by controlling the local chemical environment.

References

1. Fifth Report to the U.S. Congress and The Secretary of Energy from The Nuclear Waste Technical Review Board, NWTRB, 1100 Wilson Blvd., Suite 910, Arlington, VA 22209, June 1992.

2. Ramspott, L. Strategic Implications of Heat in a High-Level Radioactive Waste Repository, Lawrence Livermore National Laboratory, as presented to the NWTRB October 8, 1991.

3. Buscheck, T.A. and Nitau, J.J., The Impact of Thermal Loading on Repository Performance at Yucca Mountain, Proceedings of the Third Annual International Conference on High Level waste Management, April 2-6, 1992, Las Vegas, NV.

Brian M. Ikeda,[1] M. Grant Bailey,[1] Michael J. Quinn[1] and David W. Shoesmith[1]

THE DEVELOPMENT OF AN EXPERIMENTAL DATA BASE FOR THE LIFETIME PREDICTIONS OF TITANIUM NUCLEAR WASTE CONTAINERS

REFERENCE: Ikeda, B. M., Bailey, M. G., Quinn, M. J., and Shoesmith, D. W., "The Development of an Experimental Data Base for the Lifetime Predictions of Titanium Nuclear Waste Containers," Application of Accelerated Corrosion Tests to Service Life Prediction of Materials, ASTM STP 1194, Gustavo Cragnolino and Narasi Sridhar, Eds., American Society for Testing and Materials, Philadelphia, 1994.

ABSTRACT: To predict the expected lifetimes of nuclear waste containers under Canadian disposal vault conditions, specific criteria upon which a predictive model can be based must be developed. The anticipated evolution of the corrosion of titanium waste containers is described. Failure is most likely to occur by a combination of crevice corrosion, hydrogen-induced cracking and general corrosion and, for long lifetimes, the duration of crevice corrosion must be limited. The crevice corrosion of Grade-2 and Grade-12 titanium has been studied using a galvanic coupling technique. The propagation of crevice corrosion on the Grade-2 material is dependent on both temperature and oxygen concentration and repassivation occurs once the oxygen is consumed. The propagation of crevice corrosion on the Grade-12 alloy appears almost independent of temperature, and repassivation occurs for T ≤73°C even in the presence of copious amounts of oxygen. The implications for predicting the lifetimes of containers under disposal conditions are discussed.

KEYWORDS: nuclear waste containers, crevice corrosion, Grade-2 titanium, Grade-12 titanium, hydrogen-induced cracking, general corrosion, crevice repassivation.

The need to predict the lifetimes of nuclear waste containers requires the development of models capable of predicting corrosion propagation to failure [1]. Commonly this involves either the statistical analysis of penetration depths [2-4] or the development of mechanistic models requiring substantial information on the electrochemical properties of the material and the solution properties

[1]AECL Research, Whiteshell Laboratories, Pinawa, Manitoba, ROE 1LO, CANADA.

of the dissolved metal cations [5-7]. A major problem with this approach is the accumulation of a meaningful data base upon which to construct a predictive model. If such predictions are to be justified, then it is necessary to know both the geometric mode and the depth of penetration by a particular corrosion process as a function of the controlling parameters. The situation is complicated by the expectation that waste vault conditions (e.g., temperature, redox potential) will change over the lifetime of the container.

In this paper we present: (i) a description of the corrosion processes expected on waste containers fabricated from titanium alloys, especially Grades-2 (Ti-2) and -12 (Ti-12) under Canadian waste vault conditions, and how they are expected to change as vault conditions evolve; and (ii) a description of our approach to establishing a data base upon which to predict the lifetimes of these containers. Emphasis is placed on understanding the effects of temperature and oxygen concentration and on developing modelling criteria.

Although not discussed here, copper is an alternative candidate material in the Canadian Nuclear Fuel Waste Management Program (CNFWMP) [1].

Corrosion under Canadian Waste Vault Conditions

In the reference disposal concept it is proposed that titanium containers be placed in granitic rock at a depth of 500 to 1000 m. These containers would be embedded in a compacted mixture (1:1) of sodium bentonite clay and silica sand, which will eventually become saturated with saline groundwater ($[Cl^-]$ ~0.1 to 1.0 $mol \cdot L^{-1}$). The processes most likely to lead to failure of these containers are crevice corrosion, hydrogen-induced cracking (HIC) and general corrosion. For low hydrogen concentrations the occurrence of HIC by either slow crack growth processes, such as sustained load cracking, or fast crack growth processes, such as overload, is avoidable by prudent engineering of the container because high stress intensities are required at low hydrogen concentrations [8,9]. Only when hydrogen contents in the metal are $\gtrsim 500$ $\mu g \cdot g^{-1}$ does HIC appear to be a feasible failure process, and then only by the fast cracking associated with embrittlement [8]. Since hydrogen levels in as-received materials are generally in the order of tens of $\mu g \cdot g^{-1}$, significant absorption of hydrogen appears a prerequisite before HIC becomes a potential failure mechanism.

Recently, we published a model to predict the lifetimes of Ti-2 containers [10]. In this model, either crevice corrosion or HIC is predicted to fail the container. To avoid underestimating failure times, this model contains a number of pessimistic assumptions. It is assumed that crevice corrosion initiates rapidly on all containers (the vault could contain up to 140 000) and that a constant supply of oxidant, sufficient to sustain crevice propagation indefinitely, is maintained over the lifetime of all containers. As a consequence of these conservatisms, >97% of containers within the vault are predicted to fail by crevice corrosion between 700 and 10^4 years.

In reality, indefinite propagation of crevice corrosion is unlikely since available oxidants will be consumed by reaction with oxidizable minerals as well as by the corrosion process itself. The expected evolution of conditions within a Canadian vault is illustrated in Fig. 1 which shows: (i) the calculated change in surface temperature

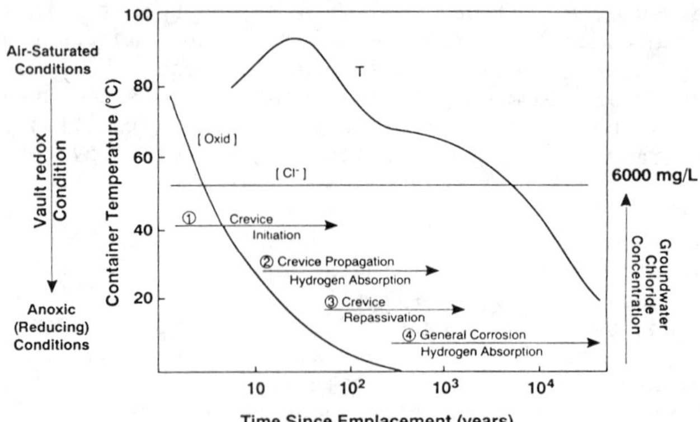

Fig. 1--Evolution of environmental conditions expected within a Canadian waste vault. The arrows marked 1 to 4 denote the stages of container corrosion expected.

with time of a container in the hottest part of the vault (i.e., in the middle); (ii) the assumed change in redox conditions expressed as a concentration of available oxidant (in this case, O_2 dissolved in groundwater); and (iii) the chloride composition of the groundwater.

The temperature of the containers can be calculated and depends on the age of the fuel, the vault geometry and the thermal properties of the materials surrounding the container. Initially, redox conditions will be established by the air and other oxidants trapped within the surrounding materials when the vault is sealed; as these oxidants are consumed, predominantly by the oxidation of Fe^{II}-containing minerals (e.g., biotite, pyrite), conditions will gradually become anoxic. The time for this to occur has been conservatively estimated to be ~300 years. The groundwaters of the Canadian Shield, the proposed location of a waste vault, are chloride-dominated [11]. Although the groundwater compositions are not expected to vary significantly with time, they may vary from site to site. The chloride concentration (6000 $mg \cdot L^{-1}$) shown in Fig. 1 is the reference groundwater composition used in the CNFWMP and is typical of Canadian Shield groundwaters, although concentrations up to 34 000 $mg \cdot L^{-1}$ also are commonly encountered [10,11].

Periods 1 to 4 in Fig. 1 denote the stages of evolution of corrosion processes expected on Ti-2 and Ti-12 under waste vault conditions. Period 1, extending over the first 100 to 200 years of emplacement, covers the interval when the container temperature and the vault oxygen concentration are high and the initiation of crevice corrosion most probable. The exact duration of this period is difficult to specify since the time of initiation of crevice corrosion is unpredictable and often irreproducible under normal corrosion test conditions. According to Schutz and Thomas [12], initiation does not occur for temperatures below ~70°C, although other authors have initiated crevice corrosion under extremely saline conditions at temperatures as low as ~40°C [13]. Unfortunately, the accumulation of a data base sufficiently extensive to predict the probability of crevice initiation would be an extremely tedious, if not impossible, task. The development of models to predict whether crevice initiation is possible [14] is impeded by the lack of data on complexation and hydrolysis of dissolved Ti^{4+}. To avoid these complications we have retained the conservative assumption that crevice corrosion initiates rapidly on all containers. Since the duration of Period 1, when initiation is favoured, is short compared with the achievable container lifetimes, little advantage is to be gained by assuming a distribution of initiation times over this period.

The expected duration of crevice propagation is indicated by Period 2, Fig. 1. The main parameters controlling the rate of penetration by crevice propagation throughout this period are: (i) the supply of oxidants to surfaces external to the crevice; (ii) the temperature; (iii) the geometrical mode of penetration within the creviced area; (iv) the composition and microstructure of the titanium alloy; and, to a lesser degree, (v) the composition of the groundwater. Substantial hydrogen absorption will also occur because of proton reduction on active metal surfaces in the acidic environment established within the crevice [14,15].

Eventually, a decrease in temperature and the consumption of available oxidant should stop propagation and induce repassivation of the crevice, Period 3 in Fig. 1. The electrochemical results of Iki and Tsujikawa [16] show that repassivation of Ti-2 should occur for temperatures of 70°C and 60°C depending on whether the groundwater concentration is 0.1 $mol \cdot L^{-1}$ or 1.0 $mol \cdot L^{-1}$ respectively. However, as we will show below, crevice propagation remains possible, but perhaps only temporarily, at even lower temperatures providing sufficient oxidant is still available.

After repassivation, corrosion will continue by the reaction of titanium with water under anoxic conditions to produce an oxide film of increasing thickness, Period 4 in Fig. 1. If film formation follows a logarithmic growth law then the corrosion rate would rapidly become negligibly slow. A more conservative possibility is that an initially formed amorphous film crystallizes into the thermodynamically stable rutile form, and the introduction of ionic transport pathways along grain boundaries will lead to a corrosion rate which is, at worst, linear with time. Whatever the form of the general corrosion kinetics, the results of Mattson and Olefjord [17] show the corrosion rate will be extremely slow in compacted clay, over 6 years of exposure at 95°C being required to produce films 12 to 23 nm in thickness. It is possible that the hydrogen produced by this reaction of Ti with water could be absorbed by the metal, thereby contributing to the formation of hydrides as the temperature falls and to the possibility of eventual container failure by HIC. Since the hydrogen absorption rate cannot exceed the general corrosion rate, this contribution is expected to be very small. Absorption of hydrogen produced by other reactions within the vault (e.g., the radiolytic decomposition of water) will be negligible.

Figure 2 summarizes schematically the expected relationships between crevice corrosion and general corrosion, the extent of hydrogen absorption compared with the amount required for HIC, and the total container wall penetration compared with the engineered corrosion allowance (4.2 mm). A combination of crevice corrosion and general corrosion will lead to failure once the wall penetration exceeds the corrosion allowance. Failure by HIC is assumed to occur once the

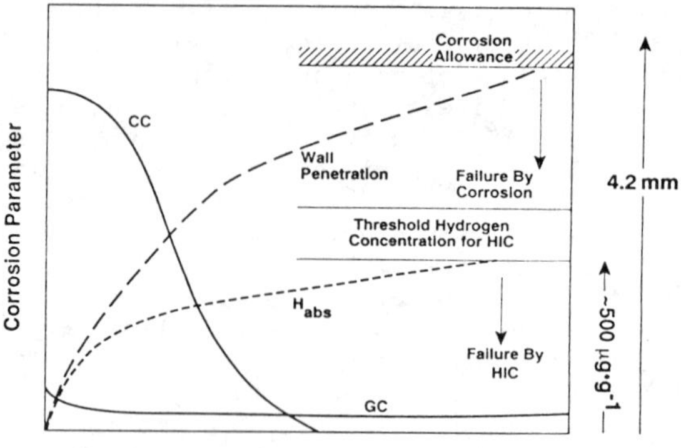

Fig. 2--Schematic summarizing the expected relationships between (i) the container corrosion rate by either crevice corrosion (CC) or general corrosion (GC), (ii) the amount of hydrogen absorbed (H_{abs}) compared with the concentration of hydrogen in the metal necessary to render it susceptible to hydrogen-induced cracking (HIC), and (iii) the total container wall penetration compared with the engineered corrosion allowance. The two vertical arrows denote the expected failure times by corrosion or HIC due to combinations of the various corrosion processes.

hydrogen content of the metal exceeds the threshold value, irrespective of the depth of wall penetration by other processes. This last assumption acknowledges that once crack growth starts, its rate will be sufficiently high that failure by HIC will be effectively instantaneous compared with container lifetimes. Clearly, it is the rate and extent of crevice propagation that will have the greatest impact on container failure, and the best way to ensure long lifetimes is to limit the extent of this process. This will occur naturally for Ti-2 as the temperature decreases and oxygen is consumed, but can be accelerated by choosing an alloy, such as Ti-12, which is designed to be resistant to crevice corrosion.

In this paper we discuss our attempts to develop an experimental data base to determine the period for which crevice corrosion may propagate. A detailed discussion of the criteria for the occurrence of HIC has been published elsewhere [8].

Experimental

A galvanic coupling technique was used to follow the progress of crevice corrosion [18]. To simulate the conditions expected for an occluded cell on the wall of a waste container, an artificially creviced working electrode was coupled to a large Ti-2 counter electrode through a zero-resistance ammeter or a potentiostat acting as a zero-resistance ammeter. The artificially creviced area was ~7 cm^2, and the ratio of this area to the cathodically coupled area on the counter electrode was approximately 1:40. The coupled current (I_c) flowing through the ammeter is a measure of the rate of oxygen consumption on the passive counter electrode. Since crevice corrosion is also driven by proton reduction on active surfaces within the crevice - a short-circuited process producing no current through the ammeter - the total extent of crevice corrosion has to be calculated from the weight change observed on the working electrode. The ratio of the total amount of charge consumed, Q (= $\int I_c dt$), to the total weight change, q (expressed as a charge from Faraday's Law), is a measure of the fraction of the total corrosion supported by O_2 reduction.

The crevice potential (E_c) and the potential of a second non-creviced planar electrode (E_p) were measured against an internal Ag/AgCl (0.1 mol·L^{-1} KCl) reference electrode [19] using a high-input-impedance voltmeter. This reference electrode has a potential of + 200 mV vs. the standard hydrogen electrode (SHE) at 150°C (223 mV vs. SHE at 125°C and 245 mV vs. SHE at 100°C). All potentials are quoted against Ag/AgCl at the experimental temperature.

The creviced specimen consisted of two planar coupons sandwiching a PTFE crevice former, since little success was achieved when attempting to initiate crevice corrosion using metal-metal crevices. It is possible that F^- ions leached from the PTFE crevice former aided the initiation process. The materials used were commercially available Grades-2 (Fe, <0.08 wt.%; H, 0.0026; O, 0.2; Ni, <0.003; Mo, <0.002; Ti balance) and -12 (Fe, 0.1; H, 0.006; O, 0.13; Ni, 0.69; Mo, 0.27; Ti balance) titanium plate. The dimensions, preparation and assembly of the metal-PTFE creviced specimens have been described in detail [20-22]. Experiments were conducted in 500 mL of air-purged NaCl solution. For experiments conducted in a 1-L titanium pressure vessel, the vessel was sealed with an ambient atmosphere of air (at room temperature) and an overpressure of argon (to prevent boiling). Details of the pressure vessel arrangement and assembly have been published previously [18]. Experiments conducted in a glass cell were continuously purged with a gas stream of known oxygen content using ultra-high-purity gases.

Experiments to determine the effect of changes in temperature on crevice propagation were performed in the pressure vessel under conditions where the oxygen concentration was maintained almost constant. Crevice corrosion was initiated at the highest temperature employed and the coupled current was allowed to achieve an approximately constant value, i.e., a condition of steady propagation at an approximately constant creviced area. The temperature was then decreased in steps and the I_c measured as a function of temperature. These experiments were performed over a period of time short enough to avoid significant changes in oxygen concentration. A similar approach was used in experiments to determine the effect of changes in oxygen concentration. Since it was much easier to change the oxygen purge gas than the oxygen concentration in a sealed pressure vessel, these last experiments were performed in the open glass cell at temperatures below 100°C.

Following the experiments, the specimens were dried and weighed. The extent of surface corrosion as a function of depth of corrosion penetration was determined using a combination of metallographic polishing and image analysis techniques as previously described [23,24].

RESULTS AND DISCUSSION

General Crevice Corrosion Behaviour

Figure 3 shows the general form of I_c, E_c and E_p from galvanically coupled experiments on Ti-2. Examples of such curves have been published previously [9,18,21-23]. After a short induction period (generally only a few hours) initiation occurs and I_c rises as the potential falls. This increase in current is due to the onset of active corrosion on an increasingly large surface area of metal within the artificially creviced area. Beyond the peak the current falls as oxygen is consumed, eventually falling to zero when the crevice becomes starved of available oxygen. Inevitably, the crevice then repassivates and E_c converges with E_p [9,24]. The decrease in E_p is a response of the passive planar electrode to the depletion of O_2. Although the lines in Figure 3 are only schematic, generally E_p does not change markedly with O_2 concentration until significant O_2 depletion has occurred as indicated in the Figure. Apparently the potential-determining reaction in the passive region is not very dependent on O_2 concentration. The major shift in E_p on depletion of O_2 may be attributable to a change in nature of the passive film. The shaded areas in Fig. 3 show the potential ranges measured at 125°C for corrosion under various conditions. The area marked A shows the corrosion potential range measured for general corrosion under oxidizing conditions. At 125°C pitting potentials for titanium in chloride solutions are in the region +3 to +5 V [12]. Obviously, pitting is extremely unlikely under vault conditions, where temperatures will be much lower (Fig. 1), and pitting potentials even larger (+5 to +7 V around 100°C) [12].

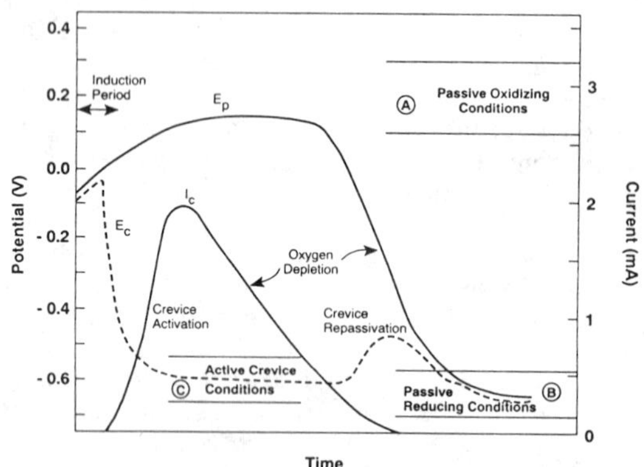

Fig. 3--The general form of the crevice current (I_c), crevice potential (E_c) and the planar potential (E_p) measured using the galvanic coupling technique: (A) range of corrosion potentials (E_p) measured for general corrosion under oxidizing conditions for T ~125°C; (B) range of corrosion potentials (E_p) measured for general corrosion under anoxic (reducing) conditions at ~125°C; and (C) range of crevice potentials (E_c) measured on fully activated crevices for Grade-2 titanium in 0.27 mol·L^{-1} NaCl solutions at ~125°C.

Area B in Fig. 3 shows the potential range for general corrosion under reducing conditions, when corrosion will involve the reaction of titanium with water. These potentials must be at, or more negative than, the potential at which the reduction of water is thermodynamically possible. At such potentials, titanium hydrides are thermodynamically stable with respect to the metal [25], and the passive film can only be considered a transport barrier, not an absolute barrier, to hydrogen absorption by the metal. As discussed above, the rate of hydrogen absorption cannot exceed the corrosion rate, which has been shown to be extremely small [17,26].

Area C in Fig. 3 indicates the crevice potentials measured for a fully activated crevice undergoing propagation. Inevitably, for Ti-2 in chloride-dominated groundwater, provided the temperature is high enough, crevice propagation continues until all the available oxidant is consumed. For Ti-12 crevice corrosion initiates but widespread activation of the artificially creviced area is suppressed. This is indicated by the much lower values of I_c (100 to 300 μA compared with 1500 to 2000 μA for Ti-2 at 125°C) (Fig. 4). The crevice potentials, E_c, are also much less negative (-0.40 to -0.47 compared with -0.55 to -0.65 V for Ti-2) consistent with only partial crevice activation in the case of Ti-12. Also, for Ti-12, repassivation of the crevice (after ~950 h, Fig. 4) occurs long before the oxygen is totally consumed.

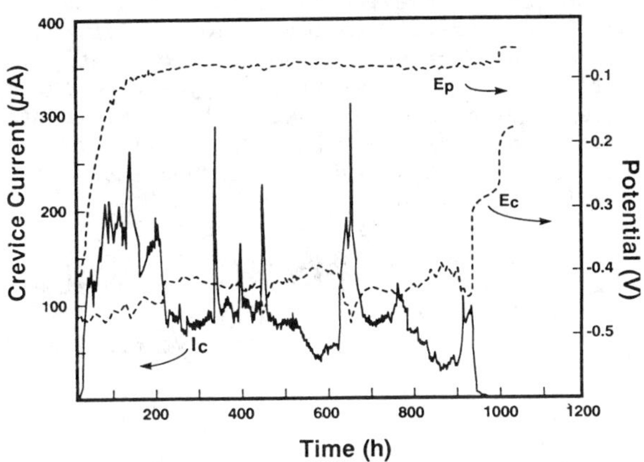

Fig. 4--Values of I_c, E_c and E_p for Grade-12 titanium in NaCl solutions ($[Cl^-] = 0.27\ mol \cdot L^{-1}$) at 125°C.

Temperature

Measurements of propagation rates as a function of temperature have been made for both Ti-2 and Ti-12. For Ti-2 a decrease in I_c with decreasing temperature is accompanied by an increase in E_c (Fig. 5). Except at the lowest temperatures (<40°C) most of this potential increase can be ascribed to changes in the potential of the reference electrode with temperature. A current for the propagation of crevice corrosion could be observed down to room temperature in the presence of substantial amounts of oxygen. The relationship between I_c and temperature is Arrhenius in form yielding an activation energy of ~55 ± 17 $kJ \cdot mol^{-1}$ [22]. Experiments on the dependence of I_c on O_2

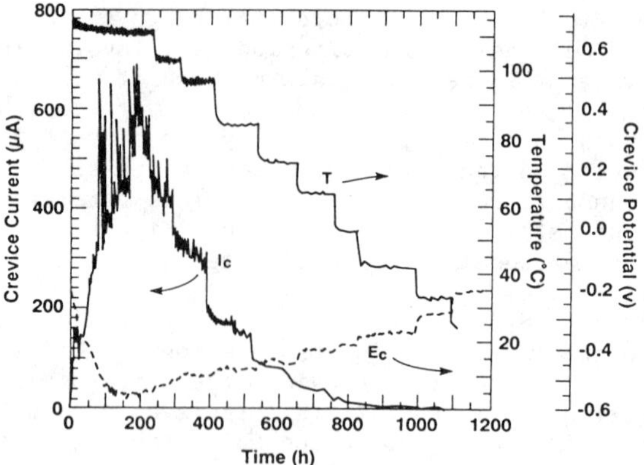

Fig. 5--Variations in I_c and E_c for Grade-2 titanium in aerated NaCl solutions ($[Cl^-] = 0.27\ mol \cdot L^{-1}$) as a function of temperature (T).

concentration (described below) suggest that the kinetics of O_2 reduction on passive surfaces external to the crevice may be rate-determining for the overall crevice propagation process.

At room temperature, crevice corrosion cannot be sustained indefinitely and I_c decays to zero after ~100 h (Fig. 5), and it is not revived by a subsequent increase in temperature to 105°C. Examination of corroded surfaces using optical microscopy and measurements of the depths of crevice penetration show that as the temperature decreases corrosion becomes more localized, i.e., propagation becomes limited to a smaller number of active sites clustered around the periphery of the artificially creviced area [24].

By contrast, a similar experiment with Ti-12 yielded values of I_c only slightly, if at all, dependent on temperatures between 73 and 110°C (Fig. 6). For a temperature $\lesssim$73°C (consistently obtained in three identical experiments), the crevice repassivates, and I_c decays to zero over a period of ~10 h as E_c shifts to positive values. Once

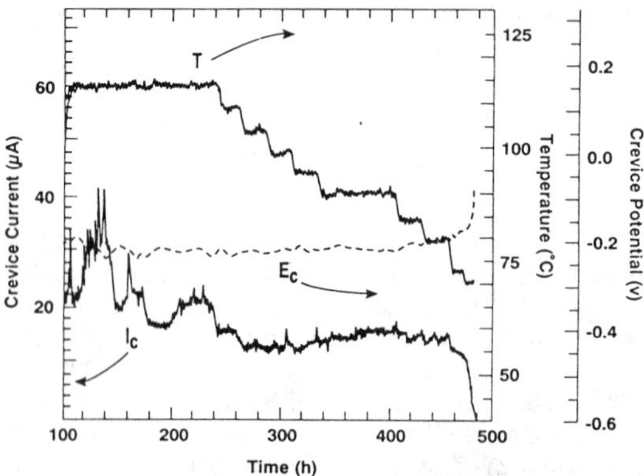

Fig. 6--Variations in I_c and E_c for Grade-12 titanium in aerated NaCl solutions ($[Cl^-] = 0.27\ mol \cdot L^{-1}$) as a function of temperature (T).

repassivation has occurred an increase in temperature does not lead to a reactivation. The amount of oxygen consumed over the 350-h period of activation was small.

Clearly, the factors controlling the repassivation of crevices, and hence the duration and extent of crevice propagation, are different for the two grades of titanium. If temperature were the only important variable for Ti-2, then an inspection of Figs. 1 and 5 indicates that crevice corrosion could continue for extensive periods of time under waste vault conditions. Obviously, to limit crevice corrosion the effect of environmental parameters other than temperature must be considered.

For Ti-12, however, repassivation is induced by a change in temperature alone. The temperature-independence of I_c indicates that the propagation rate is controlled mainly by metallurgical as opposed to environmental factors. The suppression of crevice propagation on Ti-12, and its eventual repassivation under oxidizing conditions, appear to be related to the presence of significant quantities of β-phase and the intermetallic, Ti_2Ni. The presence of 0.3 wt.% Mo may also assist the passivation process. Glass [27] has shown that Ti_2Ni is a more effective cathode for proton reduction than α-Ti and claimed that repassivation of Ti-12 is assisted by the galvanic coupling of Ti_2Ni and the metal matrix. The importance of Ni, probably in the form of Ti_2Ni, in the repassivation of Ti-12, has been demonstrated by Kido and Tsujikawa [28]. Alternative explanations have also been proposed [29].

The fact that the repassivation of Ti-12 occurs for T $\lesssim$73°C gives us a useful criterion upon which to base a prediction of the maximum duration of crevice corrosion on waste containers fabricated from this material: crevice corrosion will propagate until the container temperature falls below ~73°C. By sampling the distribution of times at which containers within a Canadian waste vault are expected to cool to 75°C, it can be shown that crevice corrosion should only last for between 10 and 140 years depending on the location of the container within the vault [10]. The containers taking the longest time to cool will be those in the centre of the vault totally surrounded by neighbouring heat-producing containers. Thus, the rapidly cooling containers will be located on the periphery of the vault. This assumption is still conservative since crevice repassivation is always observed in our experiments at all temperatures investigated up to 150°C [22].

Oxygen Concentration

McKay [30] and McKay and Mitton [20] demonstrated that the rate of crevice propagation was dependent on oxygen concentration, and we have shown previously [24] that the total amount of crevice propagation (from the total weight change) is directly proportional to the amount of oxygen consumed (Q) over the duration of the experiment.

The average rate of container wall penetration can be calculated from such weight change measurements provided the mode of penetration within the crevice is approximately uniform. However, within a creviced area there will be a distribution of penetration depths. Figure 7 shows corrosion depth profiles for the crevice corrosion of Ti-2 in 0.27 mol·L^{-1} NaCl at 150°C for various total amounts of oxygen consumed expressed as the charge Q (in coulombs). The plateau at penetration

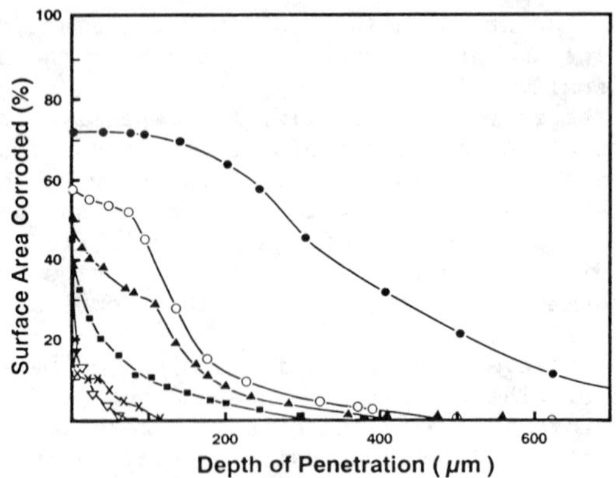

Fig. 7--Corrosion depth profiles for the crevice corrosion of Grade-2 titanium in NaCl solutions ($[Cl^-] = 0.27$ mol·L^{-1}) at 150°C as a function of the total amount of oxygen consumed expressed as the charge Q (coulombs, C): ∇ - 16 C (2h); x - 54 C (7h); ■ - 152 C (19 h); ▲ - 452 C (51 h); O - 894 C (110 h); ● - 3049 C (430 h). The bracketed terms refer to the duration of the experiment. Note that the 430-h experiment was performed in a 2-L pressure vessel and contained approximately three times the gas volume of the other experiments.

depths up to 100-200 μm, observed for the more extensively corroded specimens, indicates the development of a general corrosion front within the creviced area. However, significant penetration beyond the depth of this general front is observed. Figure 8 shows the maximum penetration depths taken from these depth profiles as a function of Q. Also shown is the ratio Q/q. For short durations of crevice propagation (low values of Q), corresponding to the period of crevice activation (Fig. 3), penetration is localized and driven as much by O_2 reduction outside as by proton reduction inside the crevice (Q/q ~0.44). For more extensive crevice corrosion (large Q equivalent to long durations), the development of the general corrosion front (Fig. 7) is accompanied by a tendency for the depth of penetration to plateau (Fig. 8). The value of Q/q becomes independent of Q indicating that the total amount of corrosion (q) has become directly proportional to the quantity of oxygen consumed (Q). Its low value indicates that proton reduction within the crevice is the main cathodic reaction sustaining propagation. The constancy of Q/q with Q (and hence with the duration of the crevice reaction) suggests that an approximately constant acidity is maintained at active sites within the crevice. Under these conditions crevice corrosion tends to propagate generally across, rather than locally into, the surface. It remains to be demonstrated whether a similar geometric mode of propagation is observed at lower temperatures.

It is clear from these results that the total extent of crevice propagation and its geometric mode of penetration on a container wall will be determined predominantly by the temperature and by the total amount of oxidant. Such data will enable us to place limits on the maximum penetration depths achievable for a given amount of oxidant available within the vault.

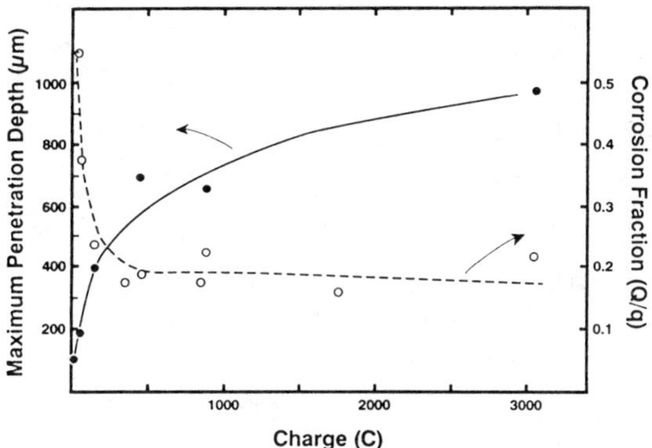

Fig. 8--Maximum penetration depths (●) due to crevice corrosion in NaCl solutions ($[Cl^-] = 0.27$ mol·L^{-1}) at 150°C as a function of the total amount of oxygen consumed expressed as the charge, Q, in coulombs. The open circles (O) show the fraction of the total amount of crevice corrosion (q) for each experiment attributable to the consumption of oxygen.

If the changes in crevice corrosion rate due to evolution of vault redox conditions are to be modelled, a more extensive data base involving a relationship between propagation rate, penetration depth and O_2 concentration is required. Figures 9 and 10 show the variations in I_c and E_c with O_2 concentration measured in a single experiment at 95°C. The response of I_c and E_c to changes in oxygen concentration was rapid (within a few minutes) and their values reproducible irrespective of whether they were recorded while the concentration was initially being decreased or subsequently being increased. The value of I_c increases linearly with the square root of O_2 concentration (Fig. 9), and it is accompanied by a negative shift in E_c, Fig. 10. The half-order concentration dependence, coupled with an activation energy of 55 kJ·mol^{-1} [22], could mean propagation is controlled by the reduction

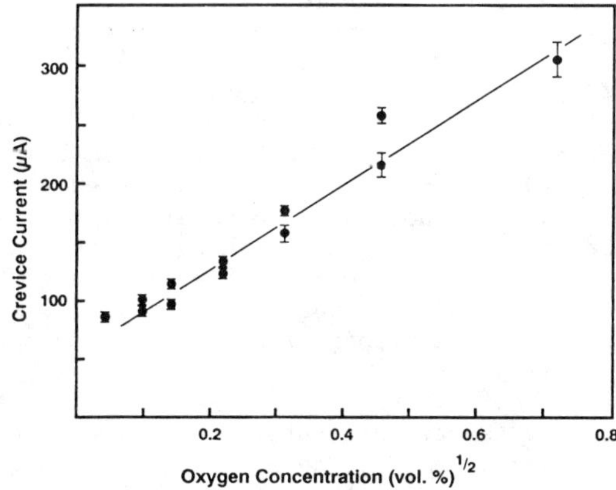

Fig. 9--Variation of I_c with oxygen concentration for Grade-2 titanium in NaCl solution ($[Cl^-] = 0.27$ mol·L^{-1}) at 95°C.

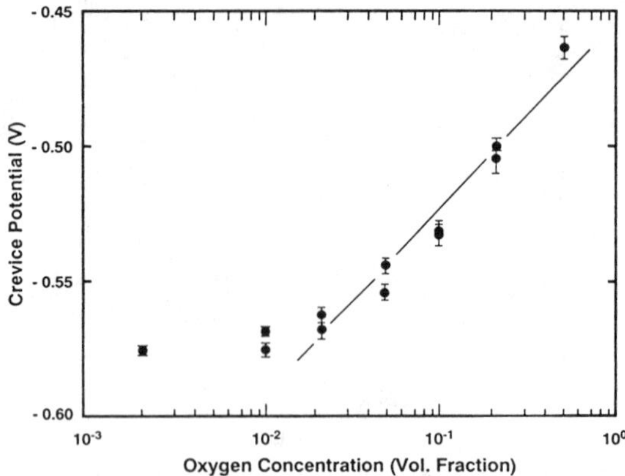

Fig. 10--Variation of E_c with the volume fraction of oxygen in the purge gas for Grade-2 titanium in NaCl solution ($[Cl^-] = 0.27$ mol·L^{-1}) at 95°C.

of O_2 on the passive surface of the titanium counter electrode in a reaction involving the prior dissociative adsorption of O_2 molecules. The accompaniment of a decrease in I_c by a negative shift in E_c indicates that the major influence of a decrease in O_2 concentration is to decrease the rate of metal dissolution at existing active sites. This is in contrast to the effect of a decrease in temperature that does not appear to affect E_c until low temperatures are reached (Fig. 5).

Extrapolation of the relationship between I_c and $[O_2]^{1/2}$ to zero O_2 concentration (Fig. 9) suggests that propagation is possible in the absence of oxygen - an unlikely situation over extensive periods of time. Conditions within the crevice should be regulated by transport processes such that the pH at actively dissolving metal sites will increase as protons diffuse out of the crevice, eventually leading to repassivation. In support of this claim, experiments in which I_c was measured at even lower O_2 concentrations than those presented here, repassivation was eventually observed [9]. That conditions within the crevice are changing at low O_2 concentrations is apparent (Fig. 10) when E_c starts to become independent of $[O_2]$ for a volume fraction of O_2 in the purge gas <1%.

Obviously, it is overly conservative for modelling purposes to accept that the propagation of crevices on Ti-2 can be sustained indefinitely at low $[O_2]$ as assumed in our published model [10]. To relax this assumption it is necessary to determine the combination of temperature and O_2 concentration for which repassivation becomes inevitable. Comparison of these critical requirements with vault temperatures and redox conditions would then enable us to specify the expected duration of crevice propagation on titanium waste containers under disposal conditions.

The results presented here suggest crevice propagation could be sustained to quite low temperatures provided oxygen is present. The results of Iki and Tsujikawa [16], determined using an electrochemical repassivation technique, suggest that, for the chloride concentrations used in our work and oxygen-degassed conditions, repassivation should occur once the temperature falls to $\lesssim$80°C. Their results were measured on metal/metal crevices, a crevice arrangement less likely to allow initiation than ours. For Ti-2, the results of Kido and Tsujikawa [28] suggest corrosion within metal/metal crevices should be impossible in oxygen-degassed solutions, at the chloride concentrations used in our work. So far we have not investigated the effect of oxygen concentration on the crevice corrosion of Ti-12. However, the ability of Ti-12 to repassivate for $T \lesssim 73°C$ in the presence of copious amounts of oxygen suggests that crevice propagation on this alloy may be seriously limited at lower O_2 concentrations.

SUMMARY

The process most likely to lead to the rapid failure of nuclear waste containers emplaced in a Canadian disposal vault is either crevice corrosion or HIC as a consequence of hydrogen absorbed into the metal during crevice corrosion. If long container lifetimes are to be achieved, then this period of crevice corrosion must be of relatively short duration. If these lifetimes are to be predicted with acceptable assurance then justifiable criteria upon which a model can be based must be developed.

For Ti-2 the duration of crevice corrosion will be limited by the decrease in temperature and O_2-concentration within the vault. The rate of crevice propagation appears to be controlled by the reduction of oxygen on passive surfaces external to the crevice and is possible even at quite low temperatures ($\lesssim 30°C$) providing sufficient oxygen is present. A prior dissociation of O_2 into adsorbed O species may be involved. At high temperatures (150°C), crevice corrosion tends to propagate across the surface, and localized deep penetrations are not far ahead of the general penetration front within the activated area. Whether a similar geometric mode of propagation occurs at lower temperatures remains to be discovered. Repassivation of crevices only occurs when the great majority of available oxidant has been consumed. The critical combinations of temperature and O_2 concentration at which repassivation occurs remain to be determined.

The propagation of crevices on Ti-12 surfaces is much slower than for the Grade-2 material. The process is almost independent of temperature and crevices repassivate for $T \lesssim 73°C$ even in the presence of substantial remaining oxygen concentrations. Clearly, the addition of the alloying elements, molybdenum and nickel has a major beneficial effect on crevice corrosion behaviour. By sampling the distribution of emplacement times at which containers in a waste vault cool to less than this temperature, it can be shown that crevice propagation should continue for no longer than 140 years even on the slowest cooling Ti-12 container in the vault [31].

Clearly, crevice corrosion will not propagate indefinitely on either Ti-2 or Ti-12 under Canadian waste vault conditions. Consequently, the lifetimes predicted using our previously published model [10] can be considered unduly pessimistic. Presently, we are redefining our failure model to include a limited period of crevice corrosion based on the results presented in this paper. The use of Ti-12 rather than Ti-2 in the fabrication of waste containers will severely restrict the period of crevice corrosion.

ACKNOWLEDGEMENTS

The Canadian Nuclear Fuel Waste Management Program is jointly funded by AECL Research and Ontario Hydro under the auspices of the CANDU Owners Group (COG).

REFERENCES

[1] King F., Ikeda, B.M., and Shoesmith, D.W., Chemtech, Vol. 22, 1992, p. 214, (also AECL-10619).

[2] Marsh, G.P., Taylor, K.J. and Harker, A.H., SKB Technical Report, Stockholm, SKB 91-62, 1991.

[3] Henshall, G.A., Proceedings, Nuclear Waste Packaging, FOCUS '91, American Nuclear Society Inc, La Grange Park, 1992, p. 225.

[4] Asano, H., Wakamatsu, H. and Akashi, M., Proceedings of the Third International Conference on High Level Radioactive Waste Management, American Society of Civil Engineers, New York, and American Nuclear Society Inc, La Grange Park, 1992, p. 1658.

[5] Sharland, S.M. and Tasker, P.W., Corrosion Science, Vol. 28, 1988, p. 603.

[6] Sharland, S.M., Jackson, C.P. and Diver, A.J., Corrosion Science, Vol. 29, 1989, p. 1149.

[7] Macdonald, D.D. and Urquidi-Macdonald, M., Corrosion, Vol. 46, 1990, p. 380.

[8] Clarke, C.F., Hardie, D. and Ikeda, B.M., Atomic Energy of Canada Limited Report, AECL-10477, COG-92-162, 1992.

[9] Shoesmith, D.W., Ikeda, B.M., Hardie, D., Bailey, M.G. and Clarke, C.F., Proceedings, Nuclear Waste Packaging, FOCUS '91, American Nuclear Society, La Grange Park, 1992, p. 193.

[10] Shoesmith, D.W., Ikeda, B.M. and LeNeveu, D.M. Proceedings of the International Conference on Life Prediction of Corrodible Structures, National Association of Corrosion Engineers, Houston, 1992, to be published.

[11] Gascoyne, M., Atomic Energy of Canada Limited Technical Record, TR-463, 1988.*

[12] Schutz, R.W. and Thomas, D.E. in Metals Handbook, 9th Edition, ASM International, Metals Park, 1987, Vol. 13, p. 669.

[13] Kobayashi, M., Araya, S., Fujiyana, S., Sunayama, Y. and Uno., H., Proceedings of the Fourth International Conference on Titanium, H. Kimura and O. Izumi (Eds.) The Metallurgical Society, AIME, Warrendale, 1980, Vol. 4, p. 2613.

[14] Sharland, S.M., Corrosion Science, Vol. 33, 1992, p. 183.

[15] Clarke, C.F., Ikeda, B.M. and Hardie, D., Proceedings of the International Conference on Environment-Induced Cracking of Metals, National Association of Corrosion Engineers, Houston, 1988, NACE-10, p. 419.

[16] Iki, F. and Tsujikawa, S., Tetsu to Hagane, Vol. 72(2), 1986, p. 292.

[17] Mattson, H. and Olefjord, I., Werkstoffe and Korrosion, Vol. 41, 1990, p. 383 and p. 578.

[18] Ikeda, B.M., Bailey, M.G., Clarke, C.F. and Shoesmith, D.W., Atomic Energy of Canada Limited Report, AECL-9568, 1989.

[19] King, F., Bailey, M.G., Clarke, C.F., Ikeda, B.M., Litke, C.D. and Ryan, S.R., Atomic Energy of Canada Limited Report, AECL-9890, 1989.

[20] McKay, P. and Mitton, D.B., Corrosion, Vol. 41, 1985, p. 52.

[21] Ikeda, B.M., Bailey, M.G., Cann, D.C. and Shoesmith, D.W. in Advances in Localized Corrosion, National Association of Corrosion Engineers, Houston, TX, 1990, NACE-9 p. 439.

[22] Ikeda, B.M., Bailey, M.G., Quinn, M.J. and Shoesmith, D.W., Proceedings of The Eleventh International Corrosion Congress, Associazone Italiana di Metallurgia, Milan, 1990, Vol. 5, p. 5.371.

[23] Quinn, M.J., Ikeda, B.M. and Shoesmith, D.W., Atomic Energy of Canada Limited Report, to be published.

[24] Ikeda, B.M., Bailey, M.G., Quinn, M.J. and Shoesmith, D.W., in Mechanical Behaviour of Materials - VI, Pergamon Press, Oxford, 1991, Vol. 2, p. 619, (also AECL-10360).

[25] Beck, T.R., Electrochimica Acta, Vol. 18, 1973, p. 807.

[26] Ikeda, B.M. and Clarke, C.F., Proceedings of the Second International Conference on Radioactive Waste Management, Canadian Nuclear Society, Toronto, 1986, p. 605.

[27] Glass, R.S., Electrochimica Acta, Vol. 28, 1983, p. 1507.

[28] Kido, T. and Tsujikawa, S., Tetsu to Hagane, Vol. 75, 1989, p. 1332.

[29] Sedricks, A.J., Green, J.A.S. and Novak, D.L., Corrosion, Vol. 28, 1972, p. 137.

[30] McKay, P., in Corrosion Chemistry within Pits, Crevices and Cracks, A. Turnbull (Ed.) Her Majesty's Stationary Office, London, 1984, p. 107.

[31] Ikeda, B.M. and Shoesmith, D.W., unpublished results.

* International report, available from SDDO, AECL Research, Chalk River Laboratories, Chalk River, ON, K0J 1J0.

Digby D. Macdonald[1], Mirna Urquidi-Macdonald[1], and Jim Lolcama[2]

DETERMINISTIC PREDICTIONS OF CORROSION DAMAGE TO HIGH LEVEL NUCLEAR WASTE CANISTERS

REFERENCE: Macdonald, D. D., Urquidi-Macdonald, M., and Lolcama, J., **"Deterministic Predictions of Corrosion Damage to High Level Nuclear Waste Containers,"** Application of Accelerated Corrosion Tests to Service Life Prediction of Materials, ASTM STP 1194, Gustavo Cragnolino and Narasi Sridhar, Eds., American Society for Testing and Materials, Philadelphia, 1994.

ABSTRACT: The corrosion of titanium alloy canisters in a plutonic rock repository, as envisaged in the AECL/Ontario Hydro HLNW disposal concept, has been explored using four deterministic models. A Mixed Potential Model (MPM) was used to calculate the corrosion potential of the canister as a function of time after resaturation, and it is predicted that the corrosion potential will shift in the negative direction as the ambient oxygen is consumed coupled with a much smaller positive shift due to the decay in temperature. Maximum general corrosion rates have been estimated using the Maximum Rate Model (MRM) and the Radiolytic Maximum Rate Model (RMRM), both of which equate the corrosion rate to the maximum possible flux of oxygen (MRM) and oxygen plus oxidizing radiolysis products (RMRM) to the canister surface. These calculations suggest that the radiolysis of water is not of major significance in establishing the general corrosion rate. Furthermore, general corrosion is predicted not to be a threat to the integrity of a thin wall (6.35mm) titanium container within the 500 year design life. Finally, corrosion rates have also been estimated using the WAPPA code and the container is predicted to withstand crevice corrosion for up to 1200 years and to be unbreached by general corrosion for 2000-2500 years after resaturation.

KEYWORDS: high level nuclear waste canisters, life prediction, deterministic models, plutonic rock repositories, titanium.

The disposal of high level nuclear waste (HLNW) is a major technological challenge to those nations that operate nuclear power systems. The demands on the technology are unprecedented in the annals of human history, because of the long time over which isolation from the biosphere must be assured. For unreprocessed waste from CANDU reactors, isolation must be effectively assured for at least 10,000 years, at which time many of the more dangerous nuclides will have decayed to safe levels. Even then, chemically toxic elements, such as

[1] Professor, Materials Science and Engineering and Director of the Center for Advanced Materials, and Associate Professor of Engineering Science and Mechanics, respectively, The Pennsylvania State University, University Park, PA 16802.

[2] Hydrochemist, INTERA, Inc., Augusta, GA 30901.

plutonium (which has a half life of 24 x 10^3 years) remain, so that release into the far field environment, and hence eventually into the biosphere, must be maintained at a low rate (e.g. by ensuring a suitably low dissolution rate for the uranium oxide fuel matrix).

In the HLNW disposal concept currently being developed in Canada by Atomic Energy of Canada Ltd and Ontario Hydro, unreprocessed spent fuel will be contained in thin-wall (6.35mm thick) Grade-12 titanium alloy [1] canisters (Figure 1), which in turn will be placed in bore holes in a rock matrix (Figure 2). The canisters are to be surrounded by a thin layer of silica sand and by a thicker outer layer consisting of a mixture of silica sand and washed bentonite clay (Figure 3). Because the bore hole is drilled, a region of highly fractured rock, perhaps 1-2 m thick (assumed for modeling), is envisaged to exist concentrically with the canister. This region will have a porosity that is significantly greater than that of the rock in the far field (estimated to be ~0.003) but less than that of the near field bentonite/sand buffer (estimated at 0.35). For our purposes, we assume that the near field (sand and bentonite/sand layers) can be regarded as a homogeneous porous phase of porosity 0.35, whereas that at greater distances is also homogeneous but of much lower porosity (0.003). Because of the finite porosity, oxygen may be transported through the far-field and near-field matrices and subsequently corrode the canister wall. Accordingly, it is of great interest to determine whether corrosion poses a threat to the integrity of the canister over the design life (in this case 500 years). This question cannot be addressed satisfactorily by experimental methods alone, because any data base that might be established within the foreseeable future would cover only a small fraction of the design life. Accordingly, any empirical method for estimating corrosion damage functions and failure probabilities (e.g. extreme value statistical techniques) are unlikely to yield predictions that are sufficiently accurate to satisfy engineering needs. An alternative procedure, which is employed here, is to develop deterministic models that are based on sound physico-chemical laws, such as Fick's laws of diffusion and the conservation of charge. These models may then be used to predict future behavior for various scenarios that encompass the conditions that are likely to exist over the design life of the canister. Ideally, any experimental data base that might be developed in the short term (compared with the time scale of the repository) could be used to test or at least calibrate the models to further enhance their reliability, and to provide values for various model parameters.

In this paper, we describe four deterministic models that have been used to estimate general (uniform) corrosion rates for HLNW canisters emplaced in repositories of the type shown in Figures 1-3. These models address different scenarios, including unrestricted supply of oxygen to the canister surface, diffusion limited supply of oxygen, and diffusion limited supply of oxidizing species (including oxygen) as produced by the radiolysis of water in the near-field environment. The predictions of the models are compared and used to construct a likely scenario for the fate of the canister with respect to uniform attack and with regards to crevice corrosion.

THE PROBLEM

The current work was carried out to address a number of important corrosion-related issues [2-7] in the design and specification of HLNW repositories, particularly within the Canadian concept of disposal in a

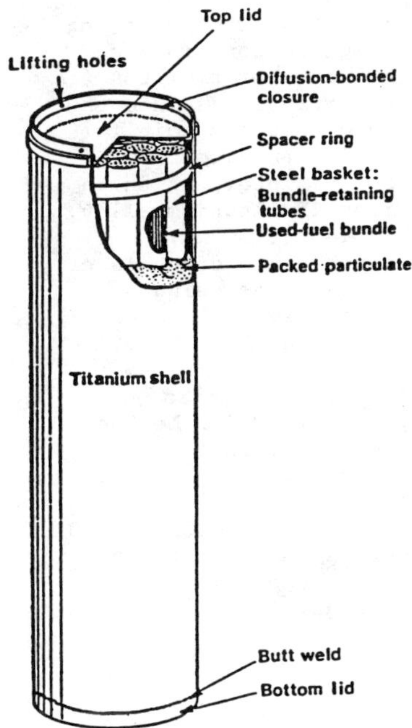

FIG. 1--AECL particulate-packed HLNW container

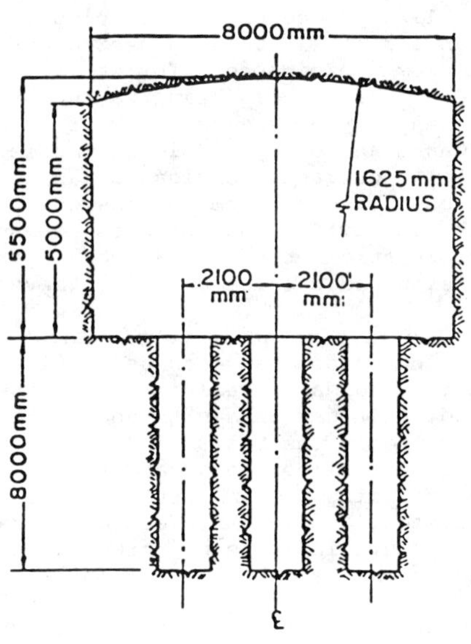

FIG. 2--Schematic of HLNW repository according to the AECL/Ontario Hydro disposal concept.

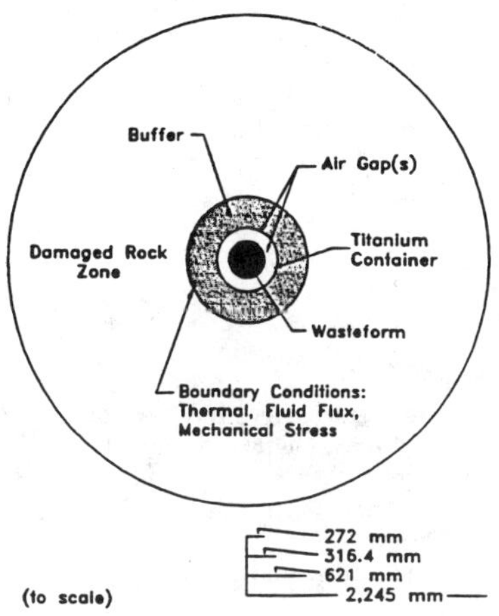

FIG. 3--Modeled waste package geometry

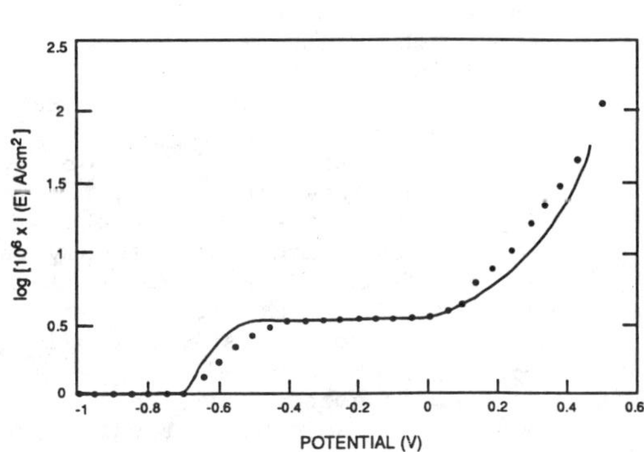

FIG. 4--Comparison of the anodic curve for Type 304 stainless steel in pure water at 250°C, as calculated from the data of Lee

plutonic rock environment. In doing so, we believe that the following factors and issues must be recognized in any attempt to construct a "most likely" scenario describing the fate of the canister over the design life: (i) The engineered barriers must effectively isolate the HLNW from the biosphere for at least 10,000 years although credit for the canister is only being taken for 500 years. (ii) An important component of the engineered barrier concept is the canister, which in the Canadian concept would be fabricated from Grade-12 titanium alloy. (iii) The currently envisaged canister is thin walled (6.35 mm). Accordingly, corrosion is assumed to be the ultimate cause of canister rupture. (iv) Our approach is to estimate corrosion rates using deterministic methods under conservative conditions. (v) Both general corrosion and crevice corrosion have been dealt with in this study.

The last point is particularly important because localized corrosion phenomena, such as crevice corrosion, are likely to be the principal modes of failure due to their very high penetration rates. Indeed, AECL has chosen to concentrate on crevice corrosion as the mode of failure on the basis of literature data that indicates that pitting corrosion will not be a factor. However, great care must be exercised when excluding pitting attack because it is a nucleation and growth phenomenon, and long nucleation times (compared with experimental times) could indicate no threat when, in fact, a significant threat actually exists.

MODELING OF THE CORROSION PROCESSES

The calculations reported here were carried out using four deterministic models. The general features of the models are outlined below:

- MIXED POTENTIAL MODEL (MPM) - The MPM [8-10]calculates the corrosion potential and corrosion rate assuming that the oxygen concentration in the near field environment and the current/voltage characteristics of the canister alloy are known. In a sense, this model explores the scenario where mass transport through the rock and buffer are sufficiently rapid that the corrosion rate is determined by conditions existing at the canister/near field environment interface. This scenario contrasts those outlined below, where we explore the maximum rate at which corrosion can occur due to limited mass transfer of oxidizing species (e.g. O_2) to the canister surface.
- MAXIMUM RATE MODEL (MRM) - The MRM calculates the corrosion rate and hence the damage function (depth of attack) by assuming that every molecule of oxygen transported to the canister surface immediately reacts with the metal. This corresponds to the maximum rate at which corrosion can occur, provided that oxygen is the only reactant.
- RADIOLYTIC MAXIMUM RATE MODEL (RMRM) - The RMRM is similar in concept to the MRM, except that the model takes into account the radiolytic generation of oxidizing (O_2, H_2O_2, H^+, O_2^-, HO_2, HO_2^-, OH) and reducing (e^-, H, H_2) species in the near field environment. This model was developed to determine whether the radiolysis of water is a significant factor in determining the general corrosion rate.
- WAPPA - WAPPA [11] was developed for the Office of Nuclear Waste Isolation (ONWI) SCEPTER program, during the period 1982-1983. This early (mainframe) version is available for public

use through the U.S. DOE software library. The PC version used for these scoping calculations has been enhanced with an accumulation-type boundary condition. WAPPA was selected because of its availability and because of the complete list of process models within the code, and therefore its abilities to perform the deterministic calculations required for the waste package scoping calculations.

We have emphasized determinism in developing these models, in order to preserve their predictive capabilities. Accordingly, the models are based on sound physico-chemical laws (e.g. Fick's law of diffusion and Faraday's law). However, as with all deterministic models, the ultimate limitations on predictability lie with the completeness of the models and the quality of the input data. Furthermore, it is important to note that determinism does not imply that all of the canisters will fail simultaneously, because a deterministic model can (and should) contain distributed parameters provided that the distributions can be derived from fundamental laws or can be measured directly (i.e. in a separate experiment).

At the initiation of this work, a search of the literature was carried out to locate and assess input data for the models. Data on the geometry of the canister as well as on the materials used in its construction were taken from AECL sources [1]. Likewise, information on the expected operating conditions, including temperatures, energy dose rates, etc., were also taken from AECL publications or were supplied by AECL personnel. However, data for corrosion rates and for fundamental electrochemical parameters required to estimate degradation rates and mechanisms are non-existent or at best very sparse. Specifically:

- No reliable steady-state current/voltage data for titanium and titanium alloys in neutral or near-neutral brine environments at temperatures over the range of interest (50°C - 100°C) could be found.
- No reliable exchange current density or Tafel constant data for the reduction of H^+, O_2, and H_2O_2 or the oxidation of H_2 on titanium and titanium alloys in neutral or near neutral brines could be located.
- Some data on pitting corrosion and crevice corrosion of titanium in brines are available [2-6]. Short term laboratory data indicate that pitting will not be a significant threat to canister integrity, but one must be cautious of this conclusion because of the very long time that is available for pit nucleation. Crevice corrosion is likely to be the principal form of corrosion if preexisting crevices are present.
- The lack of experimental data is a serious impediment to qualifying titanium alloys for the fabrication of HLNW canisters.

In light of these shortcomings, we estimated values for certain parameters, or we adopted values for other alloys in similar environments where the corrosion phenomena were similar. A key recommendation of this study is that work should be initiated to establish a data base that can be used in calculations of the type described here. In our view, this is a particularly pressing matter because of the need to predict phenomena over periods of time (into the future) where direct experimental information will not be available.

MIXED POTENTIAL MODEL

This model, which is a variant of a model developed some time ago to explore corrosion phenomena in the heat transport circuits of light water nuclear reactors [8-10], seeks to calculate corrosion potentials and corrosion rates for canisters in HLNW repositories of the type described in the AECL/OH concept. The principal features of the model are as follows: (i) The model is based on Wagner-Traud hypothesis that anodic and cathodic processes are separable. (ii) The MPM has been previously used to calculate corrosion potentials of HLNW canisters in Tuff repositories [12]. (iii) The model has been tested experimentally for stainless steel in high temperature water [13].

Calculation of the corrosion potential is important because it is known that specific corrosion processes occur only at potentials above or below "critical" values, which depend on temperature, the composition of the environment in contact with the metal, and the properties of the substrate. For example, in the case of titanium, pitting corrosion has been reported [14] to occur only at potentials more positive than

$$V_c = 1.4 - 0.10 \log M_{Cl} (\text{at } 200\,^\circ\text{C}) \tag{1}$$

where M_{Cl} is the molar concentration of chloride ion. Although the critical potential for pitting of titanium is known to shift strongly in the positive direction as the temperature is lowered [15], a knowledge of how the corrosion potential evolves with time is essential for predicting whether pitting corrosion poses a threat to the canister at some future time.

Briefly, the mixed potential model finds the potential at which the total current due to all charge transfer reactions at the surface is equal to zero. This condition, which is simply a statement of charge conservation, may be expressed in the form

$$\sum_{j=1}^{n} i_{O/R,j} + i_{corr} = 0 \tag{2}$$

assuming that the redox reactions may be written in the form

$$R \Leftrightarrow O + ne^- \tag{3}$$

The quantity $i_{O/R,j}$ is the current density for j-th reaction, expressed in the form of a generalized Butler-Volmer equation

$$i_{O/R,j} = \frac{e^{(E-E^e)/b_a} - e^{-(E-E^e)/b_c}}{\frac{1}{i_o} + \frac{1}{i_{\ell,f}} e^{(E-E^e)/b_a} - \frac{1}{i_{\ell,r}} e^{-(E-E^e)/b_c}} \tag{4}$$

where E^e is the equilibrium potential, i_o is the exchange current density, b_a and b_c are the Tafel constants for the forward and reverse directions, respectively, and $i_{\ell,f}$ and $i_{\ell,r}$ are the corresponding mass transport-limited currents for the j-th redox couple in the environment. In the present calculations, we consider only one redox species (oxygen), although the model is sufficiently general that any number can be included if appropriate data are available.

Kinetic data for the reduction of oxygen on titanium in neutral-pH or near neutral-pH brine under the conditions of interest in this work could not be located. Consequently, we have assumed data for the same parameters for oxygen reduction on stainless steel in sulfate solutions at temperatures ranging to 250°C (see [10]). We consider this assumption to be reasonably valid, because the electrochemical characteristics of stainless steel and titanium frequently are similar. However, we cannot justify this procedure on quantitative grounds, so that it should be regarded with caution.

Although polarization data are available for titanium in highly acidified brines, we could not locate any data that correspond to the conditions of interest here. Accordingly, we have assumed polarization data for Type 304SS, which are available from Lee (see [10]) for sulfate-containing solutions at temperatures up to 250°C. This assumption was made on the basis that both materials are "corrosion resistant" and have similar passive current densities (~0.1μA/cm^2) under ambient conditions where comparisons can be made. However, the passive range for titanium is greater than that for Type 304SS and, theoretically, the passive current is not a function of voltage for the former whereas it is for the latter [12]. This difference is due to the fact that the passive film on titanium is an oxygen ion conductor whereas that on stainless steel is thought to be a cation conductor. In the latter case, the Point Defect Model [16] for passivity predicts that

$$\log [i_{ss}] = \log [i^{o}_{ss}] + aE \tag{5}$$

where i_{ss} is the passive current density, E is the voltage, and

$$a = 0.5 - 1.0 \quad (V^{-1}) \tag{6}$$

depending on the conditions.

Typical data from Lee are plotted in Figure 4. The solid line corresponds to a function used to fit the experimental data (points) in the algorithm. Because Lee (see [10]) measured steady state polarization data over a wide temperature range (150°C to 250°C), we were able to express the passive current as a function of temperature and hence to extrapolate values to the temperatures of interest in this work. One feature that is notably different for titanium, in contrast to stainless steel, is that the passive current does not display the increase for E>0.2V, as shown in Figure 4. In the case of titanium, the passive current remains constant to potentials of several volts.

Corrosion potentials and corrosion rates calculated using the MPM are summarized in Table 1. In carrying out these calculations, we have assumed an exponential decay of the temperature from an initial value of 100°C with a decay constant of 0.0069yr^{-1} (corresponding to a half life of 100 years). Thus, after an exposure time of 1000 years, the temperature at the surface of the canister is calculated to be 0.1°C above the background value of 50°C. We have also carried out the calculations for oxygen concentrations in the near field environment ranging from 10 ppm to 0.001 ppm. The upper value is significantly greater (by about a factor of two) than that expected for air-saturated ground water at the background temperature and which may exist in the near field environment upon closure. The lower oxygen concentration of 1 ppb corresponds to an essentially deaerated environment, possibly as a result of oxygen consumption due to corrosion or due to radiolytically-catalyzed reaction of oxygen with reduced iron species in the near field (rock) matrix. To complete the calculations, we have averaged the corrosion rates and multiplied by 1000 to yield the corrosion losses listed in the last column in Table 1. As can be seen, the corrosion losses are predicted to be small, even for high oxygen concentrations.

The most probable scenario is that, upon closure and resaturation of the repository, oxygen will react rapidly with reduced iron species in the rock and with organic matter contained in the buffer and/or backfill, with both reactions being most likely catalyzed by the low (but significant) γ- fields in the vicinity of the canister. Thus, the most likely trajectory for the system is from $[O_2]$~6ppm to $[O_2] < 0.01$ ppm within the first fifty years and from then on horizontally across the table to a fully decayed state at 1000 years. Accordingly, the corrosion potential is expected to shift from about 0.2V to ~-0.35V at short times (<50 years), and then to move slowly in the positive direction (by about 50 mV) as the repository ages. This potential is sufficiently low that, ordinarily, pitting corrosion would not be considered a threat. However, recent theoretical work, which is supported by experiment [17-19], shows that pit nucleation is expected at potentials more negative than the "pitting potential," because of the distributed nature of the breakdown sites at the surface. Furthermore, pitting potentials are normally measured potentiodynamically and, because of the existence of an induction time for pit nucleation, the measured pitting potential can be greatly in error. This factor, coupled with the extraordinarily long time available for pit nucleation (hundreds of years), renders the possibility of pitting attack on the canister important and a matter that needs to be explored in a thorough manner.

A second scenario is that, due to the decaying γ-field adjacent to the canister, the concentration of oxygen in the near-field will slowly increase from <0.01 ppm at 50 years to perhaps >1 ppm at 1000 years after closure (or resaturation) as the result of continued diffusion from the far-field environment. In this case, the corrosion potential is predicted to slowly increase from about -0.35V to ~0.2V. Such an increase would render the canister more susceptible to both pitting attack and crevice corrosion and, again, the possibility of pit nucleation and growth needs to be explored.

In both scenarios, the corrosion loss due to general attack is predicted to be less than 0.1mm, which is negligible compared with the thickness of the canister wall (6.35mm). Even if the passive currents employed in our calculations were in error by a factor of ten (which is possible), the loss after 1000 years of operation due to general corrosion would still be less than 1 mm, which again provides sufficient margin to ensure that general corrosion is not a threat to canister integrity over the 500 year design lifetime. In this case, however, the corrosion potential will decrease (at least for stainless

TABLE 1 -- Corrosion potential and corrosion rate data estimated using the mixed potential model.

$[O_2]$ ppm	Time After Closure (Years)						
	O	50	100	200	500	1000	L (mm) (c)
	T(°C) 100	85.4	75.1	62.6	51.6	50.1	
10	0.301 (a)	0.302	0.302	0.298	0.294	0.293	0.08
	1.76×10^{-7} (b)	1.08×10^{-7}	7.41×10^{-8}	4.58×10^{-8}	2.91×10^{-8}	2.72×10^{-8}	
1	0.189	0.234	0.197	0.197	0.194	0.194	0.07
	1.69×10^{-7}	1.05×10^{-7}	7.17×10^{-8}	4.46×10^{-8}	2.85×10^{-8}	2.67×10^{-8}	
0.1	-0.336	-0.296	-0.257	-0.163	-0.019	-0.009	0.05
	7.19×10^{-8}	5.88×10^{-8}	5.02×10^{-8}	4.02×10^{-8}	2.78×10^{-8}	2.61×10^{-8}	
0.01	-0.368	-0.345	-0.328	-0.308	-0.289	-0.286	0.005
	7.10×10^{-9}	5.35×10^{-9}	4.91×10^{-9}	3.87×10^{-9}	3.56×10^{-9}	3.41×10^{-9}	
0.001	-0.370	-0.348	-0.332	-0.313	-0.296	-0.294	0.0006
	1.79×10^{-9}	4.36×10^{-10}	4.04×10^{-10}	4.01×10^{-10}	3.08×10^{-10}	1.43×10^{-10}	

(a) Corrosion potential vs. Standard Hydrogen Electrode
(b) Corrosion rate in m/yr
(c) Corrosion loss in mm

steel), so that the risk of localized attack (pitting and crevice corrosion) will be lowered correspondingly.

From the above analysis, it is likely that general corrosion does not pose a significant threat to the integrity of a thin-wall titanium (or more exactly, a stainless steel) canister. It is important to note that our MPM calculations assume an infinitely high rate of mass transfer of oxygen to the canister surface, which renders the calculations highly conservative in nature. In practice, as oxygen is depleted, mass transport is expected to become an increasingly important factor in controlling the corrosion rate.

THE MAXIMUM RATE MODEL

The maximum rate model (MRM) was developed to scope the other extreme in corrosion behavior, where the corrosion rate is controlled entirely by the rate of transport of the corrodent (oxygen) to the surface. Specific assumptions made in formulating this model include the following: (i) Oxygen is the only reactant. (ii) The corrosion rate is diffusion limited - hence the concentration of oxygen at the canister surface is zero. (iii) Oxygen transport occurs through the buffer and rock matrices having different porosities. (iv) Cylindrical diffusion is assumed. (v) No convective transport is allowed. (vi) The temperature at the canister surface decays exponentially with time. (vii) The thermal and mass transport problems are coupled through the temperature-dependent diffusivity of oxygen. (viii) The buffer and rock matrix properties (e.g. porosity) remain constant over the period assumed for the calculations. (ix) All areas of the canister surface are equally accessible.

Clearly, the MRM addresses a quite different scenario than was explored with the MPM. In many respects, the maximum rate model (and the variant thereof discussed in the next section) is more useful because it permits one to estimate the maximum rate at which corrosion can occur, regardless of the kinetics of reaction between oxygen and the canister. Accordingly, the model is not sensitive to poorly known kinetic parameters, such as the passive current density of the substrate or the exchange current density and Tafel constant for the reduction of oxygen at the canister surface.

(1) Thermal Diffusion

The first step in the MRM algorithm, is to solve for the temperature versus radial distance from the canister perpendicular to the principal axis for any time after closure. For cylindrical coordinates, the appropriate differential equation is written as

$$\frac{\partial T}{\partial t} = \sigma\left[\frac{\partial^2 T}{\partial r^2} + \frac{1}{r}\left(\frac{\partial T}{\partial r}\right)\right] \tag{7}$$

where σ is the thermal diffusivity. This equation must be solved subject to the following initial and boundary conditions:

- Initial condition ($t = o$): $r = r_o$ (canister surface): $T = \Delta T_o + T_b$ (8)

- Boundary conditions ($t > o$): $r = r_o$: $T = \Delta T_o e^{-\lambda t} + T_b$ (9)

$$r_o < r < r_b : \sigma = \sigma_b \tag{10}$$

$$r_b < r < \infty : \sigma = \sigma_r \tag{11}$$

$$r \rightarrow \infty \ : \ T \rightarrow T_b \tag{12}$$

where T_b is the temperature of the rock matrix at an effectively infinite distance from the canister, ΔT_o is the temperature of the canister above T_o at zero time, r_b is the outer radius of the backfill layer, λ is the temperature decay constant, and σ_b and σ_r are the thermal diffusivities of the bentonite-sand backfill and the rock matrix, respectively.

This problem was solved using a central (finite) difference technique using up to 100 mesh points. To preserve accuracy close to the canister, the buffer region was divided into ten segments, but at greater distances the segments were scaled quadratically. This was done to permit the calculations to be carried out over sufficiently large radial distances while maintaining the matrix of coefficients to a manageable size. Likewise, time was scaled quadratically, so that the calculations could be carried out over the time scale of interest while preserving precision at short times when the corrosion rate is greatest.

(2) Mass (Oxygen) Diffusion

The second step in the MRM is to solve the diffusion equation for the transport of oxygen to the canister surface under conditions where the concentration of oxygen at the interface is zero. Accordingly, we must solve the equation,

$$\frac{\partial C}{\partial t} = D(T)\left[\frac{\partial^2 C}{\partial r^2} + \frac{1}{r}\left(\frac{\partial C}{\partial r}\right)\right] \tag{13}$$

where D(T) is the temperature-dependent oxygen diffusivity [20], for the following initial and boundary conditions:

- Initial condition (t = o): $r_o \lesseqgtr r \lesseqgtr r_b ; C = P_b C^b$:

$$r > r_b ; C = P_r C^b \tag{14}$$

where P_b and P_r are the porosities of the buffer and rock matrix, respectively.

- Boundary conditions (t > o): $r = r_o ; C = 0$:

$$r \rightarrow \infty ; C \rightarrow P_r C^b \tag{15}$$

Again, the problem was solved using a central (finite) difference technique employing graded radial distance and time scales as employed for solving the thermal diffusion problem. Once the concentration profile was obtained, the corrosion rate was calculated from the flux of oxygen at the canister surface using Faraday's law

$$CRt = (M / \rho) J_{O_2} \tag{16}$$

where M is the atomic weight of titanium, ρ is the metal density, and J_{O_2} is the oxygen flux

$$J_{O_2} = D\frac{C(1)}{h} \tag{17}$$

The quantity C(1) is the concentration of oxygen at the first mesh point and h is the increment of distance in the backfill [$h = (r_b - r_o)/10$]. The corrosion loss at time t was then obtained by the stepwise integration of the corrosion rate.

Corrosion rates and corrosion losses calculated using the parameter set given in Table 2 are summarized in Table 3. In carrying out these calculations, we assume that the concentration of oxygen in water far removed from the canister is the same as in ground water under standard surface conditions (6 ppm or 0.1875 mol/m^3); again we assume that the lithostatic and hydrostatic pressure will more than compensate for temperature in determining the solubility of oxygen.

TABLE 2 -- Parameter values for Maximum Rate Model calculations.

Parameter	Value	Units
Radius of canister, r_o	0.33	m
Radius of buffer, r_b	0.63	m
Initial canister temperature, $T_o + \Delta T$	100	°C
Background temperature, T_b	50	°C
Temperature decay constant, λ	0.0069	yr^{-1}
Thermal diffusivity of backfill, σ_b	40	m^2/yr
Thermal diffusivity of rock, σ_r	40	m^2/yr
Porosity of buffer, P_b	0.35	-
Porosity of rock matrix, P_r	0.003	-
Background oxygen concentration, C_b	0.1875	mol/m^3
Activation energy for oxygen diffusion, E_d	3490 (20)	cal/mol
Preexponential factor for oxygen diffusion, D_o	25.323 (20)	m^2/yr

TABLE 3 -- Maximum corrosion rate and corrosion loss as calculated from the Maximum Rate Model.

Time(Years)	T(°K)	Corrosion Rate (μm/yr)	Corrosion Loss (μm)
1	372.8	3.1	3.1
25	365.2	2.9	71.6
81	351.7	2.4	198.0
169	338.7	2.1	349.9
361	327.3	1.8	639.0
529	324.5	1.7	899.0
729	323.5`	1.7	1222.0
1089	323.2	1.7	1816.8
1521	323.2	1.7	2536.5
2025	323.15	1.7	3376.9
2401	323.15	1.7	4003.9

Examination of the data listed in Table 3 shows that the corrosion rate decreases from an initial value of 3.1 μm/yr to 1.7 μm/yr after 1000 years of exposure. Over the same period, the temperature decays from 100°C to 50.1°C; however, the greatest change in the corrosion rate is due to the gradual development (decrease) of the diffusion gradient for oxygen away from the canister surface. Importantly, the corrosion rates calculated using the MRM are significantly greater than those estimated using the mixed potential model (Table 1), implying that under actual operating conditions the corrosion rate will be controlled by interfacial reactions occurring at the canister surface rather than by the transport of oxygen through the buffer and rock matrix, despite the low values of the porosities for these phases. These calculations therefore reinforce the argument for experimentally measuring key kinetic parameters for the interfacial reactions and for the use of corrosion-resistant materials.

Even if the rate of corrosion was controlled by transport of oxygen through the buffer/rock matrices, the corrosion loss after 500 years of exposure is estimated to be less than 1 mm compared with a canister wall thickness of 6.35 mm. Accordingly, even in this extreme, which is not dependent on the kinetic parameters for reactions at the canister surface, general corrosion is predicted not to pose a significant threat to the integrity of the canister over the design life.

RADIOLYTIC MAXIMUM RATE MODEL

The radiolytic maximum rate model (RMRM) was developed to explore the effect of the radiolysis of water in the near-field and near far-field environments on the maximum rate at which the canister may corrode. These calculations are important because γ- and n- radiolysis produces a variety of oxidizing species that may react with the metal.

The principal features of the RMRM are as follows:

- It is similar to the MRM, except that the radiolysis of water in the near field environment is taken into account.
- The radiolytic species include: e^-, H, OH, H_2, H_2O_2, O_2^-, HO_2, and HO_2^- in addition to oxygen (O_2), which is also present in the ground water.
- The concentrations of oxidizing species (OH, O_2, H_2O_2, O_2^-, HO_2, H^+, and HO_2^-) at the canister surface are equal to zero (maximum rate condition).
- The fluxes of the reducing species (e^-, H, H_2, OH^-) at the canister surface are set equal to zero (i.e. they are neither produced nor consumed).
- The temperature and γ and n dose rates are assumed to decay exponentially with time. γ and n dose rates are also assumed to decay as $\exp[-\lambda(r-r_0)]/r$, where r is the radial distance.
- The maximum corrosion rate is calculated from the fluxes of all oxidizing species to the canister surface.

A diffusion equation is written for each of the eleven radiolysis and dissociation (H^+, OH^-) products. However, each of these species is involved in one or more of the thirty-four reactions that have been assumed in the mechanism for the radiolysis of water (Table 4). This mechanism was adapted from the work of Ishigure et al and has been

TABLE 4 -- Reaction mechanism for the radiolysis of water.

Number	Reaction	k(ℓ/mol.s)	E (kcal/mol)
(1)	$e^- + H_2O = H + OH^-$	16	3
(2)	$e^- + H^+ = H$	2.4×10^{10}	3
(3)	$e^- + OH = OH^-$	3×10^{10}	3
(4)	$e^- + H_2O_2 = OH + OH^-$	1.3×10^{10}	3
(5)	$H + H = H_2$	1.0×10^{10}	3
(6)	$e^- + HO_2 = HO_2^-$	2.0×10^{10}	3
(7)	$e^- + O_2 = O_2^-$	1.9×10^{10}	3
(8)	$e^- + e^- + 2H_2O = H_2 + 2OH^-$	5×10^{10}	3
(9)	$OH + OH = H_2O_2$	4.5×10^{9}	3
(10)	$OH^- + H = e^- + H_2O$	2×10^{7}	4.5
(11)	$e^- + H_2O + H = H_2 + OH^-$	2.5×10^{10}	3
(12)	$e^- + H_2O + HO_2^- = OH + 2OH^-$	3.5×10^{10}	3
(13)	$H + OH = H_2O$	2×10^{10}	3
(14)	$OH + H_2 = H + H_2O$	4.0×10^{7}	3.2
(15)	$H + H_2O = H_2 + OH$	1.04×10^{-4}	20.36
(16)	$H + O_2 = HO_2$	1.9×10^{10}	3
(17)	$H + HO_2 = H_2O_2$	2.0×10^{10}	3
(18)	$H + O_2^- = HO_2^-$	2.0×10^{10}	3
(19)	$e^- + O_2^- + H_2O = HO_2^- + OH^-$	1.3×10^{10}	4.5
(20)	$H + H_2O_2 = OH + H_2O$	9.0×10^{7}	3.3
(21)	$OH + H_2O_2 = HO_2 + H_2O$	2.3×10^{7}	2.0
(22)	$OH + HO_2 = O_2 + H_2O$	1.2 x 1010	3
(23)	$OH^- + H_2O_2 = HO_2^- + H_2O$	1.8×10^{8}	4.5
(24)	$HO_2^- + H_2O = OH^- + H_2O_2$	3.7×10^{5}	4.5
(25)	$H^+ + O_2^- = HO_2$	5.0×10^{10}	3
(26)	$HO_2 = H^+ + O_2^-$	1×10^{10}	3
(27)	$HO_2 + O_2^- = O_2 + HO_2^-$	1.5×10^{7}	4.5
(28)	$2O_2^- + 2H_2O = H_2O_2 + O_2 + 2OH^-$	1.7×10^{7}	4.5
(29)	$HO_2 + HO_2 = H_2O_2 + O_2$	2.7×10^{6}	3
(30)	$H^+ + OH^- = H_2O$	1.4×10^{11}	3
(31)	$H_2O = H^+ + OH^-$	$K_{30}K_{wo}/[H_2O]$	-
(32)	$OH + O_2^- = O_2 + OH^-$	1.2×10^{10}	3
(33)	$H_2O_2 = H_2O + 1/2\, O_2$	-	-
(34)	$H_2O_2 = OH + OH$	5×10^{-2}	17

successfully used to model the radiolysis of water in the heat transport circuits of boiling water nuclear reactors and to model corrosion processes on canisters in a tuff repository (see [12]).

The components of this model are listed below.

- Thermal Diffusion -- temperature vs. distance profiles were calculated as a function of time using an analytical solution of the thermal diffusion equation, assuming that the near-field and far-field environments are thermally homogeneous.
- Mass Diffusion -- an equation of the form

$$\frac{\partial C}{\partial t} = D\left[\frac{\partial^2 C}{\partial r^2} + \frac{1}{r}\left(\frac{\partial C}{\partial r}\right)\right] + \left(\frac{dC}{dt}\right)_R + \left(\frac{dC}{dt}\right)_{rad} \tag{18}$$

is written for each species in the system. This equation includes terms for diffusion (in cylindrical coordinates), chemical reaction, and radiolysis. The latter two terms are given as follows:

Radiolysis: The rate of production of species "i" due to the interaction of γ-radiation and neutrons with water is described analytically as

$$\left(\frac{dC_i}{dt}\right)_{rad} = \frac{G_i^n \Gamma^n}{100 N_v} + \frac{G_i^\gamma \Gamma^\gamma}{100 N_v}\left(mol/cm^3.s\right) \tag{19}$$

where

$$G_i = \text{Radiolytic yield (no. molecules / 100eV)}$$

$$\Gamma = \text{Dose rate } (eV/cm^3.s)$$

$$= \Gamma_o \exp[-\lambda(r - r_o)]/r$$

and $$\Gamma_o = \Gamma_o^o \exp[-k_t t]$$

The dependence of Γ on the radial distance r accounts for both absorption and geometrical dispersion.

Reaction: Because reactions occur between the various radiolysis species, the concentration of any given species may be determined by a combination of elementary reactions, each of which is characterized by a rate constant (k_{ij}) and an activation energy (Table 4). Thus, the rate of production of species "i" in the system due to the j-reactions in which it is involved may be expressed as,

$$\left(\frac{dC_i}{dt}\right)_R = \sum_{j\text{ reactions}} \delta k_{ij} C_i C_k \tag{20}$$

where $\delta = +1$ for those reactions that produce "i" and $\delta = -1$ for those reactions that consume "i".

The initial and boundary conditions assumed in solving for the concentrations of the eleven species are as follows:

$$t = o; r_o \lesseqgtr r \lesseqgtr \infty : C_i = o$$
$$C_{O_2} = P_b C_b (r_o \lesseqgtr r \lesseqgtr r_b)$$
$$C_{O_2} = P_r C_b (r_o > r_b)$$
$$t > o; r = o : C_i = 0 (i = O_2, O_2^-, OH, H^+, HO_2, HO_2^-, H_2O_2)$$
$$\frac{dC_i}{dr} = 0 (i = OH^-, H, H_2, e^-)$$
$$r \rightarrow \infty : C_i \rightarrow 0$$
$$C_{O_2} \rightarrow C_b$$

where C_b is the concentration of oxygen in the bulk environment.

The salient features of this model and our experience in solving the equations are summarized below:

- The model involves eleven coupled non-linear, second order partial differential equations, each of which is coupled to the thermal diffusion problem.
- The equations were solved using a finite (central) difference technique for specific boundary and initial conditions
- Major difficulties were experienced with:
  - -- Computer memory
  - -- Ill-conditioning of matrices because of the length and time scales
- To our knowledge, this is the first time that computations of this type have been performed for HLNW systems.

Typical corrosion rate data produced by the RMRM for times extending to 361 years are summarized in Table 5 for three sets of initial γ- and n- energy dose rates. In performing these calculations, we employed the model parameters summarized in Table 2, with the additional assumption that the energy dose rates decay with a half life of 100 years (i.e. $k_t = 0.0069\ yr^{-1}$). The highest dose rates correspond roughly to those expected at the canister surface on closure (which is assumed to coincide with resaturation). The lowest levels correspond to those of a "cold" canister; that is where negligible γ- and n- fields exist at the canister surface. By comparing the calculated corrosion rates and corrosion losses at equivalent times, it is possible to judge the impact of radiolysis at the assumed energy dose rates on the integrity of the canister. The corrosion rates, in most, cases, differed among the three cases only in the third decimal place (not shown), but some influence of radiolysis is noted in the corrosion loss. However, the impact of radiolysis is predicted to be negligible and its omission in modeling the corrosion of HLNW canisters within the AECL/Ontario Hydro concept appears to be justified.

WAPPA

WAPPA utilizes a one-dimensional radial set of equations in analytic and algebraic form for simulating the physicochemical processes of metal corrosion and wasteform leaching, which can operate simultaneously on the waste package [11]. These processes, which

TABLE 5 -- Calculate maximum corrosion rate and corrosion loss as a function of γ- and n- energy deposition rates.

Time (years)	$\Gamma^{\gamma}(eV/m^3.s)$ $\Gamma^{n}(eV/m^3.s)$	10^{17} 10^{14}		10^{19} 10^{16}	
		(a)	(b)	(a)	(b)
1		3.1	3.1	3.1	3.1
25		2.9	74.0	2.9	74.0
81		2.5	198	2.5	198
169		2.1	350	2.1	350
361		1.8	639	1.8	639
529		1.7	900	1.7	900
729		1.7	1222	1.7	1222
1089		1.7	1817	1.7	1817
1521		1.7	2537	1.7	2537
2025		1.7	3377	1.7	3377
2401		1.7	4004	1.7	4004

(a) Corrosion rate (μm/yr)
(b) Corrosion loss (μm)

gradually degrade the nuclear waste package components with time, are driven by the modeled processes of heat flux, thermal stress, radiation damage from the decaying nuclear fuel, and water penetrating the waste package from the surrounding host rock. Flux to the host rock beyond the buffer layer and concentrations in the surrounding repository are also considered.

The various degradation processes are modeled individually within separate process models and the coupled effects are described at the system level within a time step. Each process is allowed to modify within a single time step the entire waste package. The approach taken is therefore described as barrier-integrated and process-sequential.

The waste package model recognizes the following physical and chemical processes:

- Time-dependent radionuclide inventories and radiation doses, radiation damage and radiolysis effects on the wasteform
- Heat distribution and heat flux across the layers of the waste package
- Mechanical stress applied to the repository in the horizontal and vertical direction, compounded by thermal stress
- Temperature-dependent uniform (general) and crevice corrosion of metallic barriers and enhancement of the corrosion rate in the presence of radiation
- Temperature-dependent leaching of the wasteform by ground water
- Transport of radionuclides from the waste package through breached barriers by diffusion, and through the buffer by advection and diffusion with retardation
- Mass accumulation at the waste package/host rock boundary

WAPPA predicts uniform thinning of the titanium container by general corrosion, and localized thinning by crevicing [21]. The accumulation of a corrosion film on the container surface is neglected resulting in a maximum rate of thinning. Only the corrosion aspects of

WAPPA are discussed in this paper; the reader is referred elsewhere [21] for an analysis of waste-form leaching and subsequent transport of radionuclides into the near-field environment. Uniform corrosion is described by the following equation,

$$X = X_0 - CR\,(f_0) \bullet \Delta t \tag{21}$$

where,

X = Container wall thickness (m)
f_0 = Radiolysis-induced corrosion enhancement factor
X_0 = Initial wall thickness (m)
CR = Corrosion rate (m/yr)

Uniform corrosion rates are expressed as a function of temperature and are summarized in Table 6. For this study, these rates are theoretical values calculated for a stainless steel analog in a thermal and radiation field. The radiolysis of water is modeled in the rate calculation; enhancement factors (above) are therefore 1.

TABLE 6 -- Corrosion properties database.

Parameter	Units	74°C	81°C	94°C	100°C
General Corrosion Rate(1)	m/yr	2.24E-06	2.44E-06	2.87E-06	3.11E-06
Parameter	Units	75°C	85°C	90°C	95°C
Crevice Corrosion Rate(2)	m/yr	5.0E-06	5.0E-06	5.0E-06	1.0E-05
Average Crevice Area(3)	m^2	2.0E-04	2.0E-04	2.0E-04	2.0E-04
Average Crevice Density(4)	$1/m^2$	1	1	1	1

(1) from Macdonald (1992); Radiolytic Maximum Rate Model results.
(2) 95°C data from Ikeda et al. [6].
(3) based on an anticipated 1cm X 2 cm surface dimension.
(4) arbitrary selection.

Crevice corrosion can be initiated under favorable electrochemical potential conditions at the container surface and in certain temperature-chemistry domains [6]. The empirical crevicing model in WAPPA can be used to predict the time of penetration of the container wall with pre-initiated crevices or crevices initiated after wetting. Areal degradation of the container can increase over time. Following an initial breach, the container wall is further degraded according to the following equation

$$AF_{c_2} = AF_{c_1} + \left[a_c d_c f_0\right] / A \tag{22}$$

where AF_c is the areal degradation factor, a_c is average crevice area (m^2), d_c is average crevice density ($/m^2$), f_o is the radiolysis corrosion enhancement factor, and A is the surface area of barrier (m^2). The parameters a_c, d_c, f_o, and A are user-specified quantities. The areal degradation factor is employed in the calculation of a diffusion coefficient for transport through the breached container. Thus, it is assumed that for all time following initial breach, the crevicing rate is sufficient to breach the container during each timestep. By inspection, we see that the increase in creviced area over time is directly proportional to timestep frequency.

The depth of crevice penetration is calculated in the same manner as for uniform corrosion. Crevices propagate ahead of the uniform corrosion. Crevicing rates have been adopted from the laboratory corrosion data of AECL [6], which show that the rate declines significantly at metal temperatures below 100°C. Minimum crevicing rates were conservatively adopted at lower temperatures, where crevicing was not observed during short term experiments. Crevicing rates are summarized in Table 6. An arbitrary initial crevice density of 1 per m^2 was specified, and the area per crevice is taken as 2.0 cm^2. The number of crevices and thus the creviced area on the container surface increases with each timestep. The rate of propagation of the crevices is a function of the container temperature at the current timestep. Empirical corrosion models are employed throughout the code largely because the present data would not support theoretical calculation of reliable rates for all relevant anodic and cathodic corrosion reactions on all relevant barrier materials in waste package modeling. The theoretical calculation of uniform corrosion rates for this work has been made possible only by very recent developments in mathematical models that consider cathodic reactions and radiolysis of water simultaneously.

The thermal process model of WAPPA computes a radial temperature profile within the waste package using a standard analytical expression for conductive and radiative heat transport in cylindrical co-ordinates. A heat production rate from the wasteform determined by the ORIGEN-S code, and a modeled repository temperature-boundary condition, were applied at the interior and exterior of the waste package, respectively. The thermal model results are used in the mechanical stress model (when applicable), and in the metal corrosion and wasteform leaching models [21] The metal container surface is initially at 99°C following emplacement, and cools with decay of the fuel.

AECL has reported a waste package temperature of less than 100°C at emplacement in the vault. From the heat production rates of the CANDU fuel determined by ORIGEN-S, we have simulated an initial wasteform temperature of 99°C. Cooling of the waste package is simulated through reduced heat production from the fuel. The simulated container surface temperature is about 37°C upon emplacement and reaches a maximum temperature of 95°C after 60 years of emplacement. After 1,200 years, the surface temperature is 84°C, which is also the specified waste package boundary temperature [1].

Degradation of the titanium container is assumed to be due to general corrosion and crevicing. General corrosion is applied at a maximum rate of 3 x 10^{-6} m/yr at 95°C, and at a minimum rate of 2.3 x 10^{-6} m/yr at 76°C. Crevice propagation has been specified at 1 x 10^{-5} m/yr at 95°C and decreasing to 5 x 10^{-6} m/yr at 76°C. The source of these rates has been discussed previously. General corrosion and crevicing are initiated upon wetting of the container. General corrosion continuously thins the entire container wall over time. Initial breach by crevicing is simulated at about 1,220 years after wetting. The container wall, originally 6.35 mm thick, has thinned to 3.2 mm at the time of initial breach by crevices. The surface area of the container is approximately 4.5 m^2, thus the crevice breach is characterized in this study as four or five crevices with a width of 1 cm and length of several centimeters. The combined area of the crevices upon initial breach is 1,100 cm^2 (about 33cm per dimension) has formed by year 2,500 after emplacement, at which time general corrosion is predicted to have consumed the container.

DISCUSSION

The corrosion of HLNW canisters in plutonic rock repositories, as envisaged in the AECL/Ontario Hydro disposal concept, has been addressed using four deterministic models:

- Mixed potential model
- Maximum rate model
- Radiolytic maximum rate model, and
- WAPPA

Each of these models emphasizes a different aspect of the corrosion problem, with the result that the mathematical complexities are kept to a minimum while maintaining a high level of physical realism. Thus, the MPM is used to estimate the corrosion potential of the container as a function of repository conditions, with the principal goal of determining whether various localized forms of corrosion are likely to occur. Contrariwise, the MRM and RMRM were used to estimate the maximum rate at which (general) corrosion could occur, with little information being obtained as to the variation of the corrosion potential with time. Clearly, more general models, such as the thin layer mixed potential model (TLMPM) developed by Macdonald and Urquidi-Macdonald [12], can be devised to provide a wider range of information, but this can only be done at the expense of greater mathematical complexity. The alternative approach is to use a "spread-sheet" algorithm, such as that employed in WAPPA, to input corrosion data from various sources. While such an approach is convenient and is capable of assimilating a wide range of information in a single code, it lacks some of the determinism afforded by those calculations that are based on well-established physical laws. Without doubt, one of the major challenges that we face in attempting to model HLNW repositories is to strike a reasonable balance between physical realism and mathematical tractability.

Recognizing the limitations outlined above and that the calculations are of a "scoping" nature, the most important findings of this study are as follows:

1. General corrosion is not a significant threat to the thin-walled (6.35 mm) titanium canister for exposure times up to 2000-2500 years.
2. The radiolytic generation of oxidizing species in the near-field environment does not significantly increase the general corrosion rate.
3. The rate of canister corrosion is predicted to be controlled by the reactions occurring at the canister surface, at least over the 500 year design life of the canister.

4. The titanium container might withstand crevice corrosion for about 1200 years after wetting first occurs.

These conclusions are tentative in that titanium was assumed to behave electrochemically like stainless steel.

Our study has identified a number of important deficiencies in our ability to develop deterministic models for the corrosion of HLNW canisters, as envisaged in the Canadian plutonic rock repository concept. These deficiencies include:

- The sensitivity of the corrosion rate to important model parameters (e.g. buffer and rock porosities) needs to be explored, not only to assess the reliability of a model but also to explore how the corrosion rate may be controlled by modifying the near-field environment (e.g. by reducing the porosity).
- Important kinetic parameters for the corrosion of titanium alloy in repository environments must be measured. These parameters include the steady-state polarization curves and the exchange current densities and Tafel constants for the reduction of oxygen and hydrogen peroxide, and for the oxidation of hydrogen on the alloy substrate.
- The radiolytic model needs to be modified to estimate the actual, rather than maximum, corrosion rate as has been done in the TLMPM that was previously used to model canisters in a tuff repository [12].
- More consistent and realistic models for the radiolysis of water under repository conditions need to be developed.
- The possibility that pitting corrosion can occur needs to be explored by measuring pit nucleation and growth data under "steady-state" (as opposed to potentiodynamic) conditions in simulated repository environments.

Finally, no amount of mathematical complexity can compensate for the lack of experimental data for key model parameters. Accordingly, a concerted and sustained effort is required to generate the required data in order that the predictions of physical models of the type described in this paper can be employed with confidence.

ACKNOWLEDGMENTS

The support of this work by Environment Canada through a contract to INTERA, Inc. is gratefully acknowledged.

REFERENCES

[1] Radioactive Waste Management and the Nuclear Fuel Cycle. Special Issue: Canadian Nuclear Fuel Waste Management Program (K.W. Dormuth, Guest Editor), 8 (2-3), 93-272 (1987).

[2] J. Kruger and K. Rhyne, Nucl. Chem. Waste Man., 3, 204 (1982).

[3] G. Marsh, Nucl. Energy, 21, 353 (1982).

[4] P. McKay and D.B. Mitton, Corrosion, 41, 52 (1985).

[5] P. McKay, Proc. 9th Int. Metal. Cong. 3, 288 (1984).

[6] B. M. Ikeda, M. G. Bailey, C. F. Clarke, and D. W. Shoesmith, "Crevice Corrosion of Titanium Under Nuclear Fuel Waste Conditions., " AECL Report, AECL-9568, (1989).

[7] D. D. Macdonald, "Modeling Corrosion Processes on Canadian HLNW Canisters," Technical Appendix-TA3 in Summary Report of the SAT to Environment Canada on the Canadian Nuclear Fuel Waste Disposal Concept, AUSCAN, Inc., Austin, TX, (1992).

[8] D.D. Macdonald, A.C. Scott, and P. Wentrcek, J. Electrochem. Soc., 128, 250 (1981).

[9] D.D. Macdonald, "Calculation of Corrosion Potentials in Boiling Water Reactors," in Proc. 5th Int. Symp. Environ. Degrad. Mat. Nucl. Power Systs.-Water Reactors," ANS/NACE, in press (1992).

[10] D. D. Macdonald, Corrosion, 48, 194 (1992).

[11] INTERA Environmental Consultants, Inc., WAPPA: A Waste Package Performance Assessment Code. ONWI-452, Office of Nuclear Waste Isolation Battelle Memorial Institute, Columbus, OH, (1983).

[12] D.D. Macdonald and M. Urquidi-Macdonald, Corrosion, 46, 380 (1990).

[13] D.D. Macdonald, H. Song, K. Makela, and K. Yoshida, Corrosion, in press (1992).

[14] T. Koizumi and S. Furuya, Proc. 2nd Int. Conf. Titanium Sci. Technol., 4, 2383 (1973).

[15] F.A. Posey and E.G. Bohlmann, Desalination, 3, 269 (1967).

[16] D.D. Macdonald, S. Biaggio, and H.-K. Song, J. Electrochem. Soc., 139, 170 (1992).

[17] L.F. Liu, C.-Y. Chao, and D.D. Macdonald, J. Electrochem. Soc., 128, 1194 (1981).

[18] D.D. Macdonald and M. Urquidi-Macdonald, Electrochim. Acta., 31, 1070 (1986).

[19] M. Urquidi-Macdonald and D.D. Macdonald, J. Electrochem. Soc., 134, 41 (1987).

[20] B. Case, Electrochim. Acta., 18, 293 (1973).

[21] J. L. Lolcama and D. D. Macdonald, Proc. 45th Canadian Geotechnical Conf., Toronto, October, (1992).

John A. Beavers[1], Neil G. Thompson[2], and Carolyn L. Durr[3]

APPROACHES TO LIFE PREDICTION FOR HIGH-LEVEL NUCLEAR WASTE CONTAINERS IN THE TUFF REPOSITORY

REFERENCE: Beavers, J. A., Thompson, N. G., and Durr, C. L., **"Approaches to Life Prediction for High-Level Nuclear Waste Containers in the Tuff Repository,"** Application of Accelerated Corrosion Tests to Service Life Prediction of Materials, ASTM STP 1194, Gustavo Cragnolino and Narasi Sridhar, Eds., American Society for Testing and Materials, Philadelphia, 1994.

ABSTRACT: An assessment was performed of approaches to life prediction for high-level nuclear waste containers. It was concluded that the commonly used remaining life assessment techniques are inadequate for the prediction of performance of the waste canisters primarily because of the reliance on empirical damage accumulation laws. The best hope for prediction of performance of the repository relies on utilizing mechanistic models based on first principles. It was further concluded that prediction of performance of the waste containers should rely primarily on propagation phenomenon, because of the uncertainties in proving that initiation will not occur.

The above philosophy imposes significant restrictions on the selection of container materials, and selection of the repository site. The materials selection for the waste canister should focus on corrosion allowance materials because the general corrosion rate can be bounded and the rates of pitting corrosion are expected to be much lower than for corrosion resistant materials. Corrosion resistant materials should be avoided because the probability of pit initiation is high for the range of possible environments and the rates of pitting corrosion are expected to be high, as a result of hydrolysis reactions. In order to reduce possible rates of localized corrosion, an anoxic repository should be selected. Because of the uncertainties in the predictive capability of the mechanistic models, corrosion monitoring, and a multiple barrier design for the waste package should be used.

KEYWORDS: life prediction, corrosion, waste containers, nuclear waste disposal, Tuff Repository

---

[1]Vice President of Research, Cortest Columbus Technologies, Inc., 2704 Sawbury Boulevard, Columbus, OH 43235.

[2]President, Cortest Columbus Technologies, Inc., 2704 Sawbury Boulevard, Columbus, OH 43235.

[3]Chemist, Cortest Columbus Technologies, Inc., 2704 Sawbury Boulevard, Columbus, OH 43235.

INTRODUCTION

The Department of Energy (DOE) is conducting a program for the disposal of high-level radioactive waste in a deep-mined geologic repository. The Nuclear Regulatory Commission (NRC), which is responsible for regulating high-level radioactive waste disposal, will review DOE's application for the construction and operation of the repository. To assist in evaluating DOE's application, the NRC's Office of Nuclear Regulatory Research is developing an understanding of the long-term performance of the geologic repository. As part of this effort, Cortest Columbus Technologies, Inc. (CC Technologies) conducted a program to investigate the long-term performance of container materials used for high-level waste packages [1-4].

The prediction of the long term performance of container materials is undoubtedly the most ambitious undertaking ever attempted in the area of predictive modeling. The required container life is greater than 1 000 years, which is over an order of magnitude longer than required in most industrial applications, where 20 to 50 year lives are common. Approaches to life prediction for general corrosion, pitting, and SCC are described below.

THE REPOSITORY SITE

Currently, the sole candidate repository site in the U.S. is the Tuff Repository, which will be located under Yucca Mountain, 100 miles northwest of Las Vegas, Nevada in the Nevada Test Site (NTS). The site is located in an extremely arid zone with about 15 cm/year annual precipitation. The evaporation-transpiration rates also are very high so the net water percolating down from the surface is of the order of a few millimeters per year [5].

The repository will be located in the Topopah Spring Member of the Paintbrush Tuff under Yucca Mountain. Tuff is an igneous rock of volcanic origin and is composed of volcanic rock fragments (shards) and ash. The potential repository horizon is in the lower, densely welded and devitrified portion of the Topopah Spring Member located 120 to 425 meters above the static water table [1].

The location of the repository above the static water table has a major impact on the anticipated environment. First of all, the environment will be aerated; the J-13 well water contains 5.7 ppm dissolved oxygen which probably represents a lower limit for oxygen. This condition is unique in that the plans for all other repositories, either in the United States or elsewhere, have called for locations below the static water table where conditions are deaerated (anoxic).

A second feature of the location of the repository above the water table is the elimination of the hydrostatic head on the waste container. At the repository elevation, the boiling point for water is about 95°C, and thus the environment at the waste package surface will be steam and air during the early life of the repository. As the temperature drops below atmospheric boiling, periodic contact of the waste containers with vadose water will occur [6]. The composition of this groundwater will be modified by radiation, the elevated temperature, interaction with prior precipitated salts around the containers, and interaction with corrosion products or possibly species from failed canisters.

CANDIDATE CONTAINER MATERIALS

Two classes of materials have been considered for construction of the waste container for the Tuff repository; corrosion resistant alloys and corrosion allowance alloys [1-4]. Corrosion resistant alloys are defined as alloys that rely on tenacious oxide films for their corrosion resistance. They exhibit negligible rates of general corrosion and are resistant to localized forms of corrosion. If selected, they would be used with a thin-wall design which relies on resistance to localized corrosion initiation for the required design life. No allowance in the design is given for general or localized corrosion. The Fe-Cr-Ni alloys, such as stainless steels (304L,316L) and nickel-base alloys (Alloy 825 and Alloy C276), are the primary candidate corrosion resistant alloys.

Corrosion allowance alloys are defined as alloys that exhibit measurable rates of general corrosion, necessitating the use of a thicker wall in the canister design than that used with the corrosion resistant alloys. The corrosion allowance alloys are generally less expensive that the corrosion resistant alloys, on a weight basis, but the overall cost of a canister constructed of a corrosion allowance alloy may be comparable to that of one made of a corrosion resistant alloy because of the thicker canister required. The thicker design has the added advantage that the canister provides some shielding from the radiation. The copper base alloys (CDA 102, CDA 613, CDA 715) are the primary candidate corrosion allowance alloys.

BACKGROUND ON LIFE PREDICTION

In recent years, there has been significant attention given to life prediction methodology in response to aging of major industrial equipment. Because life prediction is being applied to existing equipment, the term *Remaining Life Assessment* (RLA) generally has been applied to the analysis. As an example of one approach, the Electric Power Research Institute (EPRI) has developed and refined a three level procedure to assess the remaining life of equipment and facilities in nuclear and fossil power plants. Each successive level consists of greater detail in non-destructive examination, sampling, modeling, testing and analysis.

In related work, Jaske and Viswanathan [7] describe the seven major steps involved in remaining life assessment. These items are summarized in Figure 1 for failure of high temperature/high pressure equipment. For a component in service, the amount of degradation or damage must be measured by some means to quantify the current damage state or condition of the material from which the component is fabricated (Step 1). The measurements are made by non-destructive monitoring or by destructive examination, where several components or a generic class of components are of interest. In some instances, it is also possible to destructively examine sub-sized specimens from an operating component. The information from this analysis along with information from previous analyses of similar components are then used to identify the mechanisms of degradation and the root causes of any failures that have occurred (Step 2).

Equipment and component configurations are determined from the original design drawings and specifications, the records of as-built dimensions, and the measurements made in the field (Step 3). This effort is generally focused on key areas of the component where degradation is most likely to compromise the design life of the component.

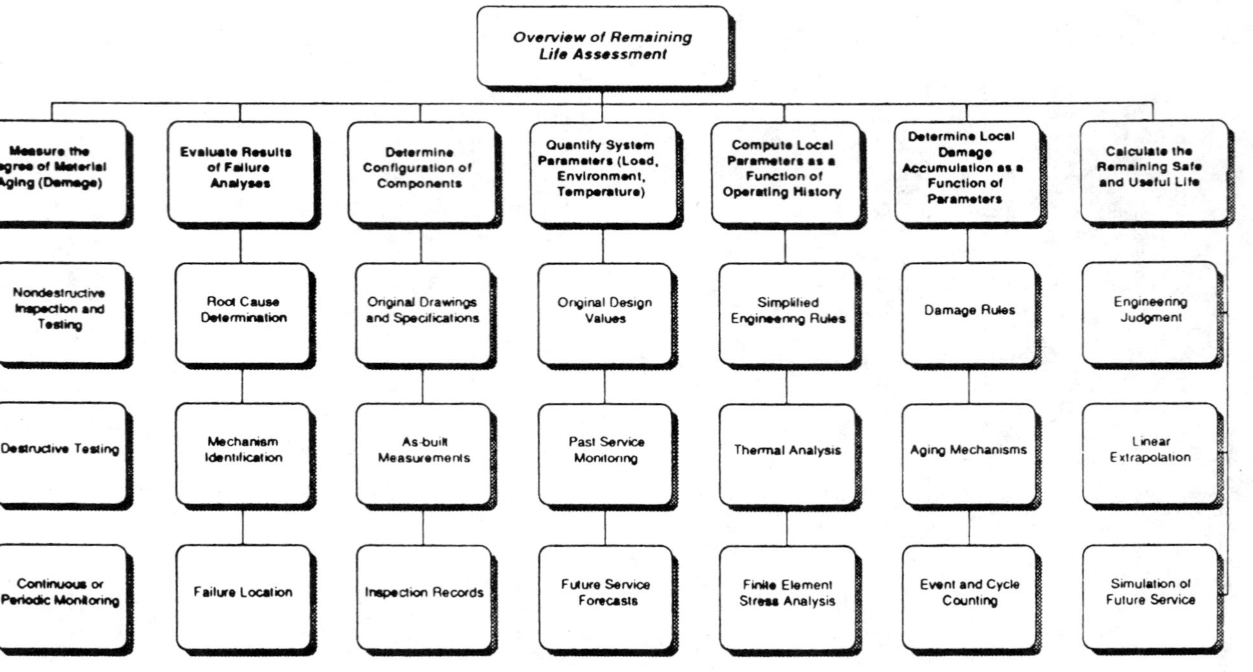

FIG. 1--Summary of major steps in remaining life assessment of engineering equipment (After Jaske and Viswanathan [6]).

The system operating parameters are then quantified. These parameters include the operating environment and temperature, the mechanical loads on the component and any other parameter that may affect service performance. Included in this quantification are upset conditions as well as standard operating conditions (Step 4). The system parameters and component configuration are then used to calculate the local parameters such as stress, stress intensity, and strain, at the critical locations in the component being assessed (Step 5). For corrosion related processes, local chemistry changes may be included in the analysis. These computations are performed using simple engineering rules, finite element analyses, thermal analysis or other analytical methods.

These calculated values of local stress, strain and environment are then used in appropriate models of material damage accumulation (Step 6). The remaining life is then predicted from a knowledge of the current condition and the rate of damage accumulation (Step 7).

APPLICATION OF RLA TO NUCLEAR WASTE DISPOSAL

The application of these existing prediction methodologies to the estimation of life of containers for high-level nuclear waste is difficult, at best. Steps 1 and 2 above focus on the identification of possible modes of degradation for the component. In operating equipment, the prior service performance provides specific information on the most probable degradation modes, as well as rates of degradation for the modes. For the waste repository, it is difficult to reasonably eliminate many possible failure mode because of the imprecise definition of the environment and the absence of any prior operating data for the repository.

The major goal of the program performed by CC Technologies for the Nuclear Regulatory Commission was the identification of possible corrosion related failure modes for the Tuff Repository [1-4]. Corrosion failure modes considered included common failure modes such as general corrosion, pitting and crevice corrosion, galvanic corrosion (with the borehole liner) and Stress Corrosion Cracking (SCC), as well as more arcane possible modes of failure such as thermogalvanic corrosion.

A further complication in the identification of degradation modes through laboratory testing is the fact that many of the alloys considered for construction of the repository are thermodynamically unstable and may change their structure over the life of the repository. Therefore, tests performed on new material may not reflect performance characteristics of aged materials.

The determination of the equipment and component configurations from the repository designs (Step 3) should be one of the easiest tasks in a life assessment. However, the final selection of a material of construction, and the container and overall repository design, that are based on the material selection, have not been finalized at this time. Therefore, the researcher must select the most current alternative candidate designs, and identify the weakest aspects of each design, where degradation is most likely to compromise the design life of the component. This approach tends to dilute the effort and reduce the likelihood of a successful analysis.

As described above, Steps 4 and 5 involve the identification of the overall operating parameters and the values of these parameters at the local sites, identified in Step 3, where failure is most likely to occur. Again, these steps are very difficult for the proposed repository, and exacerbate the task of identifying the mechanisms of degradation (Task 2). Most of the prior corrosion testing has been performed on actual or simulated J-13 well water. A typical composition is given in Table 1. The actual J-13 well water is taken from a well located several miles from the proposed repository site, which produces water from the Topopah Spring Member. This water may not be a good approximation of the actual water that will be present in the repository. The vadose water present in the repository will be altered by radiation, boiling, and interaction with corrosion products and possibly leaking radioactive waste. The influence of heating the groundwater and radiation on composition, as found in several studies, also is shown in Table 1.

In the recently completed program [1-4], a statistical modeling approach was used to identify potential failure modes in the complex repository environment. Corrosion performance was assessed primarily by means of the cyclic potentiodynamic polarization (CPP) technique. This is a screening technique for assessing localized corrosion behavior. Forty-two synthetic environments were developed using ranges of solution variables selected from data in the literature for the Tuff well water, as influenced by heating, interaction with Tuff rock and radiation. The ranges of environmental species considered are given in Table 2. Most of the species considered in the solutions evaluated were concentrated by a factor of 200, which may be too small a factor for a situation where heat transfer and drying will occur. The compositions of the actual test solutions are given in Table 3. One of the most significant findings of this work was that all of the candidate alloys considered exhibited failures from one or more corrosion failure modes in the limited factor space considered. Results for the most corrosion resistant candidate container material considered, Incoloy Alloy 825, are given in Table 4.

The final two Steps, Application of Damage Models (Step 6) and Remaining Life Calculation (Step 7) have their own inherent limitations as well as the limitations imposed by the uncertainties in the prior five steps. For example, most of the damage models used for power plant equipment are empirical deterministic models that have been developed through years of operating experience. These models generally cannot be extrapolated more than an order of magnitude past their range of verification without introducing significant uncertainty.

The extrapolations required for the repository are much greater. The required lifetimes for the waste canister are generally greater than 1 000 years, while data are rarely available for more than ten years, giving a minimum extrapolation of two orders of magnitude. Figure 2 is an example of the error introduced by such long extrapolations. In this example, a power law of the form:

$$D = At^n \tag{1}$$

is assumed for the rate of degradation, where D = maximum depth of attack in µm, A and n are constants and t is time in years. Such power laws are frequently used to described atmospheric corrosion and pitting corrosion of some alloys. In the example given in Figure 2, a value of 50 µm is assumed for A, and a value of 0.7 is assumed for n. As shown in the figure, a variation of ± 0.2 in the exponent causes an uncertainty of 23 mm in the predicted wall loss after 1 000 years. This uncertainty in the predicted wall loss is greater than the actual planned canister thickness for most corrosion resistant material designs being considered. Since the exponents are rarely known to a better accuracy, such extrapolation uncertainties can be expected. Figure 2 also graphically shows the large extrapolations needed for life prediction based on empirical relationships.

TABLE 1--Concentration ranges for environmental species in Tuff groundwater [3].

Environmental Variable	Nominal Concentration Of Well Water,[a] mg/l	Concentration Range Of Groundwater Heated To 90-250°C With Tuff Rock,[a, b] mg/l	Concentration Range Of Groundwater Heated To 90-150°C In Presence Of Radiation[c], mg/l
pH	7.6	5.8 - 8.3	7.6 - 9.0
SiO_2	58	29 - 394	1.05 - 3.10
HCO_3^-	125	45 - 195	8.1 - 21
F^-	2.2	2.2 - 4.4	0.38 - 1.4
Cl^-	6.9	6.5 - 8.9	2.5 - 5.8
NO_3^-	9.6	8.5 - 16.8	2.5 - 18.8
SO_4^{2-}	18.7	13.3 - 22	1.8 - 6.2
NO_2^-	...	0.7 - 1.5	1.2 - 3.8
H_2O_2	...	...	0 - 4.8[d]
Al^{3+}	.012	0.016 - 4.8	<0.15 - 0.18
Fe^{2+}	0.006	...	<0.02 - 0.04
Ca^{2+}	12.5	0.21 - 13.2	2.7 - 9.8
Mg^{2+}	1.9	0.009 - 2.0	0.6 - 1.1
K^+	5.1	3.2 - 19.4	2.4 - 4.8
Na^+	44	35 - 74	2.8 - 36

(a) Knauss, 1985 [8].

(b) Oversby, 1983 [9].

(c) Yunker, 1986 [10].

(d) Glass, 1985 [11].

TABLE 2--List of variables included in the matrix of potentiodynamic polarization tests [3].

Variable Number	Variable Name	Origin	Test Matrix High Concentration mg/l	Test Matrix Low Concentration mg/l	Nominal Concentration Of J-13, mg/l
1	SiO_2	J-13	215	2.2	58
2	HCO_3^-	J-13	2000	0.4	125
3	F^-	J-13	200	0.04	2.2
4	Cl^-	J-13	1000	0.2	6.9
5	NO_3^-	J-13	1000	0.2	9.6
6	NO_2^-	Radiolysis	200	0	...
7	H_2O_2	Radiolysis	200	0	...
8	Ca^{2+}	J-13	20	0.004	12
9	Mg^{2+}	J-13	20	0.004	1.9
10	Al^{3+}	J-13	0.8	0.0004	0.01
11	PO_4^{3-}	J-13	2.0	0.01	0.12*
12	Oxalic Acid	Radiolysis	172	0	...
13	O_2	Open Repository and Radiolysis	30**	5**	...
14	Temp	J-13	90°C	50°C	...
15	pH	J-13	10	5	7.6

* McCright, 1985 [12].

** Volume %.

TABLE 3--Test environments used in statistical matrix of polarization tests [3].

Test No.	SiO_2	HCO_3^-	F^-	Cl^-	NO_3^-	NO_2^-	H_2O_2	Ca^{2+}	Mg^{2+}	Al^{3+}	PO_4^{3-}	Oxalic Acid	O_2	Temp °C	pH
1	2.2	0.4	200	1000	1000	200	200	0.8	0.004	0.0004	0.01	0	5	90	5
2	2.2	2000	200	1000	1000	0	0	0.004	0.004	0.8	2.0	172	5	50	5
3	215	0.4	0.04	0.2	1000	200	0	0.8	0.004	0.8	2.0	0	30	50	5
4	215	2000	0.04	0.2	1000	0	200	0.004	0.004	0.0004	0.01	172	30	90	5
5	2.2	0.4	200	0.2	0.2	0	200	0.8	0.8	0.8	0.01	172	30	50	5
6	2.2	2000	200	0.2	0.2	200	0	0.004	0.8	0.0004	2.0	0	30	90	5
7	215	0.4	0.04	1000	0.2	0	0	0.8	0.8	0.0004	2.0	172	5	90	5
8	215	2000	0.04	1000	0.2	200	200	0.004	0.8	0.8	0.01	0	5	50	5
9	2.2	0.4	0.04	1000	0.2	200	200	0.004	0.004	0.0004	2.0	172	30	50	10
10	2.2	2000	0.04	1000	0.2	0	0	0.8	0.004	0.8	0.01	0	30	90	10
11	215	0.4	200	0.2	0.2	200	0	0.004	0.004	0.8	0.01	172	5	90	10
12	215	2000	200	0.2	0.2	0	200	0.8	0.004	0.0004	2.0	0	5	50	10
13	2.2	0.4	0.04	0.2	1000	0	200	0.004	0.8	0.8	2.0	0	5	90	10
14	2.2	2000	0.04	0.2	1000	200	0	0.8	0.8	0.0004	0.01	172	5	50	10
15	215	0.4	200	1000	1000	0	0	0.004	0.8	0.0004	0.01	0	30	50	10
16	215	2000	200	1000	1000	200	200	0.8	0.8	0.8	2.0	172	30	90	10
17	215	2000	0.04	0.2	0.2	0	0	0.004	0.8	0.8	2.0	172	30	50	10
18	215	0.4	0.04	0.2	0.2	200	200	0.8	0.8	0.0004	0.01	0	30	90	10
19	2.2	2000	200	1000	0.2	0	200	0.004	0.8	0.0004	0.01	172	5	90	10
20	2.2	0.4	200	1000	0.2	200	0	0.8	0.8	0.8	2.0	0	5	50	10
21	215	2000	0.04	1000	1000	200	0	0.004	0.004	0.0004	2.0	0	5	90	10
22	215	0.4	0.04	1000	1000	0	200	0.8	0.004	0.8	0.01	172	5	50	10
23	2.2	2000	200	0.2	1000	200	200	0.004	0.004	0.8	0.01	0	30	50	10
24	2.2	0.4	200	0.2	1000	0	0	0.8	0.004	0.0004	2.0	172	30	90	10
25	215	2000	200	0.2	1000	0	0	0.8	0.8	0.8	0.01	0	5	90	5
26	2.2	0.4	200	0.2	1000	200	200	0.004	0.8	0.0004	2.0	172	5	50	5
27	2.2	2000	0.04	1000	1000	0	200	0.8	0.8	0.0004	2.0	0	30	50	5
28	2.2	0.4	0.04	1000	1000	200	0	0.004	0.8	0.8	0.01	172	30	90	5
29	215	2000	200	1000	0.2	200	0	0.8	0.004	0.0004	0.01	172	30	50	5
30	215	0.4	200	1000	0.2	0	200	0.004	0.004	0.8	2.0	0	30	90	5
31	2.2	2000	0.04	0.2	0.2	200	200	0.8	0.004	0.8	2.0	172	5	90	5
32	2.2	0.4	0.04	0.2	0.2	0	0	0.004	0.8	0.0004	0.01	0	5	50	5
33*	108	500	50	250	250	50	50	0.2	0.2	0.2	1.3	43	15	70	7.5
34*	108	500	50	250	250	50	50	0.2	0.2	0.2	1.3	43	15	70	7.5
35*	108	500	50	250	250	50	50	0.2	0.2	0.2	1.3	43	15	70	7.5
36*	108	500	50	250	250	50	50	0.2	0.2	0.2	1.3	43	15	70	7.5
37±	64.2	121	1.7	6.4	12.4	0	0	12	1.7	0	0	0	0	90	7.0
38+	64.2	121	1.7	1000	12.4	0	0	12	1.7	0	0	0	0	90	7.0
39	108	500	50	250	250	50	50	0.1	0.1	0.5	1.3	50	15	70	5
40	108	500	50	250	250	50	50	20	20	0.5	1.3	50	15	70	5
41	108	500	50	250	250	50	50	20	0.1	0.5	1.3	50	15	70	10
42	108	500	50	250	250	50	50	0.1	20	0.5	1.3	50	15	70	10

* Tests 33 through 36 are quadruplicate midpoint tests which help to establish the degree of reproducibility of the CPP Tests.

± Simulated J-13 well water.

\+ Simulated J-13 well water containing 1000 ppm Cl^-.

TABLE 4--Polarization parameters for the test solutions in the statistically designed matrix of experiments for alloy 825 [3].

Test Solution	E_{cor}, V, SCE	i_{cor}, μA/cm² × 10⁻⁶	i_{pas}, μA/cm² × 10⁻⁶	E_b, V, SCE	E_{rp}, V, SCE	Comments
1	+0.225	0.40	16	+0.72	+0.48	No pitting/tarnished
2	-0.325	0.24	2.6	+0.78	+0.68	Very shallow pitting
3	-0.005	0.10	1.7	+0.94	+0.80	Pitting
4	+0.030	0.37				No pitting/slight tarnishing
5	+0.225	0.50				No pitting/slight tarnishing
6	-0.400	0.13	1.7	+0.65	+0.53	No pitting/tarnished
7	-0.090	0.27	1.3	+0.60	+0.15	Pitting
8	+0.105	0.14		+0.35	+0.12	Pitting/crevice/intergranular attack
8 repeat	+0.100	0.41		+0.28	+0.09	Pitting/crevice corrosion
8 repeat	-0.070	0.55	1.2	+0.24	+0.09	Pitting/crevice corrosion
9	+0.260	0.30	4.8	+0.58	+0.10	Pitting/crevice/intergranular attack
10	-0.320	0.14	2.0	+0.68	+0.63	No pitting/tarnished
11	-0.465	0.17	2.0	+0.67	+0.54	No pitting/tarnished
12	-0.010	0.12				No pitting/tarnished
13	+0.180	0.31		+0.36	+0.33	Pitting
14	-0.250	0.24	3.0	+0.69	+0.67	No pitting
15	-0.205	0.13	1.6	+0.75	+0.57	No pitting/tarnished/(possible very slight pitting)
16	-0.070	0.22				No pitting/local tarnishing/slight crevice crack
17	-0.190	0.12	2.0	+0.73	+0.70	No pitting
18	+0.135	0.32		+0.29	+0.14	Pitting/crevice attack
19	-0.070	0.10	10	+0.72	+0.15	Pitting/crevice attack
20	-0.320	0.10	1.5	+0.69	+0.09	Pitting/crevice attack
21	-0.450	0.14	2.0	+0.67	+0.60	No pitting
22	-0.040	0.16				No pitting/local tarnishing/slight etching
23	+0.020	0.26				No pitting/local tarnishing/slight etching
24	-0.180	0.16	2.6	+0.64	+0.50	No pitting/slight tarnishing
25	-0.420	0.11	2.5	+0.70	+0.60	No pitting/tarnishing/local active attack
26	+0.200	0.14				No pitting/tarnishing/very shallow local active attack
27	+0.120	0.22		+0.27	+0.11	Pitting/local active attack
28	-0.020	0.16	3.0	+0.64	+0.54	No pitting/tarnishing
28 repeat	-0.125	0.15	2.0	+0.64	+0.46	No pitting/tarnishing
29	-0.070	0.08	1.8	+0.80	+0.70	Very shallow pitting
30	+0.180	0.68	50	+0.67	+0.18	Pitting
31	+0.080	0.39				No pitting/local tarnishing/slight active attack
32	-0.270	<0.08	1.3	+0.76	+0.76	No pitting
33	+0.095	0.15				No pitting/heavy tarnishing
34	+0.100	0.33				No pitting/heavy tarnishing
35	+0.090	0.16				No pitting/heavy tarnishing
36	+0.170	0.32				Pitting/(shallow)/heavy tarnishing
37	-0.650	0.08	2.4	+0.70	+0.70	No Pitting
38	-0.480	0.14	2.4	+0.70	+0.16	Pitting
39	+0.267	0.14		+0.60	+0.29	Pitting
40	+0.189	0.13		+0.90	+0.90	Pitting/crevice attack
41	-0.255	0.02		+0.64	+0.62	No attack
42	+0.026	0.33		+0.75	+0.75	No pitting/etching at crevice

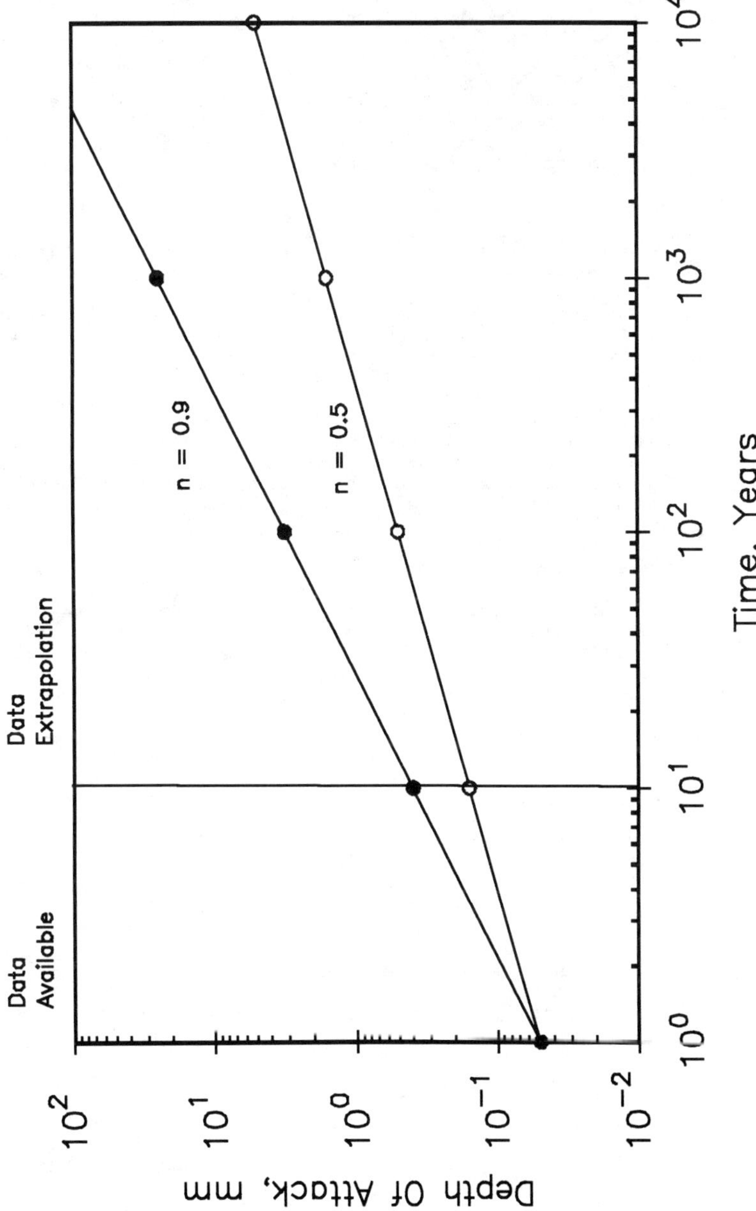

FIG. 2--Depth of attack as a function of time for two assumed exponent values in a power law relationship.

The above analysis suggests that the remaining life assessment approaches currently in use will not provide the predictive capability required for the licensing of the high-level nuclear waste repository. Alternative approaches include using a fundamental understanding of the mechanisms of failure to better estimate the rates of attack, conservatively estimating the upper bound rates from this mechanistic information, and utilizing redundant barriers to radioneuclide migration. Further considerations for the common corrosion failure modes are described below.

General Corrosion

In comparison with the localized forms of corrosion, general corrosion is not considered to be a serious problem with respect to life prediction for the waste container. The precise power law describing the rate of attack as a function of time may not be known, but the problem can be bounded since general corrosion rates normally decrease with time as protective passive films develop. Thus, assuming an exponent of 1 for the power law is usually highly conservative. Acceptable rates of attack also can be easily measured in the laboratory using standard laboratory equipment. For example, a typical weight measurement accuracy is 0.1 mg. Therefore, for a specimen with an area of 25 cm^2 and an exposure of one year, an accuracy of better than 0.01 µm/y can be achieved. This corresponds to an error in the thickness measurement of only 10 µm (0.01 mm) in 1 000 years, assuming an exponent of 1 for the power law.

The major problems with accurately bounding the rate of general corrosion are associated with the accurate identification of the environment and possible metallurgical changes in the alloys. However, where these problems exist, more serious problems are possible with the localized forms of corrosion. For example, an unexpected drop in the pH at the canister surface can promote a significant increase in the rate of general corrosion, but pitting or crevice corrosion usually occur first as the stability of protective films is compromised. Therefore, it is prudent to focus on these localized forms of corrosion in attempting to estimate container life.

LOCALIZED CORROSION (PITTING, CREVICE AND STRESS CORROSION)

Three types of approaches have been used to predict the localized corrosion performance of engineering materials; (1) the use of statistical distributions for the predictions of performance, (2) mechanistic predictions of the development of localized environments, and (3) mechanistic predictions of initiation and propagation of localized corrosion.

Statistical techniques have been applied to the initiation process as well as the combined initiation and propagation process. For example, a number of researchers have performed statistical analyses of pit initiation parameters such as current transients (Williams [13]), pitting potentials (Fratesi [14], Lemaitre [15], and Shibata [16]), and incubation time under potentiostatic control (Shibata [16], Doelling and Heusler [17], Keddam, et al. [18]). It is generally agreed that the pit initiation process is spacially and temporally probabilistic as opposed to deterministic and the data have been fitted to various distribution functions including the log normal distribution and the largest extreme value distribution.

Similar statistical approaches have been utilized for the analysis of pitting or SCC data from operating equipment (Finel and Toncre [19], Ishikawa [20], Crews [21], Strutt [22], Gallucci [23]) or laboratory experiments (Janik-Czachor [24], Baba [25], Aziz [26]).

Common probability distribution functions that are used include the log-normal, largest extreme value, and Weibull functions. Actual pit depth data may be fitted to a distribution function, as shown in Figure 3, and the shift in the distribution with time determined. An empirical model is then used to describe the time dependance of the growth rate. Alternatively, the time to failure may be analyzed directly, as was done by Finel and Toncre [19]. They applied extreme value analysis methods to leak data for a submerged pipeline. The data were then used to compare leak frequencies for different pipe diameters and thicknesses.

In a later study, Finley [27] showed that an equation of the form:

$$P_s = \left(\frac{-t}{V^k}\right) \tag{2}$$

(where P_s = probability of survival, t = time of first leak, V = characteristic age, and k = a constant) applied to his own data described above, as well as several other examples of pitting corrosion data for carbon steel.

There are two significant problems with applying this approach to life prediction for the waste canister. First of all, these are empirical analyses and, as described previously, it is dangerous to extrapolate the data to times differing in more than an order of magnitude. Secondly, the analyses assume that the environmental conditions do not change during the prediction period. As described above, the repository is not well characterized and both the environmental conditions and possibly the alloy microstructure will change over the life of the repository.

It is generally recognized that local changes in environment within pits or crevices are responsible for most corrosion related failures of operating equipment. Materials of construction are generally selected for their resistance to attack in the bulk environment. Local changes in the environment as a result of deaeration, or concentration of aggressive species may be responsible for the initiation of the corrosion process, and also occur as a result of the propagation. Accordingly, a significant effort has been made in the corrosion industry to predict the development of these aggressive local environments for important systems such as those found in nuclear power plants. The literature on this subject is quite extensive and it is not within the scope of this paper to provide an extensive review. Good reviews of the subject are given in the degradation mode surveys for the Yucca mountain project [28] and in the proceedings from a conference on the chemistry within pits, crevices, and cracks [29]. Turnbull has performed extensive research on the development of localized environments within corrosion cavities [30].

The extensive research on the development of local environments is generally useful to the problem of life prediction for the waste repository, but it is probably not reasonable to assume that the local environments adjacent to the waste canister can be well characterized since we have such a poor understanding of the bulk environments. Similarly, processes such as initiation of pitting, crevice corrosion, or stress corrosion cracking, cannot be predicted with much certainty because of the poor definition of the environment. The overall issue of predicting that a process will not occur (initiate) is also a problem from a regulatory standpoint.

From a prediction standpoint, the above analysis indicates that canister performance should be based primarily on the limiting rates of propagation of localized corrosion. This approach has been taken

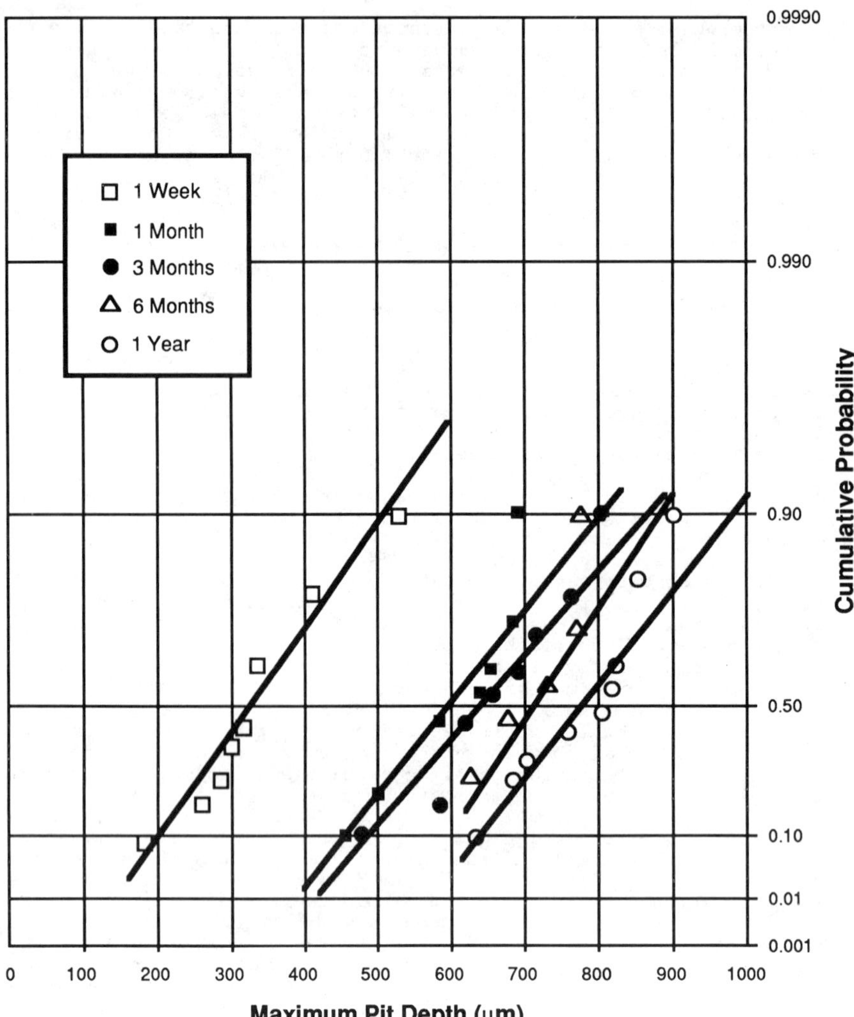

FIG. 3--Maximum pit depth data for an aluminum alloy immersed in tap water for the time periods shown plotted on extreme value probability paper against their cumulative relative frequencies (After Azis [26]).

for pitting of a carbon steel waste container in the Basalt Repository [31], [32]. It was found that the rate of pit propagation was limited by the reactivity of the pit wall near the mouth of the pit. The current generated by the corrosion reactions of the wall at the mouth created an ohmic potential drop, which greatly reduced the rate of pit propagation. The effect is shown schematically in Figure 4. The overall process limits the aspect ratio of the pits in carbon steel by reducing the rate of propagation of deep pits and opening the mouths of the pits.

While the form of the rate limiting equation was not determined in the research, it can be concluded that the rate of pit propagation approaches either the rate of general corrosion in the bulk environment or the rate of corrosion in the hydrolysis products produced in the pit, whichever is greater. In the first case, the system is self-limiting since, where the pitting rate is higher than the general corrosion rate, the aspect ratio of the pit would increase with time to such an extent that the pitting process would be stifled. In the second case, the aspect ratio of the pit can increase with time since pit propagation can be decoupled from the reduction reactions at the boldly exposed surface. The differential cell action at the mouth of the pit simply acts to maintain the aggressive environment within the pit.

It is important to point out that the hydrolysis reactions for corrosion allowance materials such as carbon steel do not produce sufficiently low pH values to drive high rates of pitting. This contention is supported by the fact that high aspect ratio pits are rarely observed in carbon steel. On the other hand, hydrolysis reactions may be more significant than differential cell action in the propagation of pits in some corrosion resistant alloys such as Fe-Cr-Ni alloys. This is supported by the low pH values possible for the hydrolysis reactions for the latter alloys, as well as the actual morphology of the pits observed. In many cases, pits in Fe-Cr-Ni alloys balloon out beneath the surface of the metal, with only a small pin-hole at the surface to indicate the presence of the pit. A typical example of this type of pitting is shown in Figure 5 [33]. The rates of attack within these pits are exceedingly high and well above an acceptable rate for a long-term waste container. For example, the pit shown in Figure 5 had a propagation rate of greater than 1 mm/year. Thus, this analysis suggests that these alloys should be avoided for construction of the waste canister.

The discussion of propagation rates for localized corrosion has focused primarily on pitting corrosion. For that failure mechanism, it appears that, through proper materials selection, the upper bound values of the rates can be quantified and the values determined may be sufficiently low that the desired design life of the container can be achieved. The process of stress corrosion cracking adds a further complication. Returning to Figure 4, we see that the rate of pit propagation, where active pit walls are present, is limited by the corrosion activity of the wall near the mouth of the pit. However, where the pit walls are passive, much higher rates of propagation are possible. Passive walls are not commonly observed with pitting, but this is generally the case with SCC. In fact, the localization of the corrosion at the crack tip by the passive walls is a distinguishing feature of SCC for all materials, including the corrosion allowance materials that are most promising from the standpoint of pit propagation.

Ford and Andresen have performed extensive modeling of stress corrosion crack propagation. A good review of the work is provided in Reference [34]. They have developed a power law to calculate the rate of crack propagation:

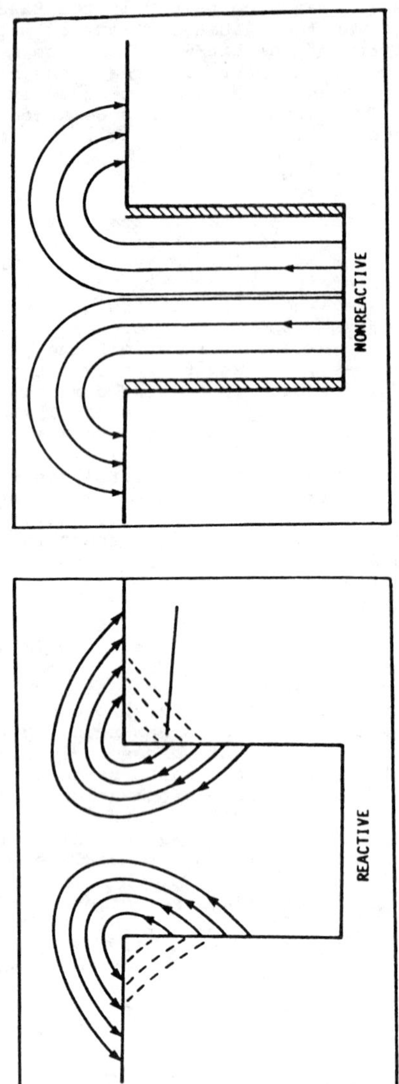

FIG. 4--Schematic showing the current behavior for nonreactive and reactive pit walls; the direction of current flow is indicated. Note that the case of the non-reactive pit wall was an experimental artifice and does not represent actual behavior (After Beavers [32]).

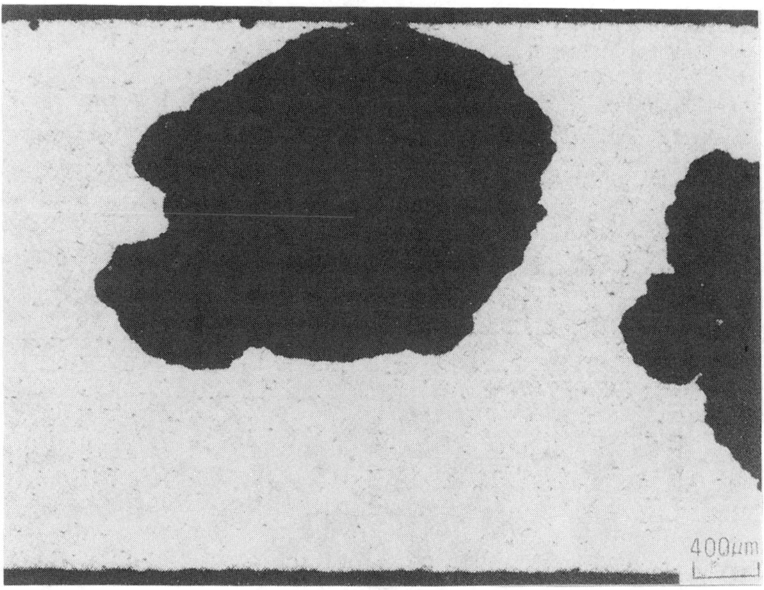

FIG. 5--Photomicrograph of pit in Hastelloy alloy G following 18-month exposure in outlet duct of a scrubber on a coal fired power plant (After Koch et al. [33]).

$$\frac{da}{dt} = \frac{M}{n\rho F} \frac{i_o t_o^n}{(1-n)\epsilon_f^n} (\dot{\epsilon})^n \tag{3}$$

where da/dt = rate of crack propagation, M = atomic weight, n = valence, ρ = density, F = Faradays constant, i_o = bare surface oxidation rate, t_o and n are constants in a power law that describes the decrease in current with time for a repassivating surface, ε_f = critical strain for rupture of the film at the crack tip, and ε = the strain rate. The power law for repassivation is given below:

$$i_t = i_o \left(\frac{t}{t_o}\right)^{-n} \tag{4}$$

where i_t = the instantaneous corrosion rate as a function of time. The constants in the equations are obtained from fitting the equations to the available data.

This model has been applied to the problem of SCC of stainless steels in light water reactor environments with good results, as shown in Figure 6. These data also show that the lowest crack propagation rate measured or predicted is of the order of 5 X 10^{-7} mm/s, which corresponds to over 15 meters of crack growth in 1 000 years.

Equation 3 suggests possible methods to reduce crack propagation rates to acceptable values. For example, the strain rate can be reduced by reducing residual stresses in the canister and eliminating notches and flaws. It may be possible to reduce the bare

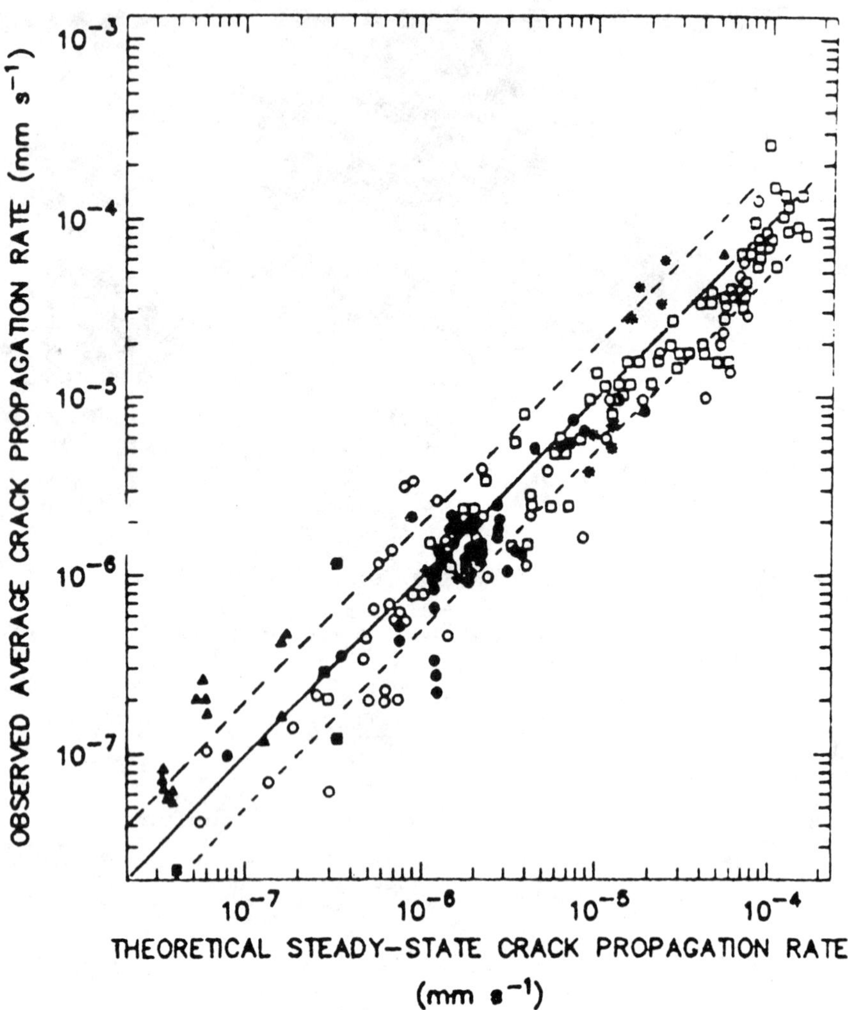

FIG. 6--Relationship between the observed and theoretical steady state crack propagation rate for stainless steel in water at 288°C. The wide range of propagation rates results from the range of degrees of sensitization of the alloy, and the range of environmental and stressing conditions (After Ford and Andresen [34]).

surface oxidation rate, i_o, by reducing the temperature or the presence of oxidants in the repository. Under activation polarization, the bare surface oxidation rate follows Butler-Volmer kinetic behavior:

$$i_o = i_e \left[\left(\exp \frac{(1-\beta)\eta F}{RT} \right) - \exp \left(\frac{-\beta\eta F}{RT} \right) \right] \tag{5}$$

where i_e = equilibrium exchange current, β is related to the Tafel constant, T = the absolute temperature, η is the difference between the equilibrium potential and the actual potential of the metal, and R = universal gas constant. Equation 5 shows that the bare surface oxidation rate is related exponentially to the temperature and overpotential.

The overpotential is, in turn, affected primarily by the presence of oxidants in the environment. There are two primary sources of oxidants in the repository; the groundwater and radiolysis products. Therefore, it appears to be prudent to select an anoxic repository, as opposed to the oxic Tuff Repository. It also appears to be prudent to have surface storage of the waste prior to permanent disposal to reduce the radiation level. This approach has the added advantage of reducing the temperature of the waste. In an anoxic repository, active metals, such as iron, will corrode due to direct reaction with water but rates of attack will be low.

Regardless of the details of the mechanistic equations selected, the method of validation and the nature of the supportive data, many uncertainties will remain in the performance prediction. Other failure modes may be important, such as hydrogen embrittlement, microbial influenced corrosion (MIC), or thermogalvanic corrosion, or unforseen changes in the environment or alloy metallurgy may occur. Therefore, a multiple barrier design for the waste package should be strongly considered. The design should utilize natural galvanic relationships to enhance performance. For example, utilize a carbon steel overpack with a copper canister. If and when failure of the overpack occurs, the carbon steel will act as a sacrificial anode to cathodically protect the copper canister from corrosion. Because initial failure of the overpack will occur locally, the area ratio between the exposed copper at the failure site and the remainder of the overpack will be optimum for the extended protection of the copper canister.

Because of the many uncertainties in the overall life prediction process, corrosion monitoring in the repository also will be necessary to assess the predictive capability of the models and whether any unforseen mechanisms of corrosion have initiated.

SUMMARY AND CONCLUSIONS

The following are the conclusions of the analysis described above:

(1) The commonly used remaining life assessment techniques are inadequate for the prediction of performance of the waste canisters because of the poor definition of the environment and the reliance on empirical damage accumulation laws.

(2) It is not prudent to extrapolate any empirical prediction equation by more than one order of magnitude, which is far short of the extrapolation needed to predict performance of the waste containers in the repository.

(3) The localized forms of corrosion (pitting and stress corrosion) pose the greatest problem with respect to limiting the life of the container and estimating that life.

(4) The best hope for prediction of performance in the repository relies on utilizing mechanistic models based on first principles and bounding the problem.

(5) Prediction of performance of the waste containers should not be based on initiation phenomenon, but should rely primarily on propagation phenomenon, because of the uncertainties in the definition of the environment and the difficulty in licensing a repository based on proving that initiation will not occur.

(6) The above philosophy imposes significant restrictions on the selection of container materials, and the actual repository site.

(7) The materials selection for the waste canister should focus on corrosion allowance materials because the general corrosion rate can be bounded and the rates of pitting corrosion are expected to be much lower than for corrosion resistant materials.

(8) Corrosion resistant materials should be avoided because the probability of pit initiation is high for the range of possible environments and the rates of pitting corrosion are expected to be high, as a result of hydrolysis reactions.

(9) In order to reduce possible rates of stress corrosion crack propagation, notches and residual stresses in the canister should be minimized and an anoxic repository should be selected. Further, the waste should be given surface storage prior to permanent burial to reduce the temperature and radiation level.

(10) Because of the uncertainties in the predictive capability of the mechanistic models, a multiple barrier design for the waste package should be strongly considered. The design should utilize natural galvanic relationships to enhance performance.

(11) Corrosion monitoring in the repository will be necessary to assess the predictive capability of the models and whether any unforseen mechanisms of corrosion have initiated.

REFERENCES

[1] Beavers, J. A. and Thompson, N. G., "Environmental Effects on Corrosion in the Tuff Repository," NUREG/CR-5435, U. S. Nuclear Regulatory Commission, February, 1990.

[2] Beavers, J. A. and Durr, C. L., "Immersion Studies on Candidate Container Alloys for the Tuff Repository," NUREG/CR-5598, U. S. Nuclear Regulatory Commission, May, 1991.

[3] Thompson, N. G., Beavers, J. A., and Durr, C. L., "Potentiodynamic Polarization Studies on Candidate Container Alloys for the Tuff Repository," NUREG/CR-5708, U. S. Nuclear Regulatory Commission, January, 1992.

[4] Beavers, J. A., Thompson, N. G., and Durr, C. L., "Pitting, Galvanic, and Long-Term Corrosion Studies on Candidate Container Alloys for the Tuff Repository," NUREG/CR-5709, U. S. Nuclear Regulatory Commission, January, 1992.

[5] Montazer, P. and Wilson, W. E., "Conceptual Hydrologic Model of Flow in the Unsaturated Zone, Yucca Mountain, Nevada," U. S. Geological Survey - Water Resources Investigation Report, USGS-WRI-84-4345, 1984.

[6] Halsey, W. G., "Selection Criteria for Container Materials at the Proposed Yucca Mountain High Level Nuclear Waste Repository", Corrosion/90, Paper Number 509, National Association of Corrosion Engineers, 1990

[7] Jaske, C.E., and Viswanathan, R., "Predict Remaining Life of Equipment for High Temperature/Pressure Service," Corrosion/90, Paper Number 213, National Association of Corrosion Engineers, 1990

[8] Knauss, K. G., Beiriger, W. J., and Peifer, D. W., "Hydrothermal Interaction Of Crushed Topopah Spring Tuff and J-13 Water At 90, 150, and 250°C Using Dickson-Type, Gold-Bag Rocking Autoclaves," Lawrence Livermore National Laboratory, Livermore, CA, UCRL-53630, DE86-014752, May, 1985.

[9] Oversby, V. M., Knauss, K. G., "Reaction Of Bullfrog Tuff With J-13 Well Water At 90°C and 150°C," Lawrence Livermore National Laboratory, Livermore, CA, UCRL-53442, DE84-000659, September, 1983.

[10] Yunker, W. H., Westinghouse Hanford Company, and Glass, R. S., Lawrence Livermore National Laboratory, "Long-Term Corrosion Behavior Of Copper-Base Materials In A Gamma-Irradiated Environment," UCRL-94500, DE87-007098, December, 1986.

[11] Glass, R. S., Overturf, G. E., Van Konynenburg, R. A., and McCright, R. D., "Gamma Radiation Effects On Corrosion: I - Electrochemical Mechanisms For The Aqueous Corrosion Processes Of Austenitic Stainless Steels," Lawrence Livermore National Laboratory, Livermore, CA, UCRL-92311, February, 1985.

[12] McCright, R. D., FY 1985 Status Report, UCID-20509, September 30, 1985.

[13] Williams, D. E., "A Statistical Approach to the Study of Localized Corrosion," AERE-R-10832, Atomic Energy Research Establishment, Harwell, England, 1983.

[14] Fratesi, R., "Statistical Estimate of the Pitting Potential of AISI 316L Stainless Steel in 3.5% NaCl Measured by Means of Two Electrochemical Methods," Corrosion, Vol. 41, No. 2, 114, 1985.

[15] Lemaitre, C. et al., "Chromate as a Pitting Corrosion Inhibitor: Stochastic Study," Werkstoffe und Korrosion, Vol. 40, No. 4, 229, April, 1989.

[16] Shibata, T. "Stochastic Studies of Passivity Breakdown," Corrosion Science, Vol. 31, 413, 1990.

[17] Doelling, R. and Heusler, K. E., "A Statistical Analysis of Process Resulting in Pitting of Passive Iron," International Congress on Metallic Corrosion, Vol. 2, 129, 1984.

[18] Keddam, M., Krarti, M., and Pallotta, C., "Some Aspects of the Fluctuations of the Passive Current on Stainless Steel in the Presence of Chlorides - Their Relation to the Probabilistic Approach of Pitting Corrosion," Corrosion, Vol. 43, No. 8, 454, August 1987.

[19] Finley, H. F. and Toncre, A. C., Materials Protection, Vol 3, 29, 1964.

[20] Ishikawa, Y. et al., "Application of Extreme Value Statistical Analysis to Actual Localized Corrosion Data From Some Operating Machine Components," Corrosion Engineering Japan, Vol. 29, 502, 1980.

[21] Crews, D. L., "Interpretation of Pitting Corrosion Data From Statistical Prediction Interval Calculations," Galvanic and Pitting Corrosion - Field and Laboratory Studies, ASTM STP 576, American Society for Testing and Materials, 217, 1976.

[22] Strutt, J. E. et al., "The Prediction of Corrosion By Statistical Analysis of Corrosion Profiles," Corrosion Science, Vol. 25, No. 5, 305, 1985.

[23] Gallucci, R. H. V. et al. "Statistical Prediction of the Numbers of Degraded Tubes in Nuclear Power Plant Steam Generators," CSNI-CEC Specialist Meeting on Trends and Pattern Analyses of Operational Data From Nuclear Power Plants, Rome Italy, June 1989, DE89793177, Avail NTIS.

[24] Janik-Czachor, M. and Ives, M. B., "Use of Pit Size Distribution for the Study of the Localized Breakdown of Passive Films," Passivity of Metals, Ed. R. P. Frankethal and J. Kruger, The Electrochemical Society, 369, 1979.

[25] Baba, H., "Pitting Corrosion of Copper Tubing and Prevention Effect Due to the Addition of Phytic Acid in Hot Water," Corrosion Engineering, Japan, Vol. 34, No. 1, 10, 1985.

[26] Aziz, P. M., "Application of the Statistical Theory of Extreme Values to the Analysis of Maximum Pit Depth Data for Aluminum," Corrosion, Vol. 12, No. 10, 495t, 1956.

[27] Finley, H. F., Corrosion, Vol. 23, 83, 1967.

[28] Bullen, D. B., Farmer, J. C., Gdowski, G. D., McCright, R, D., VanKonynenburg, R. A., and Weiss, H., Survey of Degradation Modes of Candidate Materials for High-Level Radioactive Waste Disposal Containers, 8 Volumes, Lawrence Livermore National Laboratory, Livermore, California, UICD-21362

[29] Corrosion Chemistry Within Pits, Crevices, and Cracks, Proceedings of a Conference, National Physical Laboratory, Teddington, Middlesex, U.K., October, 1984, HMSO, London, U.K., 1987.

[30] Turnbull, A. "Chemistry Within Localized Corrosion Cavities," Advances in Localized Corrosion, Proceedings of the Second

International Conference on Localized Corrosion, June, 1987, Eds. H. S. Isaacs, U. Bertocci, J. Kruger, and S. Smialowska, NACE-9, National Association of Corrosion Engineers, Houston, TX, 1990.

[31] Beavers, J. A., Thompson, N. G., and Markworth, A. J., "Pit Propagation of Carbon Steel in Groundwater," Advances in Localized Corrosion, Proceedings of the Second International Conference on Localized Corrosion, June, 1987, Eds. H. S. Isaacs, U. Bertocci, J. Kruger, and S. Smialowska, NACE-9, National Association of Corrosion Engineers, Houston, TX, 1990.

[32] Beavers, J. A. and Thompson, N. G., "Effect of Pit Wall Reactivity on Pit Propagation of Carbon Steel," Corrosion, Vol. 43, No. 3, 185, 1987.

[33] Koch, G. H., Kistler, C., and Mirick, W., Evaluation of Flue Gas Desulfurization Materials in the Mixing Zone: R. D. Morrow Sr. Generating Station, EPRI CS-5476, Electric Power Research Institute, Oct. 1987.

[34] Ford, F. P. and Andresen, P. L., "Unresolved Modeling Issues and Their Effect on Quantitative Predictions of Environmental Cracking," Corrosion/89, Paper Number 559, National Association of Corrosion Engineers, 1989.

J. Y. Park[1], W. J. Shack[2], and D. R. Diercks[3]

CRACK GROWTH BEHAVIOR OF CANDIDATE WASTE CONTAINER MATERIALS IN SIMULATED UNDERGROUND WATER

REFERENCE: Park, J. J., Shack, W. J., and Diercks, D. R., **"Crack Growth Behavior of Candidate Waste Container Materials in Simulated Underground Water,"** Application of Accelerated Corrosion Tests to Service Life Prediction of Materials, ASTM STP 1194, Gustavo Cragnolino and Narasi Sridhar, Eds., American Society for Testing and Materials, Philadelphia, 1994.

ABSTRACT: Fracture-mechanics crack growth tests were conducted on 25.4-mm-thick compact tension specimens of Types 304L and 316L stainless steel and Incoloy 825 at 93°C and 1 atmosphere of pressure in simulated J-13 well water, which is representative of the groundwater at the Yucca Mountain site in Nevada that is proposed for a high-level nuclear waste repository. Crack growth rates were measured under various load conditions: load ratios of 0.2-1.0, frequencies of 2 x 10^{-4}-1 Hz, rise times of 1-5000 s, and peak stress intensities of 25-40 MPa·$m^{1/2}$. The measured crack growth rates are bounded by the predicted rates from the current ASME Section XI correlation for fatigue crack growth rates of austenitic stainless steel in air. Environmentally accelerated crack growth was not evident in any of the three materials under the test conditions investigated.

KEYWORDS: crack growth, nuclear waste, waste container, metallic canister, stress corrosion cracking, corrosion fatigue

---

INTRODUCTION

The tuff formations at the Yucca Mountain site in Nevada have been selected as a potential repository for high-level nuclear waste. The geologic horizon at the proposed location consists of compacted volcanic

---

[1]Metallurgist, Materials and Components Technology Division, Argonne National Laboratory, Argonne, IL 60439 USA.

[2]Associate Director, Materials and Components Technology Division, Argonne National Laboratory, Argonne, Illinois 60439 USA.

[3]Section Manager, Materials and Components Technology Division, Argonne National Laboratory, Argonne, Illinois 60439 USA.

ash, or tuff, situated above the permanent water table. The rock consists of quartz, christobalite, alkali feldspar, and other minor phases [1]. Interaction between groundwater and the waste packages is expected to be minimal.

The high-level waste packages consist of three major components: the metallic containment barriers (the container or canisters), the waste itself, and other materials, including packing materials and emplacement-hole liners. The metallic containers are intended to provide essentially complete containment of the nuclear wastes for 300-1,000 years after emplacement. Types 304L and 316L stainless steel (SS), Incoloy 825, deoxidized copper (CDA 122), Cu-30% Nickel (CDA 715), and Cu-7% Aluminum (CDA 614) were selected as initial candidate container materials. These materials were chosen because of the favorable corrosion properties they exhibit in extensive marine, nuclear, and process industries applications.

The present investigation considers only the three austenitic alloys, with the objective of determining whether any of them can survive for 300 years without developing through-wall cracks. Fracture-mechanics crack growth tests were conducted in simulated J-13 well water, which is representative of the groundwater present at the Yucca Mountain site. The temperature of the waste package will be sufficiently high for much of the time so that the limited groundwater in the vicinity of the package will be in the form of steam. However, aqueous environments at ≈93°C are expected to provide "worst case" conditions for environmentally assisted cracking; hence they were the focus of the present investigation. Some of the results from this investigation were reported previously [2, 3].

EXPERIMENTAL PROCEDURE

Compact tension (CT) specimens, 25.4 mm thick were machined from 25.4-mm-thick plates of Type 304L SS (Heat No. V70200), Type 316L SS (Heat No. 16650), and Incoloy 825 (Heat No. HH2125F). The materials were tested in the as-received mill-annealed condition without additional heat treatment. The elemental composition and mechanical properties of the materials are shown in Tables 1 and 2, respectively.

Side grooves with a semicircular cross section were cut into both sides of the specimen to a depth of 1.27 mm to restrict crack growth to a single plane. Except for the side grooves and six small threaded holes on the front face for instrumentation, the design of the CT specimens is in accordance with ASTM Standard Test Method for Plane-Strain Fracture Toughness of Metallic Materials (E 399) [4]. The direction of crack extension was perpendicular to the short transverse thickness direction of the plates. The specimens were fatigue-precracked in air at room temperature to introduce a sharp starter crack for a length of 1.91 mm. A triangular loading waveform at a frequency of 1-2 Hz, a load ratio (algebraic ratio of minimum to maximum load) (R) of 0.25, and a maximum stress intensity of 17.5 MPa·m$^{1/2}$ was used for precracking. This maximum stress intensity value is 70% of initial peak stress intensity for subsequent crack growth tests in simulated J-13 well water.

The initial peak stress intensity for the tests in the simulated J-13 well water was ≈25 MPa·m$^{1/2}$. Welding residual stresses are the

TABLE 1-- Elemental composition of test materials (%) [a]

Element	304L SS	316L SS		Incoloy 825
	Vendor Analysis	Vendor Analysis	Indep. Analysis	Vendor Analysis
C	0.023	0.018	0.018	0.02
Mn	1.79	1.78	1.86	0.39
P	0.030	0.026	0.022	b
S	0.018	0.013	0.016	0.003
Si	0.39	0.43	0.40	0.14
Cr	18.15	16.5	16.84	21.98
Ni	8.25	10.39	11.1	43.89
Cu	0.26	0.19	0.17	2.03
Mo	0.30	2.09	2.1	2.79
Co	0.14	0.1	b	b
N	0.079	0.054	0.058	b
O	b	b	0.010	b
Al	b	b	b	0.03
Ti	b	b	b	0.75
Fe	b	b	b	27.98

[a]Vendors were Eastern Stainless Steel Co. for the Type 304L SS, Jessop Steel Co. for the Type 316L SS, and Inco Alloy International for the Incoloy 825. The Independent analysis was performed by Anamet Laboratories, Inc.

[b]Not analyzed.

TABLE 2-- Mechanical properties of test materials

Mechanical Properties	Type 304L SS	Type 316L SS	Incoloy 825
Yield (MPa)	273	231	298
Tensile (MPa)	594	530	676
Elongation (%)	59	53.9	43
RA (%)	66	72.0	62
Hardness	BHN156	BHN137	R_B085

most important driving force for cracks. Extensive calculations performed to analyze the growth of cracks in boiling water reactor piping [5] suggest that for the type of welding processes expected to be used to fabricate waste canisters and for the largest defects that would be expected to be marginally detectable, the stress intensity values chosen for the tests should be fairly conservative. However, measurements of the residual stress distributions produced by the actual welding processes on the actual container geometries will be needed to verify this.

The crack-growth-rate tests were performed in a 5-L nickel (Ni) vessel with a once-through flow system at a flow rate of 3 mL/min, under 1 atm pressure at 92-94°C. The simulated J-13 well water was prepared from deionized high-purity water and reagent-grade-purity salts of calcium sulfate ($CaSO_4$), calcium nitrate ($Ca(NO_3)_2$), calcium chloride ($CaCl_2$), ferrous chloride ($FeCl_2$), lithium sulfate (Li_2SO_4), magnesium sulfate ($MgSO_4$), manganese sulfate ($MnSO_4$), aluminum chloride ($AlCl_3$), sodium carbonate (Na_2CO_3), sodium bicarbonate ($NaHCO_3$), potassium bicarbonate ($KHCO_3$), sodium silicate (Na_2SiO_3) and hydrofluoric acid (HF). High-purity mixed gas containing 20% oxygen (O_2), 12% carbon dioxide (CO_2), and 68% nitrogen (N_2) was used as a cover gas at 3-5 psig to maintain the desired dissolved O_2 and bicarbonate (HCO_3^-) concentrations. Table 3 shows a typical composition of J-13 well water [6] and analysis results for the simulated test solution used for the crack growth tests. The test solution is a good simulation of the ionic and dissolved gaseous species in the J-13 well water, with the exception of the silicon (Si) concentration. Silicon was not added to the test solution to avoid precipitation of silicate. Silicates are commonly used as corrosion inhibitors for low-temperature applications [7,8], and it is considered unlikely that silicates would contribute to environmentally assisted crack growth of the test materials. The higher $pH_{25°C}$ values for the effluent test solution are due to some loss of CO_2 to the atmosphere during the tests. Ni in the effluent is present primarily in the form of corrosion products from the Ni test vessel.

TABLE 3--Chemical composition of J-13 well water and test solution.

Species	J-13 Analysis (mg/L)	Test Solution (mg/L)	
		Feed	Effluent
Ca	11.5	10.0 - 11.3	10.1 - 11.5
Mg	1.76	1.71 - 1.78	1.72 - 1.88
Na	45.0	48.3 - 48.8	50.2 - 50.9
K	5.3	5.21 - 5.26	5.32 - 5.44
Li	0.06	0.057 - 0.058	0.059 - 0.060
Fe	0.04	< 0.01	< 0.01
Mn	0.001	< 0.01	< 0.01
Al	0.03	< 0.1	< 0.1
Si	30.0	a, b	a, b
Cu	b	< 0.01	< 0.01
Ni	b	< 0.02	1.61 - 4.66
F^-	2.1	2.02 - 2.10	2.03 - 2.16
Cl^-	6.4	6.68 - 6.94	6.75 - 7.21
SO_4^{2-}	18.1	19.3 - 20.5	20.4 - 20.9
NO_3^-	10.1	11.7 - 12.7	12.7 - 13.1
HCO_3^-	143.0	127.0 - 129.5	132.2 - 141.0
Diss. O	5.7	7.0 - 9.0	3.5 - 7.0
$pH_{25°C}$	6.9	6.21 - 6.81	7.71 - 9.31

[a]Not added.
[b]Not analyzed.

The specimens and the platinum (Pt) electrode were electrically insulated from the test vessel and load train. The corrosion potential of the specimens and a Pt reference electrode with respect to an external 2×10^{-4} M KCl/AgCl/Ag reference electrode were measured continually during the crack growth tests. The potential of the reference electrode at 93°C was previously determined to be +378 mV SHE [9]. The corrosion potential varied from 376 to 431, 345 to 433, 337 to

403, and 399 to 497 mV SHE for Type 304L SS, Type 316L SS, and Incoloy 825 specimens and the Pt electrode, respectively.

TABLE 4--Loading conditions for crack-growth-rate tests on candidate waste container materials

Alloy	K_{max} ($MPa \cdot m^{1/2}$)	Load Ratio	Rise Time (s)	Unload Time (s)
High-load-Ratio Tests				
All	25	1	-	-
All	25	0.95	12	2
All	25	0.9	12	2
Low-load-Ratio Tests				
304 SS	35.7 - 36.2	0.7	1 - 500	1
	27.4 - 28.5	0.5	1 - 1000	1
	28.5 - 29.7	0.2	1 - 5000	1
316 SS	39.2 - 39.8	0.7	1 - 500	1
	26.8 - 31.2	0.5	1 - 1000	1
	34.7 - 39.4	0.2	1 - 5000	1
Incoloy 825	36.4 - 36.9	0.7	1 - 500	1
	27.4 - 28.3	0.5	1 - 500	1
	29.5 - 31.0	0.2	1 - 5000	1
	31.5 - 31.9	0.2	10	1 - 2000

Tension-to-tension loads were applied to specimens with either MTS servohydraulic equipment under load-control or deadload testing machines. A sawtooth cyclic load waveform with variable rise and unloading time was used at a fixed R. Both high-load-ratio ($R \geq 0.9$) and low-load-ratio ($R = 0.2$, 0.5 or 0.7) tests were conducted; the test conditions are summarized in Table 4. Three precracked specimens were loaded simultaneously in a daisy-chain serial loading arrangement. The crack length in each specimen was continually monitored by means of an electric potential drop method [10]. The resolution of the crack length measurement was 50 µm. A schematic representation of the experimental setup and the CT specimen is shown in Fig. 1.

RESULTS AND DISCUSSION

High-Load-Ratio Tests

Specimens of all three alloys were initially loaded under a constant ($R = 1$) stress intensity of 25 $MPa \cdot m^{1/2}$ in the simulated J-13 well water environment, but no crack growth was observed after 315 h. The load ratio was then decreased to $R = 0.95$, thus imposing a slight "wiggle" on the loading waveform, but no crack growth was observed after an additional 334 h under these conditions. Following this, R was further decreased to 0.9 and the tests were continued for an additional 18,300 h (more than two years), and again no crack growth was observed. Figures 2-4 show crack length as a function of time for test times from 9,300 to 19,000 h for the three alloys under $R = 0.9$ loading.

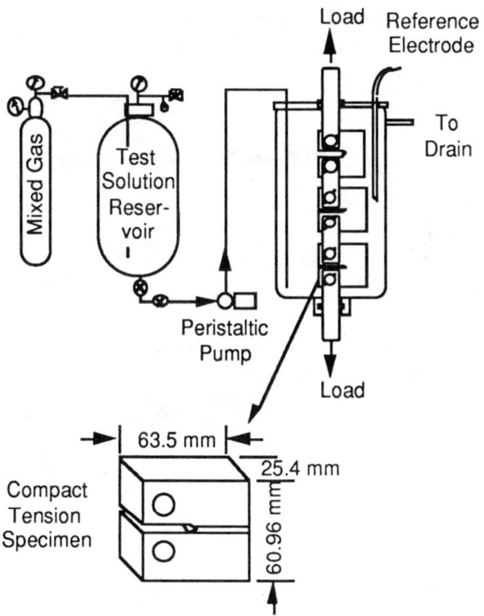

Fig. 1. Schematic representation of experimental setup and test specimen.

The maximum average crack growth rate in these tests can be estimated from the resolution of the crack-growth-measurement technique (50 μm) and the time over which no crack extension was detected. Dividing the overall resolution of 50 μm by the test duration of 19,000 h results in an estimated maximum crack growth rate of $\approx 8 \times 10^{-13}$ m/s under the test conditions. This compares with a maximum allowable crack growth rate of $\approx 1 \times 10^{-12}$ m/s for the waste package canister, based on a wall thickness of 1 cm and a target life of 300 years.

Low-Load-Ratio Tests

Crack-growth-rate tests in the same environment, conducted under more severe conditions of greater stress intensities and lower R ratios, resulted in observable crack extension. In Fig. 5, Type 304L SS under conditions of R = 0.7, K = 36 $MPa \cdot m^{1/2}$, and frequency of 0.5 Hz, shows a crack growth rate of 4.0×10^{-9} $m \cdot s^{-1}$. The results of cyclic load tests on Types 304L and 316L SS and Incoloy 825 at R = 0.2, 0.5 and 0.7 and frequencies of 10^{-4} to 1 Hz are summarized in Table 5. For comparison, fatigue crack growth rates of austenitic stainless steel in air at 93°C computed with the current ASME Section XI correlation, which is based on the work of James and Jones [11], are also included.

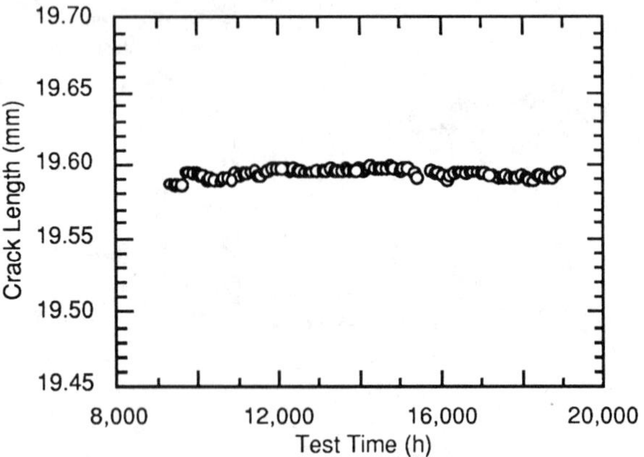

Fig. 2. Crack length vs. test time for Type 304L SS at K_{max} = 25 MPa·m$^{1/2}$ and R = 0.9.

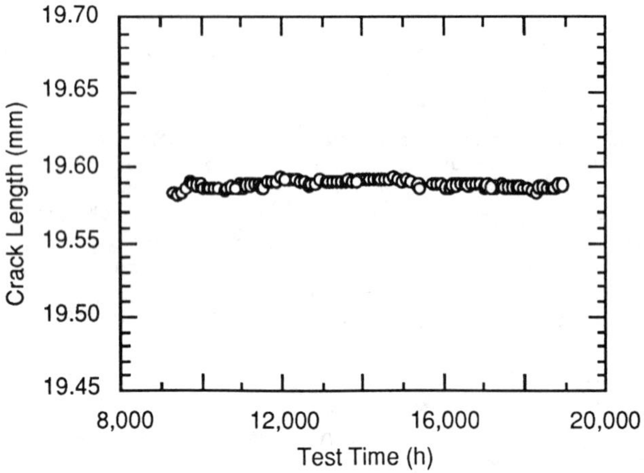

Fig. 3. Crack length vs. test time for Type 316L SS at K_{max} = 25 MPa·m$^{1/2}$ and R = 0.9.

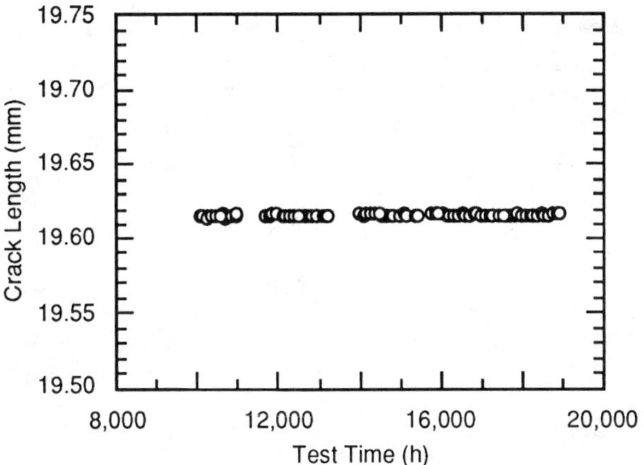

Fig. 4. Crack length vs. test time for Incoloy 825 at K_{max} = 25 MPa·m$^{1/2}$ and R = 0.9.

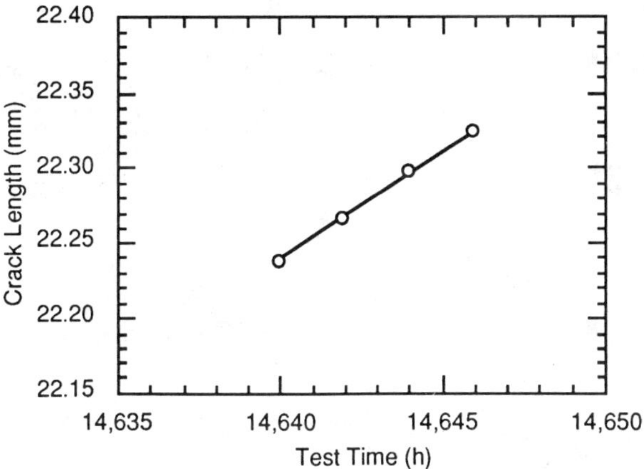

Fig. 5. Crack length vs. test time for 304L SS at K_{max} = 36 MPa·m$^{1/2}$, R = 0.7 and 0.5 Hz.

Figures 6 and 7 show a comparison between the crack growth rates observed for the test materials in the simulated J-13 well water and the predicted crack growth rates of austenitic stainless steel in air, based on the ASME Section XI correlation. In the figures, growth rates on the solid line (with slope of unity) would be equal to those predicted by

Table 5--Observed cyclic crack growth rates and rates calculated with ASME Section XI correlation

Load Ratio	Freq. (Hz)	Rise Time (s)	Unload Time (s)	K_{max} (MPa·$m^{1/2}$)	Observed Rate ($m·s^{-1}$)	ASME XI Rate ($m·s^{-1}$)	Ratio (ASME / observed)
304L SS							
0.7	0.500	1	1	35.7	3.96×10^{-9}	8.37×10^{-9}	2.11
0.7	0.167	5	1	35.9	1.62×10^{-9}	2.83×10^{-9}	1.75
0.7	0.091	10	1	36.0	7.64×10^{-10}	1.56×10^{-9}	2.04
0.7	0.048	20	1	36.0	4.41×10^{-10}	8.19×10^{-10}	1.86
0.7	0.020	50	1	36.1	2.31×10^{-10}	3.39×10^{-10}	1.47
0.7	0.010	100	1	36.1	9.73×10^{-11}	1.72×10^{-10}	1.76
0.7	0.005	200	1	36.1	4.91×10^{-11}	8.64×10^{-11}	1.76
0.7	0.002	500	1	36.2	2.29×10^{-11}	3.48×10^{-11}	1.52
0.5	0.500	1	1	27.4	1.55×10^{-8}	1.59×10^{-8}	1.03
0.5	0.091	10	1	27.8	2.15×10^{-9}	3.01×10^{-9}	1.40
0.5	0.048	20	1	28.0	7.89×10^{-10}	1.61×10^{-9}	2.04
0.5	0.020	50	1	28.1	3.96×10^{-10}	6.76×10^{-10}	1.71
0.5	0.010	100	1	28.2	1.78×10^{-10}	3.46×10^{-10}	1.94
0.5	0.005	200	1	28.3	9.40×10^{-11}	1.76×10^{-10}	1.87
0.5	0.002	500	1	28.4	3.94×10^{-11}	7.14×10^{-11}	1.81
0.5	0.001	1000	1	28.5	1.95×10^{-11}	3.59×10^{-11}	1.84
0.2	0.500	1	1	28.8	2.73×10^{-8}	6.29×10^{-8}	2.30
0.2	0.091	10	1	29.6	7.21×10^{-9}	1.25×10^{-8}	1.74
0.2	0.010	100	1	29.7	4.49×10^{-10}	1.38×10^{-10}	3.07
0.2	0.001	1000	1	29.7	3.38×10^{-11}	1.39×10^{-10}	4.12
0.2	0.0002	5000	1	29.7	7.99×10^{-12}	2.79×10^{-11}	3.49
316L SS							
0.7	0.500	1	1	39.2	4.09×10^{-9}	1.14×10^{-8}	2.79
0.7	0.167	5	1	39.4	1.93×10^{-9}	3.87×10^{-9}	2.00
0.7	0.091	10	1	39.6	9.66×10^{-10}	2.13×10^{-9}	2.20
0.7	0.048	20	1	39.6	5.25×10^{-10}	1.12×10^{-9}	2.14
0.7	0.020	50	1	39.7	2.06×10^{-10}	4.64×10^{-10}	2.25
0.7	0.010	100	1	39.7	6.81×10^{-11}	2.35×10^{-10}	3.45
0.7	0.005	200	1	39.7	3.10×10^{-11}	1.18×10^{-10}	3.82
0.7	0.002	500	1	39.8	1.82×10^{-11}	4.76×10^{-11}	2.62
0.5	0.500	1	1	26.8	2.33×10^{-8}	2.02×10^{-8}	0.86
0.5	0.091	10	1	30.0	3.97×10^{-9}	3.88×10^{-9}	0.98
0.5	0.048	20	1	30.3	1.12×10^{-9}	2.10×10^{-9}	1.88
0.5	0.020	50	1	30.6	5.65×10^{-10}	8.91×10^{-10}	1.58
0.5	0.010	100	1	30.8	2.69×10^{-10}	4.60×10^{-10}	1.71
316L SS (cont'd)							
0.5	0.005	200	1	31.0	1.34×10^{-10}	2.36×10^{-10}	1.76
0.5	0.002	500	1	31.1	5.36×10^{-11}	9.58×10^{-11}	1.79
0.5	0.001	1000	1	31.2	2.64×10^{-11}	4.84×10^{-11}	1.83
0.2	0.500	1	1	37.2	7.51×10^{-8}	1.46×10^{-7}	1.95
0.2	0.091	10	1	38.6	2.02×10^{-8}	3.01×10^{-8}	1.49
0.2	0.010	100	1	38.9	1.78×10^{-9}	3.36×10^{-9}	1.89
0.2	0.001	1000	1	39.2	1.77×10^{-10}	3.48×10^{-10}	1.96
0.2	0.0002	5000	1	39.4	4.55×10^{-11}	7.07×10^{-11}	1.56
Incoloy 825							
0.7	0.500	1	1	36.4	4.89×10^{-9}	8.86×10^{-9}	1.81
0.7	0.167	5	1	36.6	1.56×10^{-9}	3.01×10^{-9}	1.93
0.7	0.091	10	1	36.7	9.55×10^{-10}	1.66×10^{-9}	1.74
0.7	0.048	20	1	36.7	5.28×10^{-10}	8.74×10^{-10}	1.66
0.7	0.020	50	1	36.8	1.60×10^{-10}	3.61×10^{-10}	2.26
0.7	0.010	100	1	36.8	9.73×10^{-11}	1.83×10^{-10}	1.88
0.7	0.005	200	1	36.9	5.84×10^{-11}	9.23×10^{-11}	1.58
0.7	0.002	500	1	36.9	2.62×10^{-11}	3.72×10^{-11}	1.42
0.5	0.500	1	1	27.4	9.05×10^{-9}	1.58×10^{-8}	1.75
0.5	0.091	10	1	27.6	1.70×10^{-9}	2.96×10^{-9}	1.74
0.5	0.048	20	1	27.8	7.04×10^{-10}	1.58×10^{-9}	2.25
0.5	0.020	50	1	27.9	3.54×10^{-10}	6.61×10^{-10}	1.87
0.5	0.010	100	1	28.1	1.67×10^{-10}	3.39×10^{-10}	2.03
0.5	0.005	200	1	28.2	1.22×10^{-10}	1.72×10^{-10}	1.41
0.5	0.002	500	1	28.3	3.88×10^{-11}	6.99×10^{-11}	1.80
0.2	0.500	1	1	29.9	2.99×10^{-8}	7.12×10^{-8}	2.38
0.2	0.091	10	1	30.8	7.13×10^{-9}	1.43×10^{-8}	2.00
0.2	0.010	100	1	30.9	7.62×10^{-10}	1.57×10^{-9}	2.06
0.2	0.001	1000	1	31.0	7.01×10^{-11}	1.60×10^{-10}	2.29
0.2	0.0002	5000	1	31.0	1.75×10^{-11}	3.21×10^{-11}	1.83
0.2	0.091	10	1	31.5	7.24×10^{-9}	1.54×10^{-8}	2.12
0.2	0.067	10	5	31.7	4.74×10^{-9}	1.15×10^{-8}	2.43
0.2	0.033	10	20	31.8	2.55×10^{-9}	5.82×10^{-9}	2.28
0.2	0.009	10	100	31.9	7.71×10^{-10}	1.60×10^{-9}	2.08
0.2	0.002	10	500	31.9	1.53×10^{-10}	3.46×10^{-10}	2.26
0.2	0.0005	10	2000	31.9	4.32×10^{-11}	8.77×10^{-11}	2.03

the ASME correlation. The observed crack growth rates for Types 304L and 316L SS and Incoloy 825 are generally lower than the predicted rates. Since the Code correlation represents a 95% confidence limit upper bound for the observed data base, it is expected to be conservative for most heats of material in the absence of environmental effects.

Crack growth rate under cyclic loads in a corrosive environment (da/dt) may be expressed as a sum of contributions by (1) stress corrosion cracking, $(da/dt)_{SCC}$; (2) corrosion fatigue, $(da/dt)_{CF}$, representing the additional crack growth under cyclic loading due to the environment; and (3) mechanical fatigue, $(da/dt)_{air}$, representing the fatigue growth in air:

$$(da/dt) = (da/dt)_{SCC} + (da/dt)_{CF} + (da/dt)_{air} \; . \qquad (1)$$

The first two terms on the right side of the equation are environment sensitive. They depend on loading history variables, such as rise-, unload- and hold-time, as well as on frequency. In oxygenated-water environments, the environment-sensitive terms can contribute significantly to crack growth rates of austenitic stainless steels [12-14]. Under low-R and high-frequency loading, mechanical fatigue dominates. Environmental contributions would be expected to become more significant as the frequency decreases. Crack growth rates as a function of cyclic frequency and crack growth rates per cycle versus rise time for the current tests are shown in Figs. 9-14.

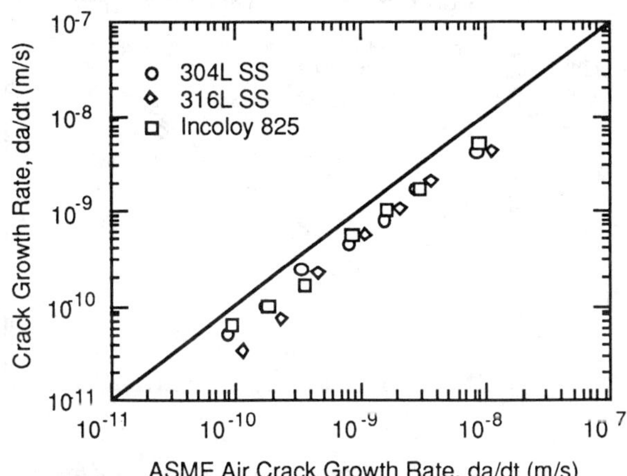

Fig. 6. Time-Based Crack Growth Rate vs. Rate for Austenitic Stainless Steels in Air from ASME Section XI Correlation for R = 0.7

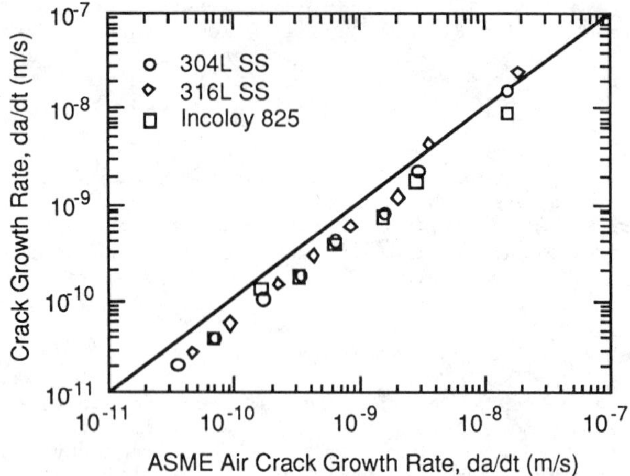

Fig. 7. Time-Based Crack Growth Rate vs. Rate for Austenitic Stainless Steels in Air from ASME Section XI Correlation for R = 0.5

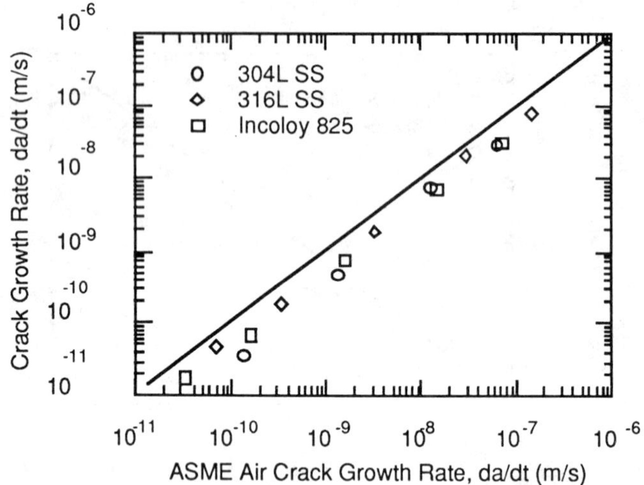

Fig. 8. Time-based crack growth rate vs. rate for austenitic stainless steels in air from ASME Section XI correlation for R = 0.2

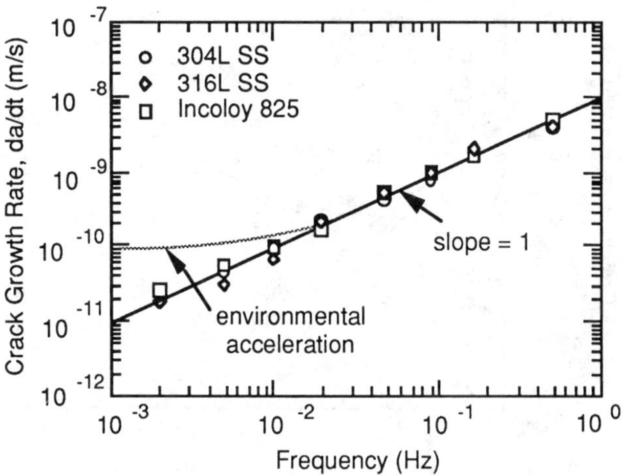

Fig. 9. Time-based crack growth rate vs. frequency at R = 0.7

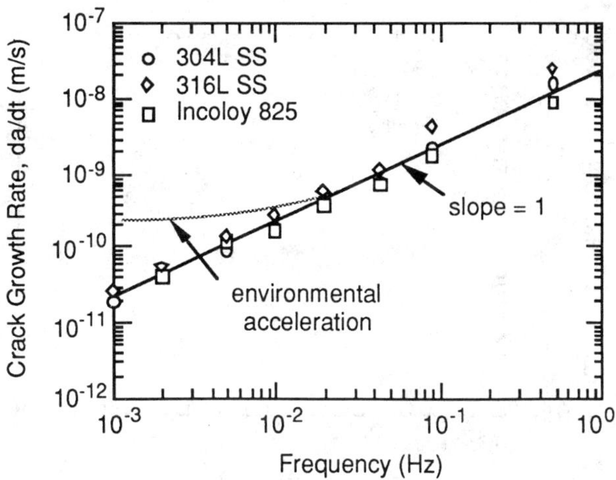

Fig. 10. Time-based crack growth rate vs. frequency at R = 0.5

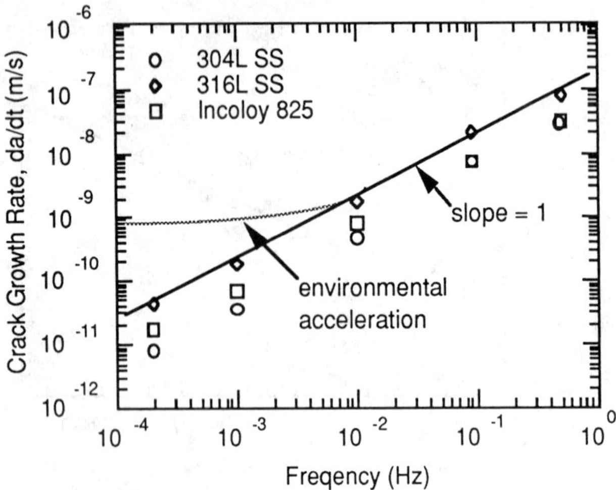

Fig. 11. Time-based crack growth rate vs. frequency at R = 0.2

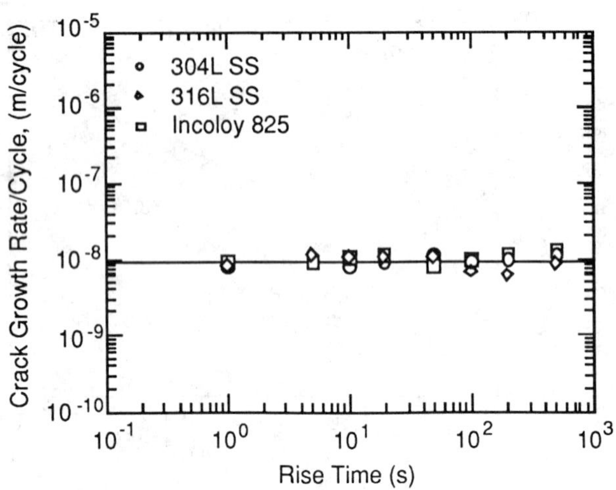

Fig. 12. Crack growth rate per cycle as a function of rise time for R = 0.7

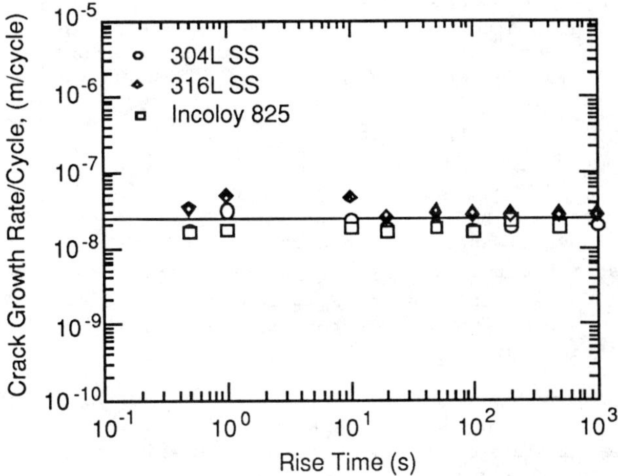

Fig. 13. Crack growth rate per cycle as a function of rise time for R = 0.5

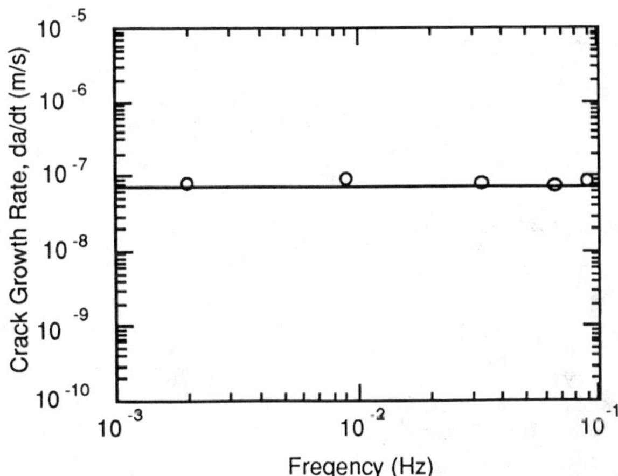

Fig. 14. Crack growth rate per cycle for Incoloy 825 as a function of frequency for R = 0.2

Figures 9-11 show that the time-based growth rates are proportional to frequency over the entire range of the frequencies used in the tests. This indicates that no environmental acceleration of crack growth is present for the test conditions considered. Figures 12-14 show that the growth rate per cycle is independent of rise time or frequency. This, too, indicates that no environmentally assisted crack growth occurred, and the crack growth mechanism is pure mechanical fatigue. For comparison, the environmentally accelerated behavior observed in high-temperature oxygenated environments [13,14] is shown schematically by the curved lines in Figs. 9-11.

SUMMARY

Types 304L and 316L SS and Incoloy 825 were tested at 93°C in simulated J-13 well water in both high- and low-load-ratio tests. In the high-load-ratio tests, K_{max} was 25 MPa·m$^{1/2}$ and R ranged from 0.9 to 1. In the low-load-ratio tests, K_{max} ranged from 26.8 to 39.8 MPa·m$^{1/2}$ and R was 0.2 - 0.7. The results of these experiments lead to the following conclusions:

1. No crack growth was detected in any of the alloys tested at K = 25 MPa·m$^{1/2}$ and R = 0.9 - 1 for test times of 19,000 h. Based upon the resolution of the crack-length-measuring technique, this indicates that the maximum average crack growth rate under these conditions was no greater than ≈8 x 10^{-13} m/s.
2. No cyclic or rise-time-dependent environmental acceleration of crack growth was observable in the simulated J-13 well water at 93°C under the test conditions.
3. Growth rates for R = 0.2-0.7 and maximum stress intensities of 26-40 MPa·m$^{1/2}$ agree with the current ASME Section XI correlation for austenitic stainless steel within a factor of 0.9-4.1.

ACKNOWLEDGMENTS

D. R. Perkins, W. F. Burke, W. E. Ruther, W. K. Soppet, S. L. Phillips, and J. C. Tezak of Argonne National Laboratory contributed to various experimental aspects of the project. The authors thank W. L. Clarke and R. D. McCright of Lawrence Livermore Laboratory for their support of this work, which was sponsored by the U.S. Department of Energy under subcontract from Lawrence Livermore National Laboratory.

REFERENCES

[1] Guzowski, R. V., Nimick, F. B., Siegel, M. D., and Finely, N. C., "Repository Site Data Report for Tuff: Yucca Mountain, Nevada," NUREG/CR-2937, SAND82-2105, Sandia National Laboratories, Albuquerque, NM (October 1983).

[2] Park, J. Y., Maiya, P. S., Soppet, W. K., Diercks, D. R., Shack, W. J., and Kassner, T. F., "Stress Corrosion Cracking of Candidate Waste Container materials - Final Report," ANL-92/28, Argonne National Laboratory (June 1992).

[3] Park, J. Y., Shack, W. J., and Diercks, D. R., "Crack-Growth-Rate Testing of Candidate Waste Container Materials," Proc. of the Topical Meeting on Nuclear Waste Packaging Focus'91, American Nuclear Society, La Grange Park, IL (1992), pp. 163-169.

[4] Designation E 399-90, "Standard Test Method for Plane-Strain Fracture Toughness of Metallic Materials," in *Annual Book of ASTM Standards, Vol. 03.01*, American Society for Testing and Materials, Philadelphia (1990) pp. 485-515.

[5] Harris, D. O., "The Influence of Crack Growth Kinetics and Inspection on the Integrety of Sensitized BWR Piping Welds," EPRI NP-1163, Electric Power Research Institute, Palo Alto, CA (1979).

[6] Ogard, A. E., and Kerrisk, J. F., "Groundwater Chemistry along Flow Path Between a Proposed Repository Site and the Accessible Environment," LA-1088-MS, Los Alamos National Laboratory (1984).

[7] Uhlig, H. H., ed., *Corrosion Handbook*, John Wiley & Sons, New York (1948), pp. 500-503.

[8] Hanlon, R. T., Steffen, A. J., Rohlich, G. A., and Kessler, L. H., "Scale and Corrosion Control in Potable Water Supplies at Army Posts," *Ind. Eng. Chem.*, *37*, 724-735 (1945).

[9] Maiya, P. S., Soppet, W. K., Park, J. Y., Kassner, T. F., Shack, W. J., and Diercks, D. R., "Stress Corrosion Cracking of Candidate Waste Container Materials," ANL-90/50, Argonne National Laboratory (November 1990), pp. 67-72.

[10] Prater, T. A., Catlin, W. R., and Coffin, L. F., "Environmental Crack Growth Measurement Techniques," EPRI NP-2641, Electric Power Research Institute, Palo Alto, CA (1982).

[11] James, L. A., and Jones, D. P., "Fatigue Crack Growth Correlation for Austenitic Stainless Steels in Air," *Proc. of the Conference on Predictive Capabilities in Environmentally-Assisted Cracking*, R. Rungta, ed., PVP Vol. 99, American Society of Mechanical Engineers, N. Y. (1985), pp. 363-414.

[12] Kawakubo, T., Hishida, M., Amano, K., and Katsuta, M., "Crack Growth Behavior of Type 304 Stainless Steel in Oxygenated 290°C Pure Water under Low Frequency Cyclic Loading," *Corrosion 36*, 638-647 (1980).

[13] Gilman, J. D., Rungta, R., Hinds, P., and Mindlan, H., "Corrosion-Fatigue Crack Growth Rates in Austenitic Stainless Steels in Light Water Reactor Environments," *Int. J. Pres. Ves. Piping 31*, 55-68 (1988).

[14] Shack, W. J., "Corrosion Fatigue Curves for Austenitic Stainless Steels in Light Water Reactor Environments," in "Environmentally Assisted Cracking in Light Water Reactors, Semiannual Report, October 1990-March 1991," NUREG/CR-4667, Vol.12, ANL-91/24 (August 1991), pp. 30-36.

Narasi Sridhar,[1] Gustavo A. Cragnolino,[1] John C. Walton,[1] and Darrell Dunn[1]

PREDICTION OF LOCALIZED CORROSION USING MODELING AND EXPERIMENTAL TECHNIQUES

REFERENCE: Sridhar, N., Cragnolino, G. A., Walton, J. C., and Dunn, D., **"Prediction of Localized Corrosion Using Modeling and Experimental Techniques,"** Application of Accelerated Corrosion Tests to Service Life Prediction of Materials, ASTM STP 1194, Gustavo Cragnolino and Narasi Sridhar, Eds., American Society for Testing and Materials, Philadelphia, 1994.

ABSTRACT: An overall approach to life prediction of a component undergoing localized corrosion involving initiation and repassivation is considered. A crevice corrosion initiation model for Fe-Ni-Cr-Mo alloys is presented in this paper. This is a finite difference model that considers the combined effects of hydrolysis and transport in predicting the evolution of the solution chemistry inside a crevice of a given geometry. Crevice corrosion initiation time is computed by equating the time it takes for the crevice-solution chemistry to evolve to a critical solution chemistry for the depassivation of a given alloy. Depassivation pH is measured on non-creviced (open) samples as a function of environmental factors such as temperature, chloride, and sulfate. The measured depassivation pH is then used as an input variable in the crevice corrosion model to calculate the initiation time. The use of crevice corrosion repassivation potential as a bounding parameter for performance prediction is also examined. The limitations of the current models and future directions are identified.

KEYWORDS: Crevice, pitting, corrosion, model, repassivation, depassivation, pH, chloride, stainless steel, alloy 825

INTRODUCTION

The U.S. Nuclear Regulatory Commission (USNRC) requirements for geologic disposal of high-level nuclear waste, as contained in 10 CFR 60.113, specify that the waste packages shall be designed such that containment of radionuclides will be substantially complete for a period of at least 300 to 1,000 years. In the U.S. geologic

[1]Center for Nuclear Waste Regulatory Analyses, Southwest Research Institute, 6220 Culebra Road, San Antonio, TX 78238-5166.

disposal program, two different types of engineered barrier system (EBS) concepts are being considered: (i) Vertical emplacement of a single-wall or double-wall container in a bore-hole with an air-gap between the container and the bore-hole wall, and (ii) Horizontal emplacement of a double-wall container surrounded by a suitable backfill in the drifts of the underground repository. The vertical emplacement of a thin, single-wall container, made of either an austenitic alloy or a copper-based alloy, is the present reference design. It has also been proposed that high thermal loading be used to create a dry-out zone around the EBS for thousands of years to minimize corrosion and radionuclide transport. In the case of the reference design, this drying out process may create deposition of solids rich in Ca and Si on the container wall [1]. However, backflow of the condensate through fractures may result in an aqueous environment around the container as indicated by the results of some field heater tests [2,3,4]. The rewetting of the dried solids can cause the formation of solutions rich in chloride and sulfate as shown experimentally by Beavers et al. [1]. Hence, a conservative approach in life prediction for the reference design is to assume the presence of aqueous conditions surrounding the containers. In this paper, life prediction of a single-walled container in an aqueous environment will be considered, with a focus on pitting and crevice corrosion.

An approach to long-term life prediction, assuming only corrosion related failure modes, is illustrated schematically in Figure 1a and 1b. While the figure illustrates the approach for pitting, the same methodology can be used for crevice corrosion. As shown in Figure 1a, the corrosion potential, E_{corr}, of the container material exposed to the repository environment changes with time in response to factors such as radiolysis, temperature, and oxygen concentration [5]. For example, this increase in E_{corr} may occur due to the generation of hydrogen peroxide by radiolysis of an aqueous environment contacting the container [5]. A simple increase in E_{corr} to a constant upper limit is indicated in Figure 1a, although a more complex behavior may be observed in reality. Additionally, significant variation in E_{corr} across containers may be expected depending upon their location and thermal output. The critical potentials, the pit initiation potential, E_p, and repassivation potential, E_{rp}, in the case illustrated in Figure 1, also will vary with time depending upon environmental changes near the waste packages. For example, concentration of chloride due to evaporation can decrease these potentials initially. After a long period of time, the decrease in temperature may result in an increase in these potentials. If E_{corr} exceeds E_p, pits initiate and propagate into the container wall. If E_{corr} drops below E_p, pits already initiated continue to grow, but no new pits initiate. Finally, if E_{corr} drops below E_{rp}, all pits repassivate and cease to grow. After repassivation, the corrosion of the container continues in a uniform manner at a low rate determined by the passive current density. Both E_p and E_{rp} can show stochastic variations [6], and hence significant variability in the performance of various containers may be expected. The stochastic nature of localized corrosion and the variations in the corrosion potential of the containers in the repository can be taken into consideration by repeating the sequence of calculations shown in Figure 1a for varying input parameters, for example by a Monte-Carlo driver, to obtain a distribution of containers with pit depths exceeding a certain value for any given time period (Figure 1b), termed a damage function [6]. Containment can then be evaluated by comparing the fraction or number of containers

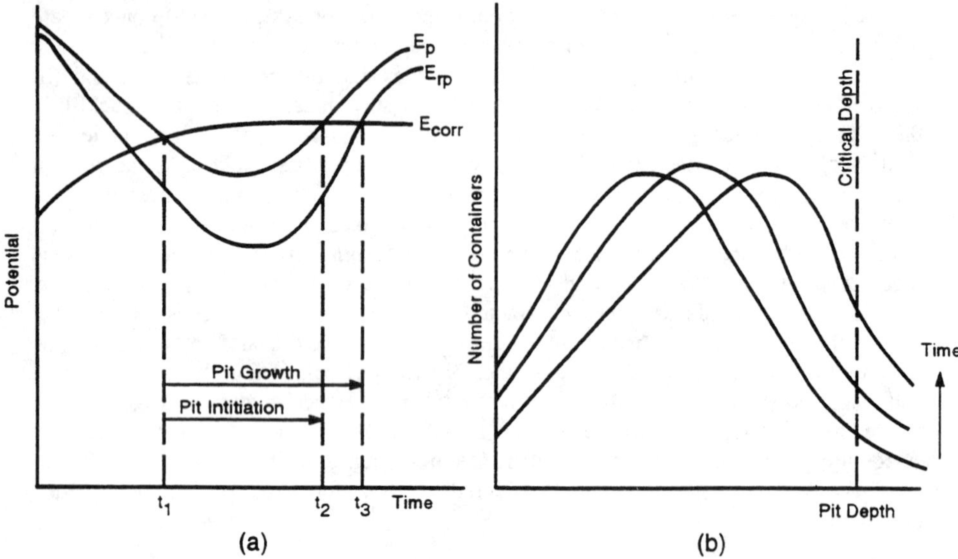

Figure 1. A schematic view of the approach for long-term life prediction of container materials assuming pitting as the failure mechanism.

with pits exceeding the critical pit depth at any given time period to a predetermined requirement. Although two potentials, E_p and E_{rp}, were defined, in long-term prediction these two potentials may coincide to one critical potential [7].

In the above approach, the initiation time for pitting or crevice corrosion, once the corrosion potential exceeds a critical value, is ignored in comparison to the required life times. Another important aspect in the above approach is the use of the repassivation potential as a lower bound parameter, which is assumed to be independent of the extent of prior pit/crevice corrosion growth, but dependent on environmental factors and the container material. Further, it is assumed that no localized corrosion occurs for very long periods of time once the corrosion potential falls below the repassivation potential. These assumptions must be verified by a suitable combination of modeling and experiments.

EXPERIMENTAL METHOD

Materials

Types 304L (UNS S30403) and 316L (UNS S31603) stainless steels and alloy 825 (UNS N08825) are considered in this paper (Table 1). For the experimental program, the materials were obtained in the form of mill-annealed, 12.5 mm thick plates. For the repassivation potential tests, cylindrical specimens (6.35 mm diameter × 48.64 mm long) were machined transverse to the rolling direction. For the cyclic polarization

Table 1. Chemical Compositions of Alloys Tested in Weight Percent

Alloy	C	Cr	Cu	Fe	Mn	Mo	Ni	Ti
825	0.01	22.1	1.8	30.4	0.35	3.2	41.1	0.8
Type 304L S.S.	0.022	18.3	0.19	Bal.	1.46	-	9.14	-
Type 316L S.S.	0.014	16.4	0.27	Bal.	1.58	2.1	10.0	-

experiments in simulated crevice/pit solutions, cylindrical specimens (6.35 mm diameter × 19.1 mm long) were used. For crevice corrosion repassivation tests, specimens (19.1 mm × 38.1 mm × 12.5 mm) were machined from the plates. A thin section, 19.1 mm × 25.4 mm × 3.1 mm thick, was machined on these specimens on which the crevice devices were mounted.

Test Techniques

Two types of tests were conducted:

a). The effect of prior pit and crevice corrosion growth on repassivation potential was measured using a decreasing potential staircase technique. The details of the technique and the data acquisition system have been described elsewhere [8]. Pits and crevice corrosion were initiated at a potential above the E_p measured by a cyclic polarization technique and then propagated potentiostatically for various time periods. The potential was then reduced in a stepwise manner until the current density dropped below 50 $\mu A/cm^2$. This potential is then considered to be the repassivation potential. For pitting tests, the specimens were partially immersed in the solution such that the gasketed region where electrical contact was made to the specimens was not immersed in the solution. This procedure was effective in eliminating crevice corrosion in the gasketed area. No preferential pitting was detected in the solution-vapor phase interface. For the crevice corrosion tests, the specimens were immersed completely. Serrated alumina crevice washers were bolted on the thin section of the crevice specimens using alloy C-276 bolts torqued to 0.28 N-m (40 in-oz). Care was taken to avoid galvanic contact with the bolts by using polytetrafluroethylene (PTFE) sleeves on the bolts. It was noted that crevice corrosion was more severe under the PTFE gasket area than under the alumina crevice washers, possibly because the PTFE conforms closer to the surfaces and forms a tighter crevice. For the crevice corrosion repassivation experiments, both these crevice areas were immersed and taken into consideration for crevice growth. All solutions were deaerated with nitrogen.

b). The electrochemical behavior of the alloys in simulated crevice solution was determined on cylindrical specimens using the ASTM G-61 procedure for cyclic polarization tests. Care was taken to prevent crevice corrosion at the electrode-gasket interface in these specimens by partial immersion of the specimens. The potentiodynamic scan was started from the corrosion potential and scanning was performed at 0.167 mV/sec. All solutions were deaerated with argon prior to and during the tests.

Solutions

The solutions for the repassivation potential tests consisted of 1000 ppm Cl^-, 85 ppm HCO_3^-, 20 ppm SO_4^{2-}, 10 ppm NO_3^-, and 2 ppm F^-, all prepared with sodium salts dissolved in high-purity (17 MΩ/cm resistivity) water. These tests were performed at 95°C. The simulated crevice solutions contained varying concentrations of chloride ranging from 0.028 M (1000 ppm) to 4 M, while maintaining the concentrations of the other anionic species at the same level as above. The pH of the simulated crevice solutions was modified by adding HCl to range from near neutral to zero. For acidic solutions, bicarbonate was not added because of its instability. The tests in simulated crevice solutions were conducted at 95°C and 30°C. The pH was measured at room temperature after the tests.

MODELING CREVICE CORROSION

A transient model for predicting the evolution of crevice solution chemistry (TWITCH) was developed [9]. The details of the model can be found elsewhere [9], but the salient points are described in the following paragraphs.

The essential approach of the model is the same as crevice corrosion models discussed in the literature [10,11,12]. A continuous liquid phase is assumed to be present in the crevice and the bulk. Crevice corrosion is assumed to initiate through three consecutive steps: a) Oxygen depletion in the crevice due to anodic dissolution of the metal, b) hydrolysis of the metal ions leading to a decrease in pH within the crevice, and c) transport processes leading to an increase in chloride concentration within the crevice. The model considers an one-dimensional crevice of gap, g, and length, L. The mass transport is assumed to occur by diffusion and electromigration:

$$N_j = -\frac{z_j D_j F}{RT} C_j \nabla \Phi_s - \frac{D_j}{\gamma_j}(\gamma_j \nabla C_j + C_j \nabla \gamma_j) \quad (1)$$

where:

N_j = Flux of species j
D_j = Diffusion coefficient of species j
C_j = Concentration of species j
z_j = Charge of species j

γ_j = Mean activity coefficient of j
Φ_s = Electrostatic potential in solution
R = Gas constant
T = Absolute temperature
F = Faraday constant

The flux of ions results in a current in the solution. Assuming that the crevice is formed between the metal and an inert material, charge conservation, for the case of a no-flux boundary at one end of the crevice, leads to:

$$i_s = F\sum_j z_j N_j = \frac{1}{g}\int_X^L i_e \, dx \qquad (2)$$

where, i_s is the net current in the solution, i_e is the current density at the metal-solution interface resulting from electrochemical reactions, and g is the crevice gap. In the above equation, electroneutrality is assumed at each point. For a passive metal, i_e is assumed to be independent of potential and equal to the passive current density. The flux of ionic species, $\bar{N}_j$, generated by anodic dissolution at the metal-solution interface is given by:

$$\bar{N}_j = \frac{\alpha_j i_e}{z_j F} \qquad (3)$$

where, α_j is a weighing factor indicating the degree of preferential dissolution of species j involved in interfacial reaction. The potential gradient in the solution is obtained by combining Eqns. 1 and 2:

$$\nabla\Phi_s = \frac{-(i_s + \sum_j \frac{z_j F D_j}{\gamma_j}(\gamma_j \nabla C_j + C_j \nabla\gamma_j))}{F^2 \sum_j \frac{z_j^2 D_j}{RT} C_j} \qquad (4)$$

Assuming the flux at the metal-solution interface is averaged over the crevice gap, the material balance at any point is given by:

$$\frac{\partial C_j}{\partial t} = -\nabla N_j + \frac{\bar{N}_j}{g} + R_j \qquad (5)$$

where R_j is the rate at which species j is produced through chemical reactions in the solution. In the simpler version of the code, the homogeneous chemical reactions, R_j, are assumed to occur much faster than the transport processes and hence are decoupled from

Eq. 5. A finite difference scheme with variable spacing mesh is used and the resulting ordinary differential equations are solved by an ODE solver [13]. The solution sequence for each time step is as follows:

1. Calculate the current in solution (i_s) at each node using Eq. 2
2. Estimate the potential gradient and potential in the solution at each node using Eq. 4
3. Iterate until the potential converges within tolerance
4. Calculate flux of ionic species through the metal-solution interface by anodic dissolution using Eq. 3
5. Calculate the flux of each ionic species at each node using Eq. 1
6. Simultaneously solve the transport equation at each node for each species to calculate C_j using Eq. 5 without the reaction term
7. Equilibrate the solution at each node through various hydrolysis and other equilibrium equations
8. Apply electroneutrality to correct for numerical round-off errors by distributing the numerical error at each time step among all ionic species according to their charge density

Chemistry Changes in a Type 304/304L Stainless Steel Crevice

The equilibrium solver in the code is quite flexible in accommodating a variety of species depending on the corroding system. Simplification of the number of species and nodes involved will result in a significant decrease in computation time. The code has been exercised [9] against the crevice corrosion experiments of Alavi and Cottis [14] on type 304 stainless steel in NaCl solution and Valdes [15] on iron in acetate buffer. The chemistry changes in a crevice on type 304L/304 stainless steel has been modeled in the following manner [9]:

Equilibrium Reactions -- Thirteen reactions were assumed to occur in the system. These, along with their equilibrium constants and diffusion coefficients at 25°C are shown in Table 2. The hydrolysis reactions for Ni and Fe are ignored as they are not found to affect the crevice pH significantly in the presence of Cr species [16].

Crevice Geometry -- The crevices were assumed to be 8.3 cm deep and two gaps were considered: 0.1 mm and 0.01 mm. The crevice was divided into 20 nodes with progressively finer nodes at the mouth of the crevice where the concentration gradients are steep.

Electrochemical Parameters -- The passive current density, i_e, was assumed to be 10^{-6} A/cm^2. The Fe, Cr, and Ni were assumed to dissolve as Fe^{2+}, Cr^{3+}, and Ni^{2+} respectively in proportion to their atomic fraction in the alloy. The simulation was run for 90 hours.

Table 2. Properties of Aqueous Species in Solution for Crevice corrosion of Type 304/304L Stainless Steel [14].

Species	ΔG_f^0 (J/mol)	D (m^2/s)
H_2O *(l)*	-237,200	—
H^+ *(aq)*	0	9.3 x 10^{-9}
OH^- *(aq)*	-157,300	5.0 x 10^{-9}
Na^+ *(aq)*	-261,900	1.3 x 10^{-9}
Cl^- *(aq)*	-131,300	2.0 x 10^{-9}
Cr^{3+} *(aq)*	-195,100	0.6 x 10^{-9}
$Cr(OH)^{2+}$ *(aq)*	-409,400	0.7 x 10^{-9}
$Cr(OH)_2^+$ *(aq)*	-614,100	0.8 x 10^{-9}
$Cr(OH)_3$ *(aq)*	-803,900	0.8 x 10^{-9}
$Cr(OH)_4^-$ *(aq)*	-987,400	0.8 x 10^{-9}
$Cr_2(OH)_2^{4+}$ *(aq)*	-789,600	0.5 x 10^{-9}
$Cr_3(OH)_4^{5+}$ *(aq)*	-1,487,400	0.4 x 10^{-9}
$CrCl^{2+}$ *(aq)*	-325,500	0.8 x 10^{-9}
$CrCl_2^+$ *(aq)*	-458,600	0.8 x 10^{-9}
Fe^{2+} *(aq)*	-91,500	0.7 x 10^{-9}
$FeCl^+$ *(aq)*	-221,900	0.8 x 10^{-9}
$FeCl_2$ *(aq)*	-340,100	0.8 x 10^{-9}
$FeCl_4^{2-}$ *(aq)*	-605,800	0.5 x 10^{-9}
Ni^{2+} *(aq)*	-45,600	0.7 x 10^{-9}
$NiCl^+$ *(aq)*	-171,200	0.8 x 10^{-9}

ΔG_f^0 = Gibbs Free Energy of Formation at 25°C
D = Diffusion coefficient at 25°C

RESULTS

Model Predictions

The predicted pH in the crevice as a function of distance into the crevice is shown in Figure 2 for two crevice gaps. As expected, the smaller gap results in a lower pH. The pH also drops relatively steeply with depth. The pH at the tip of the crevice (8.3 cm into the crevice) and near the mouth of the crevice (0.1 mm into the crevice) as a function of time is shown in Figure 3. The chloride concentration in the crevice rises with depth, with the smaller gap resulting in higher chloride concentration (Figure 4). The effect of bulk chloride concentration on the crevice chloride concentration is shown in Figure 5 for a 0.1 mm gap crevice. It can be seen that the chloride concentration at the crevice tip is approximately the same for bulk concentrations ranging from 0.028 M to 0.5 M. The pH in the crevice is also relatively unaffected by the bulk chloride concentration within this range. Finally, the predicted potential drop in the crevice is shown in Figure 6. Under the assumption of a potential-independent passive current density and without any input of the cathodic kinetics, the model can not predict the corrosion potential at the external surface of the crevice. However, significant potential drop (about 250mV) is predicted for a tight crevice, even when there is no active crevice corrosion. For the crevice with a larger gap, the predicted potential drop is less significant (about 50 mV). This can be inferred from Eq. 4 where a greater gradient in concentration due to a tighter crevice results in a higher potential drop. The total time for initiation of crevice corrosion includes the time for the pH in the crevice to drop to a depassivation pH (Figure 3) and the time for depletion of oxygen in the crevice.

Measured Electrochemical Response to Simulated Crevice Solutions

The cyclic polarization response of alloy 825 in a series of simulated crevice solutions of decreasing pH is shown in Figure 7. At pH=0 (Figure 7d), the polarization curve is representative of an actively corroding metal as indicated by the lack of a passive region and hysteresis in the cyclic polarization curve. Visual examination of the specimens showed that specimens after tests shown in Figure 7a - 7c were pitted, while the specimen after test shown in Figure 7d exhibited only uniform corrosion. No crevice corrosion was noted on any specimen. The pH at which active behavior was observed (Figure 7d) is then termed the depassivation pH, pH_D. The pH_D value was measured for alloy 825, and types 304L and 316L stainless steels under a variety of environmental conditions. The results are shown in Table 3. The pH values have not been corrected for the high chloride concentrations which can lead to significant over estimation of pH values. Another important aspect of the polarization behavior in simulated crevice solutions is the active-passive transition observed in intermediate pH solutions prior to total depassivation (Figure 7c).

The effect of chloride concentration and pH on the E_p and E_{rp} of alloy 825 from cyclic polarization tests is shown in Figure 8. It can be seen that a low chloride concentrations, below about 0.028 M, E_p is relatively independent of chloride because the potential is in the oxygen evolution region. Then a steep, logarithmic decrease in E_p

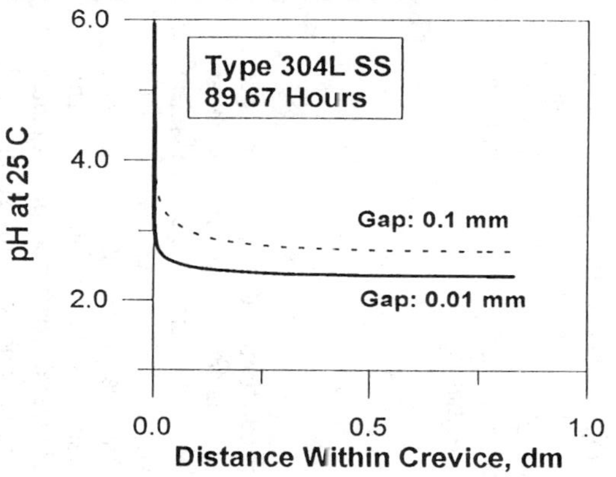

Figure 2. Predicted pH vs. Depth into the crevice

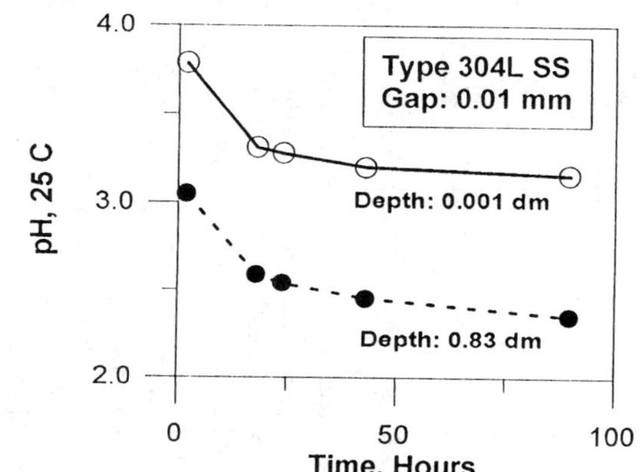

Figure 3. Predicted pH at tip and mouth of crevice vs. Time

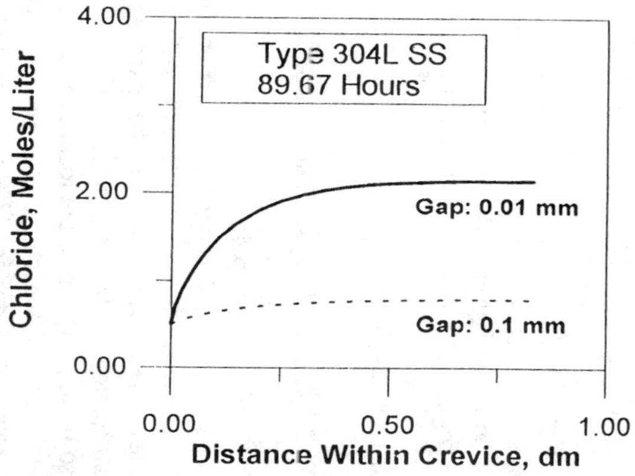

Figure 4. Predicted Cl⁻ concentration vs. depth into crevice

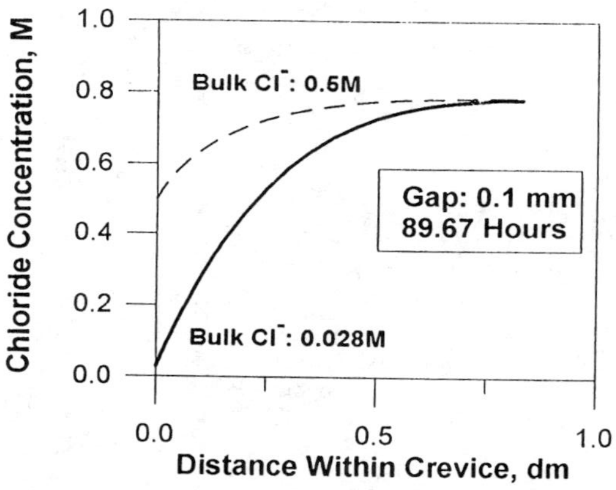

Figure 5. Effect of bulk Cl⁻ on crevice Cl⁻ concentration

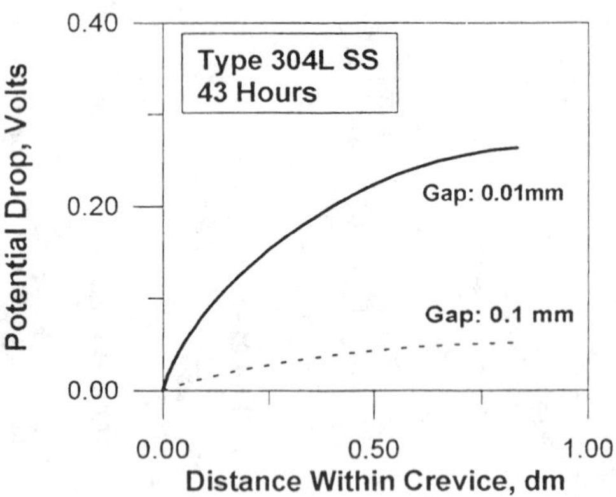

Figure 6. Predicted potential drop in the crevice

with chloride concentration is observed, followed by a more gradual decrease at high chloride concentrations. The E_{rp} shows a somewhat similar behavior, although its dependance on chloride concentration is shifted to the lower concentrations. Neither E_p nor E_{rp} show observable dependance on pH.

Repassivation Potential Measurements

The effects of prior pit growth on repassivation potentials of type 316L stainless steel and alloy 825 are shown in Figures 9 and 10 respectively. Included in these figures are results from crevice corrosion repassivation tests. It can be seen from Figures 9 and 10 that the repassivation potential attains a lower bound value beyond a certain pit and crevice corrosion growth as measured by charge density during pit growth. Interestingly, the repassivation potentials for specimens on which intentional crevices were placed (triangles in Figures 9 and 10) were within the range of repassivation potentials measured for pits. The scan rate during the backward scanning had an effect on the measured repassivation potential. Unlike cyclic polarization tests, scan rate effect can not be attributed to its impact on pit growth because pits were grown potentiostatically. Moreover, the total charge density, including the charge during the back scanning period is used in plotting the results in Figures 9 and 10. The effect of scan rate can, however, be explained by the effect of applied potential on repassivation time. The repassivation time decreases with a decrease in potential. Therefore, when the potential is scanned down rapidly, insufficient time is available for repassivation at the higher potentials and the measured repassivation potential is lower.

DISCUSSION

Overall Approach to Container Life Prediction

From the point of view of container life prediction and the performance assessment of the engineered barrier system, detailed, mechanistic models, such as TWITCH, are too cumbersome to be directly incorporated in the performance assessment codes. However, they can be used as auxiliary analyses to provide a mechanistic justification for the parameters such as initiation and repassivation potentials. They can also be used to derive the local chemistries which may determine the dissolution rate and speciation of various waste package components. Hence a container performance assessment code is likely to include various combinations of empirically and mechanistically derived parameters.

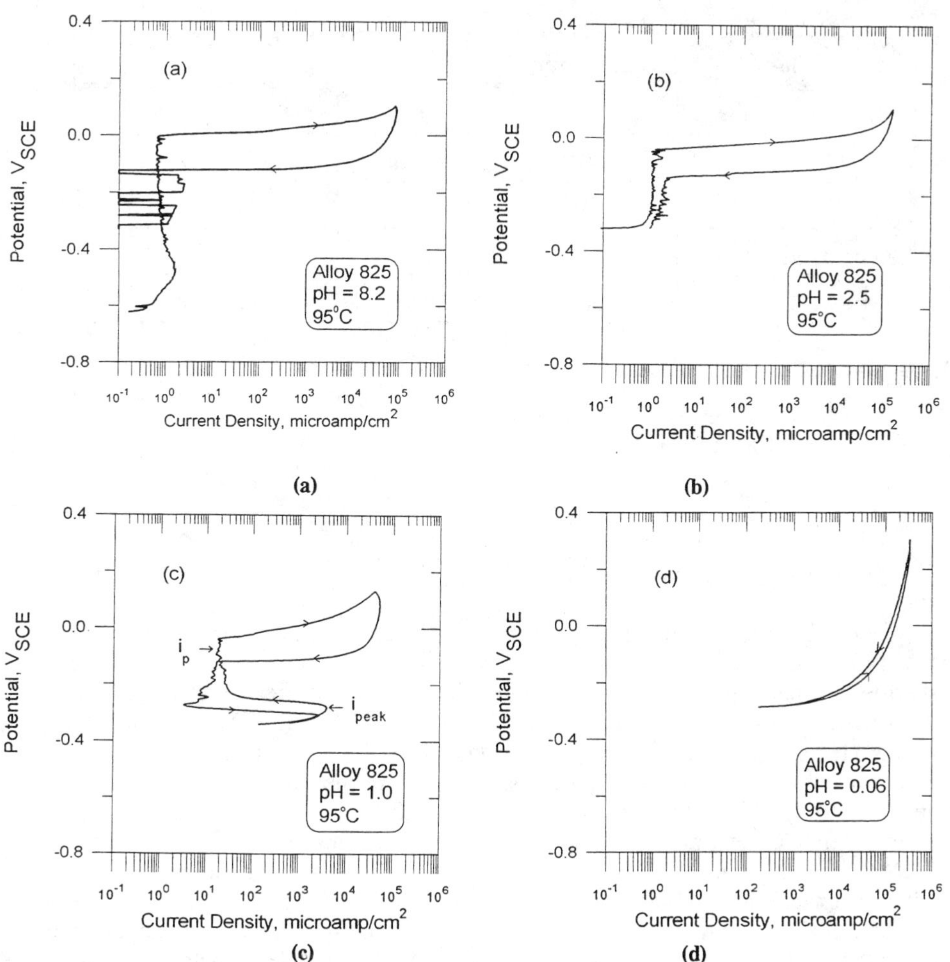

(a) (b)

(c) (d)

Figure 7. Cyclic polarization behavior of alloy 825 in simulated crevice solutions at 95°C. Scan rate: 0.167 mV/sec.

Table 3. Depassivation pH of Three Alloys under Various Simulated Crevice Solutions and Temperatures Measured by Cyclic Polarization Technique.

Alloy	Temperature °C	pH_D		
		A	B	C
Alloy 825	30	-0.36	–	–
	95	0.01-0.09	0.09	0.028
Type 316L S.S.	95	1.10	–	–
Type 304L S.S.	30	.028	0.11	–
	95	1.28	0.11	–

A: 4M Chloride
B: 0.5M Chloride
C: 3M Chloride + 0.25M Sulfate

Crevice Corrosion Initiation

The approach for long-term prediction of crevice corrosion initiation is to use the TWITCH code which requires the input criterion for depassivation, pH_D, obtained from short-term experiments. As shown in Figure 3, the pH at the crevice tip did not attain the depassivation pH (Table 3) for type 304L S.S. in 100 hrs. However, prior modeling results [16] suggest that crevice corrosion initiations time is quite short. Generally, crevice corrosion is assumed to initiate when the pH of the solution inside the crevice attains a value corresponding to pH_D of the alloy [10-12, 16]. The pH_D is assumed to be relatively independent of chloride concentration and temperature [17]. As shown in Table 3, the pH_D of alloy 825 is not dependent on the chloride and sulfate concentrations within the scatter in the data. However, pH_D decreases with a decrease in temperature. For type 304L stainless steel, the pH_D is dependent both on temperature and chloride concentration. The pH_D values for types 304L and 316L stainless steels in 4 M Cl^-, shown in Table 3, are similar to values reported by others [17]. The results differ from those in the literature in the effect of temperature and chloride concentration on pH_D, possibly because of differences in test techniques. Okayama et al. [17] used open-circuit conditions to determine the pH at which the corrosion potential fell sharply to active values. It must be assumed, from their data, that the deaeration of the solution was insufficient since rather high corrosion potentials were observed at pH values above "pH_D" while low corrosion potentials were observed at pH values below "pH_D". From Figure 7c, it can be seen that the high peak current density prior to complete

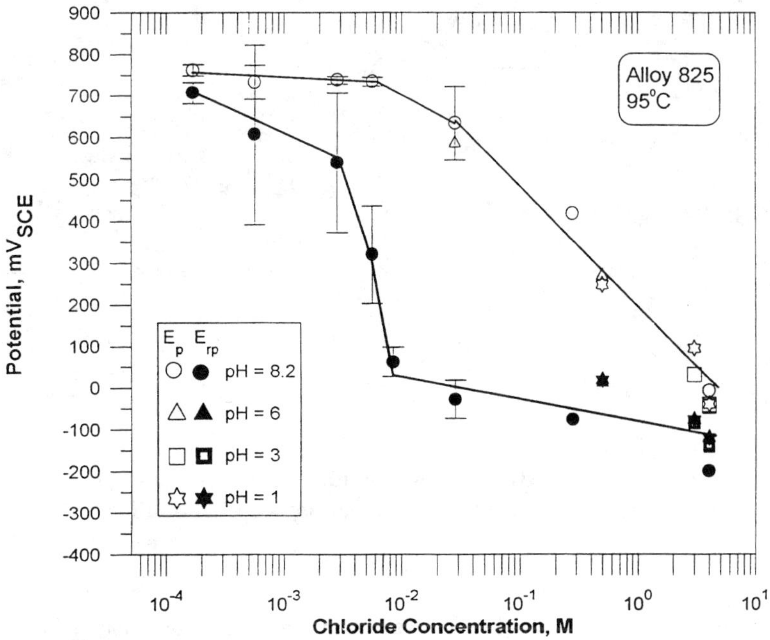

Figure 8. Effect of pH and Cl^- on pitting and repassivation potential of alloy 825.

depassivation may lead to lowering of corrosion potential or an unstable corrosion potential depending on the redox condition of the test. For example, if the cathodic polarization curve intersects the anodic curve at or below the anodic peak (Figure 7c), a lowering of corrosion potential can indicate that depassivation may have occurred at pH = 1.0. However, complete depassivation, as seen by a lack of active-passive transition did not occur until the pH was lowered further. Conducting a cyclic polarization curve, while more time-consuming, is a more reliable technique for determining depassivation.

More importantly, the assumption that crevice corrosion initiates only after complete depassivation occurs is questionable. Other events that can trigger crevice corrosion prior to complete depassivation include: (i) Pitting inside crevices, (ii) The increase in active peak current density (Figure 7c) resulting in active dissolution especially if potential drop in the crevice is of the right magnitude, (iii) The increase in passive current density with a decrease in pH accelerating dissolution inside crevices and hence the rate of change of solution chemistry, and (iv) The dissolution of MnS inclusions resulting in an acidic solution containing sulfur species that lead to destabilization of passive film.

Pits Inside Crevices-- It is generally accepted that the pitting potential is relatively unaffected by pH for a wide variety of alloys, but decreases with increase in chloride concentration and temperature [18]. The findings reported here are in agreement, as shown in Figure 8. Thus, depending upon the applied or corrosion potential, increase in chloride concentration such as that shown in Figure 4 can trigger pitting within

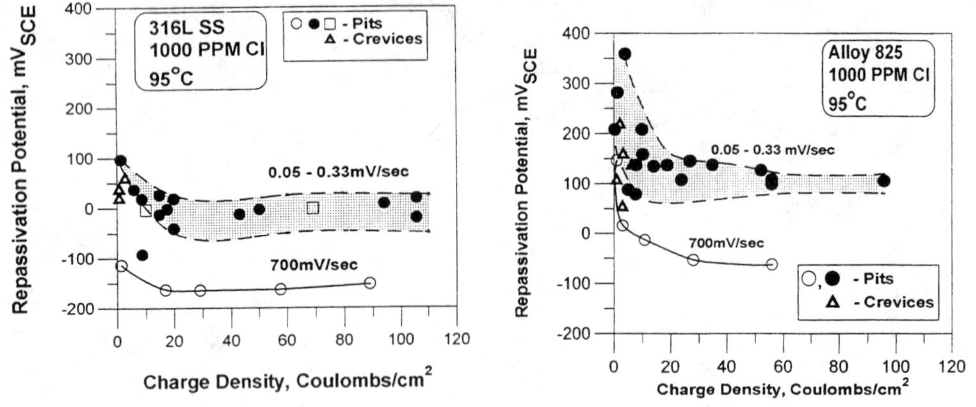

Figure 9. Effect of pit growth on repassivation potential of type 316L S.S.

Figure 10. Effect of pit growth on repassivation potential of alloy 825.

crevices which then can further acidify the environment eventually resulting in total depassivation.

Potential Dependent Current-- The predictions of the crevice corrosion model, shown in Figures 2 through 6, are based on the assumption that the anodic dissolution in the crevice prior to initiation of corrosion is the passive current density that is independent of potential, although the TWITCH code can accept more general formulation of the anodic dissolution rate. Because of this assumption, the prediction of the chemistry changes in the crevice, and hence the initiation time, is independent of the external potential. In reality, it is well established that crevice initiation time decreases with increasing potential and occurs only if the potential is above a critical potential [19]. The model prediction can be reconciled with experimental observation if a potential dependent current density in the crevice is assumed or if the criterion for crevice initiation is potential dependent instead of the simple pH_D. While potential dependent current complicates the numerical computation significantly, it has been successfully implemented in the TWITCH code for the case of iron in acetate buffer [9].

Dissolution of Inclusions-- Potential dependent current can also occur by the dissolution of MnS inclusions present in stainless steels prior to complete depassivation by Cr hydrolysis products. This mechanism was proposed by Lott and Alkire [20]. This mechanism may not be valid for alloys such as alloy 825 which do not have significant amounts of MnS inclusions.

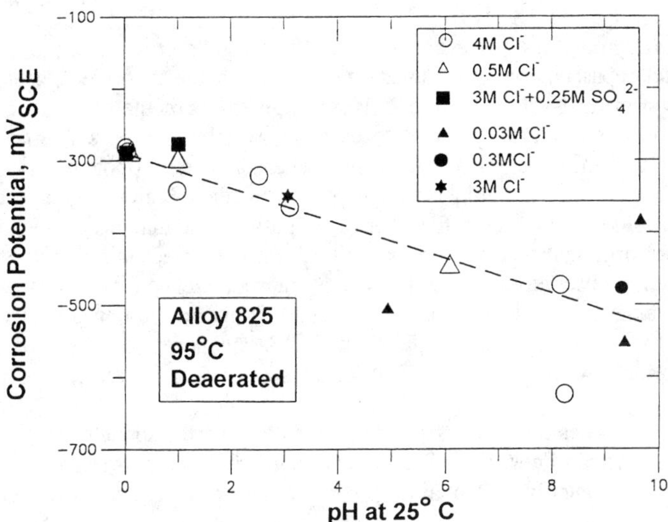

Figure 11. The effect of pH and Cl^- on corrosion potential of alloy 825 in deaerated, simulated crevice solutions

Repassivation

The use of repassivation potential for crevice corrosion prediction has been elaborated by Tsujikawa et al. [22]. In the present investigation, it has been shown that the repassivation potential for pitting can be used as a lower bound parameter for both pitting and crevice corrosion performance. Two types of mechanisms have been proposed for repassivation, although these are not mutually exclusive.

The first mechanism involves changes in the potential inside the crevice as a result of changes in external potential [21]. If one assumes that crevice corrosion is initiated by depassivation and general dissolution in the crevice, then repassivation occurs when the internal potential is below the corrosion potential of the alloy in the crevice environment. The corrosion potential of alloy 825 in various simulated crevice solutions was relatively independent of chloride and sulfate concentration, but dependent on pH, as shown in Figure 11. This indicates that the corrosion potential is essentially determined by the cathodic reaction of hydrogen evolution. Assuming a crevice solution pH of 1, the internal potential has to be below -320 mV_{SCE} to repassivate by this mechanism. The repassivation potential, under fast scanning condition, for alloy 825 is -50 mV_{SCE}, implying a potential drop of 270 mV in the crevice. This potential drop is possible (Figure 6), especially if large anodic current density is present. Because, an active crevice corrosion is being considered prior to repassivation, the pH assumed in the crevice is independent of prior crevice growth, being equal to pH_D. Because the corrosion potential is independent of chloride concentration, this mechanism predicts that repassivation potential should be relatively independent of bulk chloride concentration and prior crevice growth.

The second mechanism involves changes in solution composition inside the crevice/pit as a result of changes in external potential [22]. In this mechanism, the chloride concentration and the pH inside an actively corroding crevice/pit are assumed to be related and repassivation is assumed to occur when the chloride concentration decreases below a critical value. While a rigorous treatment of the effect of external potential on chloride concentration is yet to be done, simpler models suggest that the chloride concentration in the crevice is decreased when the external potential decreases [22,23]. Since the corrosion potential is relatively independent of chloride concentration, the effect of chloride is related to the change in pitting potential inside the crevice (Figure 8) or the formation of a chloride salt film [23]. This mechanism needs further exploration. Other observations also support a transport related mechanism. For example, Nakayama and Sasa [24] observed that E_{rp} increased considerably upon ultrasonic agitation of the specimen. The finite repassivation time observed in the present investigation as well as in others [23] also suggest a transport related mechanism.

The choice between the two types of mechanisms suggested above may be dictated by scan rate. At fast scan rates, there is insufficient time for transport related processes and E_{rp} may be determined by potential drop considerations alone. At slow scan rates or under potentiostatic conditions,. transport processes become significant and repassivation is observed at higher potentials.

SUMMARY

An overall approach to life prediction under localized corrosion conditions is outlined. This involves consideration of corrosion potential, pit initiation potential, and repassivation potential. The time for initiation once the critical potential has exceeded it is not considered to be significant. A transient transport-reaction coupled model (TWITCH) is shown to predict the changes in chemistry inside a crevice. Long-term prediction of the initiation of crevice corrosion can be accomplished by combining this model with the short-term experimental data on depassivation pH. The initiation time predicted by this model is short in comparison to anticipated container life. However, because the model assumes a potential-independent current prior to initiation of crevice corrosion, a critical potential for initiation can not be predicted. Polarization experiments in simulated crevice solutions indicate that prior to total depassivation, increased active-passive peak current density is observed leading to a potential dependent current in the crevice. Short-term experimental data of the anodic dissolution kinetics in simulated crevice solution can be used in the model to obtain the potential dependency of crevice corrosion initiation. Other approaches to predicting a potential dependence of initiation involve the assumption of the dissolution of MnS inclusions or the assumption that pitting nucleates inside crevices prior to complete depassivation.

It is shown that the repassivation potential for pitting is independent of prior pit growth beyond a certain value. The repassivation potential for crevice corrosion is within the band of repassivation potentials measured for pitting. This enables the use of a single potential for long-term life prediction under both these corrosion modes. At fast scan rates, repassivation is dictated by the corrosion potential of the active metal inside the crevice. At slow scan rates or under potentiostatic conditions, transport related changes in the crevice solution determine repassivation. Prediction of repassivation potentials using TWITCH code should be explored.

ACKNOWLEDGEMENT

The assistance of Walter Machowski and Donald Pile in conducting the experiments is acknowledged. This paper was prepared as a result of the work performed by the Center for Nuclear Waste Regulatory Analyses (CNWRA) for the U.S. Nuclear Regulatory Commission under Contract No. NRC-02-88-005. The activities reported here were performed on behalf of the Office of Nuclear Regulatory Research, Division of Regulatory Applications, Dr. M.B. McNeil project officer. The paper is an independent product of the CNWRA and does not necessarily reflect the views or regulatory position of the NRC.

REFERENCES

[1] Beavers, J. A., Thompson, N. G., and Durr, C. L., "Pitting, Galvanic, and Long-Term Corrosion Studies on Candidate Container Alloys for the Tuff Repository," NUREG/CR-5709. Cortest Columbus Technologies, Inc., Columbus, OH 43235, January 1992.

[2] Patrick, W. C., "Spent Fuel Test - Climax: An Evaluation of the Technical Feasibility of Geologic Storage of Spent Nuclear Fuel in Granite," Final Report, UCRL-53702, Lawrence Livermore National Laboratory, Livermore, CA, March 30, 1986.

[3] Ramirez, A. L., (ed.), "Prototype Engineered Barrier System Field Test (PEBSFT) Final Report," UCRL-ID-106159, Lawrence Livermore National Laboratory, Livermore, CA, August 1991.

[4] Zimmerman, R. M., Schuch, R.L., Mason, D. S., Wilson, M. L., Hall, M. E., Board, M. P., Bellman, R. P., and Blanford, M. P., "Final Report: G-Tunnel Heated Block Experiment," SAND84-2620, Sandia National Laboratories, Albuquerque, NM, May 1986.

[5] Macdonald, D. D., and Urquidi-Macdonald, M., "Thin-Layer Mixed-Potential Model for Corrosion of High-Level Nuclear Waste Canisters," Corrosion, Vol. 46, No.5, May 1990, pp. 380-390.

[6] Macdonald, D. D., and Urquidi-Macdonald, M., "Corrosion Damage Function—Interface Between Corrosion Science and Engineering," Corrosion, Vol. 48, No.5, May 1992, pp. 354-367.

[7] Thompson, N. G., and Syrett, B. C., "Relationship Between Conventional Pitting and Protection Potentials and a New, Unique Pitting Potential," Corrosion, Vol. 48, No.8, August 1992, pp. 649-659.

[8] Sridhar, N., and G. A. Cragnolino, "Long-Term Prediction of Localized Corrosion of Fe-Ni-Cr-Mo High-Level Nuclear Waste Container Materials," CORROSION/93, Paper No. 93197, National Association of Corrosion Engineers (NACE), Houston, TX, April 1993.

[9] Walton, J. C., and Kalandros, S. K., "TWITCH – A Model for Transient Diffusion, Electromigration, and Chemical Reaction in One Dimension," CNWRA 92-019, Center for Nuclear Waste Regulatory Analyses, San Antonio, TX, August 1992.

[10] Sharland, S. M., "A Mathematical Model of the Initiation of Crevice Corrosion in Metals," Corrosion Science, Vol. 33, No. 2, Feb. 1992, pp. 183-201.

[11] Watson, M., and Postlethwaite, J., "Numerical Simulation of Crevice Corrosion of Stainless Steels and Nickel Alloys in Chloride Solutions," Corrosion, Vol. 46, No. 7, July 1990, pp. 522-530.

[12] Oldfield, J. W., and Sutton, W. H., "Crevice Corrosion of Stainless Steels - I. A Mathematical Model," British Corrosion Journal., Vol.13, No.1, Jan. 1978, pp. 13-22.

[13] Hindmarsh, A. C., "ODEPACK, A systematized Collection of ODE Solvers," Scientific Computing, R. Stapleman et al. (eds.), IMACS/North-Holland Publishing Co., Amsterdam, 1983, pp. 55-64.

[14] Alavi, A., and Cottis, R. A., "The Determination of pH, Potential, and Chloride concentration in Corroding Crevices on 304 Stainless Steel and 7475 Aluminum Alloy," Corrosion Science, Vol. 27, No. 5, May 1987, pp. 443-451.

[15] Valdes-Mouldon, A., Ph.D. Thesis, Pennsylvania State University, University Park, PA, 1987.

[16] Sridhar, N., Cragnolino, G. A., Pennick, H., and Torng, T. Y., "Application of a Transient Crevice Corrosion Model to the Prediction of Performance of High-Level Nuclear Waste Container Materials," Life Prediction of Corrodible Structures, R. N. Parkins (ed.), NACE, Houston, TX, to be published.

[17] Okayama, S., Tsujikawa, S., and Kikuchi, K., "Effects of Alloying elements on Stainless-Steel Depassivation pH," Corrosion Engineering, Vol. 36, No.11, Nov. 1987, pp. 631-638.

[18] Szklarska-Smialowska, Z., Pitting Corrosion of Metals, NACE, Houston, TX, 1986, pp. 201-236.

[19] Yashiro, H., Tanno, K., Hanayama, H., and Miura, A., "Effect of Temperature on the Crevice Corrosion of Type 304 Stainless Steel in Chloride solution up to 250°C," Corrosion, Vol.46, No.9, Sep. 1990, pp. 727-733.

[20] Lott, S. E., and Alkire, R. C., "The Role of Inclusions on Initiation of Crevice Corrosion of Stainless Steel," Journal of The Electrochemical Society, Vol. 136, No. 11, 1989, pp. 3256-3262.

[21] Rosenfeld, I. L., Danilov, I. S., and Oranskaya, R. N., "Breakdown of the Passive State and Repassivation of Stainless Steels," Journal of The Electrochemical Society, Vol. 125, No.11, Nov. 1978, pp. 1729-1735.

[22] Tsujikawa, S., Sone, Y., and Hisamatsu, Y., "Analysis of Mass Transfer in a Crevice Region for a Concept of the Repassivation Potential as a Crevice Corrosion Characteristic," Corrosion Chemistry within Pits, Crevices and Cracks, A. Turnbull (ed.), Her Majesty's Stationary Office, London, 1987, pp. 171-186.

[23] Strehblow, H. -H., "Breakdown of Passivity and Localized Corrosion: Theoretical Concepts and Fundamental Experimental Results," Werkstoffe und Korrosion, Vol.35, 1984, pp. 437-448.

[24] Nakayama, T., and Sasa, K., "Effect of Ultrasonic Waves on the Pitting Potentials of 18-8 Stainless Steel in Sodium Chloride Solution," Corrosion, Vol. 32, No.7, July 1976, pp. 283-285.

Rolf Sjöblom[1]

CORROSION ASPECTS OF COPPER IN A CRYSTALLINE BEDROCK ENVIRONMENT - WITH REGARD TO LIFE PREDICTION OF A CONTAINER IN A NUCLEAR WASTE REPOSITORY.

REFERENCE: Sjöblom, R., "**Corrosion Aspects of Copper in a Crystalline Bedrock Environment - With Regard to Life Prediction of a Container in a Nuclear Waste Repository,**" Application of Accelerated Corrosion Tests to Service Life Prediction of Materials, ASTM STP 1194, Gustavo Cragnolino and Narasi Sridhar, Eds., American Society for Testing and Materials, Philadelphia, 1994.

ABSTRACT

A disposal system for high-level and long-lived radioactive waste should provide for adequate isolation of radionuclides from the biosphere.

During the work of finding such a system, and in demonstrating its long-term isolation, certain unprecedented requirements become apparent: the disposal system must function over very long times without any need for maintenance or repair. Furthermore, no experience from operation can be made available for the analysis, and there will be no results from tests with durations comparable in magnitude to the life-time of the facility.

In the systems studied in Sweden and Finland, copper containers are deposited in crystalline bedrock with bentonite as a backfill. The groundwater is reducing and the flow is very low due to the low permeability of the bentonite clay and of the rock near the deposition hole.

The chemical durability of a container in such a disposal system will depend primarily on three factors: the mechanical tensile stresses in the container, the chemical environment, and the copper material itself. The tensile stresses depend on manufacturing factors as well as external forces. The chemical environment will depend on the composition of the groundwater, the bentonite clay and possibly also phenomena that might take place in the vicinity of the container. The copper material can corrode primarily by oxidation or sulphidation. If the salt concentration is high (and a reducible specie is available), copper may also react by forming chloride complexes.

If the copper container has thick walls, penetration by general corrosion will take considerable time. It is nevertheless of interest to

---

1 The Swedish Nuclear Power Inspectorate (SKI), Sehlstedtsgatan 11, Box 27 106, S-102 52 Stockholm, Sweden. Previously at the Swedish National Board for Spent Nuclear Fuel (SKN).

study general corrosion since - at least under certain conditions - it will give rise to a protective and passivating surface layer. Such a layer may be of significance when the possibilities of localized corrosion and stress corrosion cracking are to be evaluated. It is pointed out that the development of localized corrosion is strongly dependent on the specific chemistry. Concern is expressed regarding certain treacherous aspects of stress corrosion cracking. Areas of research that may be fertile to pursue have also been identified.

A point is made that the safety analyses and scenarios should include both probable and less probable developments. Examples are given of the latter type.

A suggestion is made that samples of the container and of container material be left in the repository to enable evaluation and thus to facilitate improvements that future generations might wish to make.

KEYWORDS: corrosion, copper, metamorphic rocks, ground water, materials testing, radioactive waste storage, radioactive waste disposal, spent (nuclear) fuel

---

BACKGROUND

Currently, considerable resources are being allocated in a number of countries for research and development efforts with the purpose of finding suitable systems for the disposal of long-lived radioactive waste.

The systems studied are quite different. The major reason for this is that different types of geological formation are present, or might be made available in each particular country. There are also considerable differences in the emphasis placed on the geological barrier as compared with that of the engineered barrier.

Copper as a container material for the high-level waste in a final repository is considered and/or studied primarily in Canada, Finland, Sweden, Switzerland and the United States. The studies are made primarily in conjunction with programs for the disposal of the spent fuel without reprocessing. In Canada, Finland, Sweden and Switzerland disposal in crystalline rock is considered. In the United States, disposal in crystalline rock was in the program earlier.

The reference system used in this paper is the KBS-3 system which has been developed by the Swedish Nuclear Fuel and Waste Management Company (SKB) [1]. In this system, the spent nuclear fuel will be enclosed in copper containers (canisters) having thick walls (perhaps 50 mm). The containers will be deposited in holes drilled in the floors of tunnels with compacted bentonite clay as a buffer and backfill. The tunnels are to be excavated in good quality crystalline rock at a depth of about 500 meters.

The anticipated function of the system is as follows[2, 3]:

- The rock will provide a mechanically and chemically stable environment for the engineered barriers.

- The location of a deposition hole will be selected in such a manner that the groundwater flow in the vicinity of the engineered barriers will be very small.
- Soon after the deposition of the containers, the compacted bentonite clay will take up water, swell and provide mechanical protection to the container.
- The bentonite clay will also hinder water flow efficiently.
- The containers will be extremely corrosion resistant in the reducing hard rock environment.
- In the event that a container is penetrated, the spent fuel inside will be very limited in its solubility in the groundwater.

The Swedish strategy [1-4] has indicated, that the container is the part of such a system that is expected to provide the most long-term containment of the radionuclides.

This emphasis on the container is, however, not reflected in a consistent manner in the allocation of the R&D funds or in the safety analyses made. Considerably more effort has been spent on the geological barriers than on the engineering ones. In the most recent safety analysis made [2, 3], only a small number of mechanisms that may affect the container have been identified, and only a few of the scenarios compiled[5] deal with mechanisms for the deterioration or modes of malfunction of the copper container.

It has been estimated [3] that the life-time of a copper container may be as long as 100 million years! This estimate was based on a mechanism involving general corrosion caused by reduced sulphur that may reach the surface of the container through diffusion. The conclusion was that "it is highy unlikely that corrosion will decisively limit the life of the canister". In the "short term", the only realistic mechanism for contact between the spent fuel and the groundwater was through an initial penetration of the container (due to manufacturing factors, for example) [3].

POSSIBLE REQUIREMENTS AND PERFORMANCE ASSURANCES FOR A CONTAINER IN A REPOSITORY

A disposal system for high-level and long-lived radioactive waste should provide for isolation of radionuclides from the biosphere such that the dose to an individual in a critical group will be low. Moreover, for the long term perspective, it also might be appropriate to require that the radioactivity may reach the surface only in amounts that are small in relation to the natural flow of radionuclides from the bedrock to the biosphere.

This implies that we are actually facing certain unprecedented requirements, namely that the disposal system should function in an adequate and safe manner over essentially unlimited periods of time without any need for maintenance or repair.

In the research and development work aimed at finding such a system, in demonstrating its long-term containment of the radionuclides, and in assuring its future performance, certain characteristics particular to this issue must be adequately dealt with:

- there will be no access to any results from tests with durations comparable in magnitude to the life-time of the facility;

- there will be no access to any experience from the post-closure operation of a disposal system; and
- there will be considerable uncertainties in a number of other areas of relevance for the future performance of the repository.

Moreover, these uncertainties increase with time. On the other hand, the radioactivity in the waste will decay with time. This may imply that the assessment and assurance of the future performance should be carried out rather differently for different spans of time.

The requirement of adequate and safe functioning of a repository does not imply that all parts of the repository necessarily will have to function in the manner predicted. In a repository designed in accordance with the *multibarrier principle*, the failure of one barrier should not lead to unacceptable functioning of the repository as a whole [2, 6].

For the purpose of the development of a container, it might be practicable to consider e.g. the following time spans and possible expectations regarding its future performance:

1 *From now until a few thousand years hence.* During this time, comparatively short-lived radionuclides like strontium-90 and cesium-137 will decay to insignificant levels. For this period of time, it might be appropriate to aim for an essentially complete containment of the radionuclides inside the containers, as well as a rather comprehensive assurance regarding this containment.

2 *From a few thousand years until the next ice-age (which is expected to take place in 5 000 - 30 000 years[7, 8]).* Certain malfunction of the containers may be tolerable depending on the number of containers involved, the distribution in time, the nature of the mal-functions and the uncertainties in the assessments of the events.

3 *After the next ice-age.* Certain modes of container failure should be tolerable. It may, however, still be necessary to avoid and find assurance against common cause failures, for example [2].

FACTORS OF IMPORTANCE FOR THE CHEMICAL DURABILITY OF A CONTAINER IN A REPOSITORY

The integrity of a container in a repository depends to a large extent on three basic factors: the chemical environment, the mechanical load and the material. Other factors of importance are for example radiation and temperature. The actual combination of these basic factors encountered in a repository will depend amongst other things on rock type, repository design, container material selected, design and manufacturing processes for the container and future climate.

Different mechanisms for deterioration may appear for different combinations of these basic factors. All three factors will be of importance when the possibilities of corrosion are to be considered and analysed.

The mechanical condition of the container in a repository.

The container in a repository will be subjected to external forces from several sources. After closure, the groundwater level will be restored and the container will be subjected to a hydrostatic pressure of perhaps 6 MPa [3]. Furthermore, the bentonite will absorb water, swell and exert an additional pressure of at most 10 MPa [3] on the container. During glaciation, an additional pressure of up to perhaps 30 MPa (depending on the location of the site) may develop due to the pressure of the ice sheet.

These pressures will give rise to stresses in the walls of the container. Most of the stresses will be compressive, but some will be tensile. The externally induced stresses will be added to the stresses already present in the container originating from the manufacturing processes (e.g. moulding, working and welding) [11, 12].

One concern regarding such forces acting over a very long time is creep of the container. Such phenomena are presently being studied, e.g. in Sweden and Finland[11-15], but will not be mentioned further since they fall outside the scope of this paper.

It is, however, of interest in the present context to discuss the interaction between the tensile stresses in the container and chemical reactions (see below regarding stress corrosion). Actually, chemical reactions themselves can give rise to local mechanical stresses, e.g. if a reaction product at the tip of a crack has a volume that is larger than the metal that it was formed from.

The composition of the groundwater in crystalline rock

The composition of the groundwater in Swedish crystalline rock varies considerably with depth and also with location and distance from the sea shore. At several hundred meters depth, the salinity of the water may exceed that of the water in the Baltic or even the Atlantic. Also, the composition is different from that of the sea water [16].

The groundwater has a typical pH of 7-9 and is strongly reducing (it contains iron-II). According to thermodynamics it is possible that sulphate ions (together with iron-II), reduced sulphur or chloride ions, which are present in the water, could react with copper. Other species of interest are nitrate and ammonium ions that might be present in trace amounts.

The genesis of the groundwater is presently a matter of some debate among the geochemists. There are at least three possible explanations:

- The dissolved species in the groundwater originate primarily from the rock itself (and are thus older than the water) [17].
- Freezing of water in the rock during previous glaciations has lead to a separation of water enriched in salts (at depth) from fresh water (near the surface) [7, 8, 17, 18].
- There has been an inflow of saline marine water the composition of which has been altered through ion exchange with the rock [3, 17].

Little seems to be known about how the composition of the water in the rock may change during a glaciation cycle [7, 17] It has been suggested, however, that the flow might be very different during different phases[2, 19], and that it might even be possible to encounter situations in which directions of flow are reversed[19].

In view of these uncertainties, it has been suggested [17] that the safety analyses be made using different assumptions regarding the future composition of the groundwater.

Copper in groundwater.

Some pertinent general information about copper, its occurrence, manufacture and corrosion properties can be found in references [20-25]. General, as well as system-specific information on copper can be found in references [12, 25-46].

Although not entirely undisputed[47-49], it is generally accepted that copper is thermodynamically stable in relation to oxygen-free pure water[27, 50, 51].

Since the longevity of the container is very crucial for the Swedish program[3, 4, 52], the question of the chemical stability of copper - even in pure water - must be analysed carefully. The only risk that has been identified is if a new, and previously uncharacterised, phase were to form and if such a phase had sufficiently low free energy. At present, certain attempts are under way to resolve this matter.

Copper reacts rapidly with hydrogen sulphide to form one- or two-valenced copper sulphide (or more complex sulphides). (Actually, the electron structure of copper in sulphides is compatible only with a valency of one[53]). The concentrations of reduced sulphur found in deep groundwaters are generally low, however.

Oxidation of copper with sulphate ions is not thermodynamically possible unless some other specie participates, e.g. iron-II [27].

Reduction of sulphate ions is, however, extremely sluggish kinetically [27, 28, 54], and no evidence has been found to suggest that any such reduction takes place at conditions resembling those in a Swedish waste repository. It has been pointed out[27] that "the Quincy mineralisation [in Canada] is a good example of a geological system where native copper has been in contact with water containing both sulphate and iron(II) without any appreciable quantities of Cu_2S forming ... furnishing additional evidence for the strong kinetic inhibition of redox processes with sulphate". A recent study[28] calls for some caution, however, since "even the <u>formation</u> of native copper in the oxidation zones of sulphide ores takes place from solutions rich in SO_4^{2-}, i.e. precipitation and stability of the metal is a question of suitable Eh-pH conditions".

The reduction of sulphate ions can be relatively rapid, however, in the presence of certain microorganisms [16, 55]. One of the limiting factors for the microbial activity is the availability of nutrients, e.g. organic carbon. This feature has been utilized for estimates of upper bounds for the access of reduced sulphur to a container [12].

In water rich in chloride, significant amounts of copper may dissolve to form $CuCl_2^-$ and $CuCl_3^{2-}$ complexes [28, 45, 46, see also reference 43, page 170-171]. These complexes are quite strong and stabilize copper with valency one. Furthermore, the formation of these complexes is very strongly favored by a high chloride concentration. Thus, an increase in chloride concentration from 0.5 to 1M, may lead to an increase in the total solubility of copper by a factor of about six [28]. The solubility is also very highly dependent on pH and Eh.

POSSIBILITIES OF CORROSION OF A COPPER CONTAINER

Comprehensive compilations of the various aspects to consider for the containment of long-lived radionuclides in a container are given in references [9, 10].

General corrosion

The general corrosion of a copper container is limited by the availability of corrodants. It is usually assumed that these species appear in the groundwater in small amounts only, and that the rate of reaction is limited by diffusion (together with the flow of the groundwater).

As mentioned above, various estimates have been made[3, 12, 27, 56] concerning the durability of a container with respect to general corrosion by reduced sulphur. These estimates have indicated very long times indeed. The rate limiting mechanism assumed has been diffusion which is relatively well understood, at least by comparison to other mechanisms.

In the case of chloride complex formation, dissolution of copper will take place only if the complexes - which form in low concentrations (except under acidic and/or oxidizing conditions) - are able to diffuse away from the vicinity of the container [44, 46]. Another rate limiting factor may be the availability of a suitable reducing specie [44, 46].

Certain events might give rise to a considerably higher rate of general corrosion, for example inflow of water rich in oxygen. The oxygen might then oxidize sulphide minerals and give rise to low pH conditions [17].

It seems likely, however, that another aspect of general corrosion may be considerably more important than the rate of dissolution. It is well known, that under oxidizing conditions, a passivating layer may form on the copper and that such a layer may hinder further attack very efficiently [34]. This is probably the primary reason for the excellent corrosion properties that copper exhibits in numerous everyday applications.

If attack occurs, nevertheless, it may well be a local phenomenon such as pitting corrosion, crevice corrosion or stress corrosion.

However, the Swedish repository conditions are oxidizing only during the initial stages (after closure) whilst the remaining oxygen from the atmosphere is being consumed. After this period, the conditions will be strongly reducing and a passivating oxide type of layer will no longer be stable.

It has been reported[1, 27] that reduced sulphur may react with copper to form copper(II) sulphide. There are, however, a considerable number of different copper sulphides[28] . It is also conceivable that other species may react with copper and reduced sulphur to form a surface layer. Little appears to be known regarding the composition of any such layer, and whether or not it will be a protective or a passivating one (see for example the discussion on page 134 in Reference [43]).

Localized corrosion

In the KBS-3 report[1, 27], it was noted that the attack by reduced sulphur might not be uniform. Instead it was estimated that a pitting factor of 5 would be realistic. (The pitting factor is the ratio between the maximum pit depth and the average penetration.) In the latest safety analysis of SKB[3], the pitting factor has been estimated to have a value of 2.

The pitting factor - and perhaps even the pitting mechanism - may be altered by changes in the water chemistry. Thus, at least for oxidizing conditions[36, 37, 39, 42], pitting can be affected by for example the concentrations of chloride, sulphate and hydrogen carbonate ions.

The distribution of the pits is also known to be strongly dependent on the local conditions of the metal surface and of the bentonite clay adjacent to it [42].

In addition to the difference in penetration described by the pitting factor, there is a difference due to nonuniformity of the container and to differences in the diffusion pattern of reacting species [12 and 43 page 162].

Another possibility of localized corrosion is crevice corrosion that occurs in crevices between two metal surfaces or between a metal surface and a non-metal surface [36, 37, 39, 42]. Crevice corrosion is similar in nature to pitting corrosion and is sometimes considered to be a special case of pitting corrosion[42].

It is conceivable that localized corrosion may be due to reactions between copper, iron-III and cloride ions[28].

Stress corrosion

The general assessment by well-reputed corrosion specialists is that there is no indication that stress corrosion will take place in a copper container[27, 57] and that any such event would appear to be improbable[27].

On the other hand, when it comes to the actual evaluation and assessment of the significance of stress corrosion (if any), certain features pertinent to stress corrosion should be recognized[57].

Typically, stress corrosion is the result of specific combinations of stress, material and chemical conditions. Thus, a material may in general be rather insensitive, but for certain values of pH and Eh and/or when certain species are present in trace amounts in combination with appropriate material and stress conditions, stress corrosion may occur.

Moreover, stress corrosion often only initiates after a significant period of time.

Thus, it may be difficult to find and establish the conditions under which stress corrosion could occur. In fact, in industrial applications, stress corrosion has earned a reputation of treacherous appearance and may well appear first long after a material has been introduced in a given application.

It might be hoped that theoretical considerations and an understanding of the relevant mechanisms would be an appropriate tool to avoid surprises. However, when this is attempted, it should be kept in mind, that it is very difficult to study and understand what happens during the initiation period. Moreover, theoretical considerations have not been found to be an efficient tool for the identification of stress corrosion conditions. Instead, the approach typically used in industrial applications is empirical, and is based on testing under relatively realistic conditions for long times - often at considerable expense.

When such testing is considered, it will soon become apparent that the expected life time of a container will be orders of magnitude greater than any possible duration of a laboratory test. Furthermore, the initiation time might well exceed the entire testing time. There will also be considerable practical difficulties associated with achieving and measuring crack propagation rates that would - in a laboratory experiment - correspond even to a few thousand years of service life.

Pertinent methods for assessing the significance of stress corrosion may include: specification of the chemistries involved, identification of relevant chemical reactions, characterization of surface layers, mapping conditions for stress corrosion using accelerated tests, establishing of mechanisms and models, and modelling stress corrosion under repository conditions.

SAFETY ANALYSES AND SCENARIOS

The safety of a final repository should be evaluated in a comprehensive, integrated, coherent and transparent safety analysis. Such an analysis should consider two main aspects of the safety, namely probable developments and their consequences, as well as less probable developments together with their presumably greater consequences. These are usually described in terms of processes and scenarios [5].

Processes and scenarios are developed and analysed in order to make sure that all relevant and conceivable future evolutions of a repository are taken into account [5]. Such analyses are very useful for example when the "robustness" of a disposal system is to be assessed [2]. It is in this respect necessary to study the corrosion of a container not only for the most likely future evolution (which might be general corrosion) but also for cases that are considered less likely, but which cannot be ruled out. This point can be illustrated by the following examples of conceivable but less likely future developments in the repository:

1 During the first decades of final storage, the temperature at the surface of the container may reach a maximum value of around 70 °C, after which the temperature slowly decreases to that of the rock [3]. The changes in temperature will give rise to differences in the chemistry and the reactions will, in general, be more rapid the higher the temperature.

It is of importance that no part of the surface of the container is subjected to high temperatures and subsequent dry-out of any adjacent bentonite. In such a case, the swelling properties of the bentonite could be jeopardized. In addition, there could be locally an enrichment of salt in the water which would change the chemistry the surface of the container. Experiments carried out at Stripa have indicated, however, excellent wetting [58].

2 Another possibility for the container to be exposed to high concentrations of chlorides would be if the water underwent changes in composition during glaciation (see above).

In the analysis, it should then be possible to make some quantitative estimates of the life-time of the container based on a given mechanism for localized corrosion, for example.

3 Another possibility would be localized attack subsequent to a penetration of the container. The penetration could result from manufacturing for example, or be caused by stress corrosion.

The penetration would allow water to enter the inside of the container. This water would then be subjected to alpha radiolysis and gas would develop. Part of this gas might form a pocket inside the container which could vary in volume as a result of changes in the atmospheric pressure above the repository (and also with tidal movements inside the rock) and thus cause a pumping effect. This pumping might move matter in and out of the container. It might also affect the conditions for corrosion in the penetration.

Modelling of scenarios similar to this has recently been carried out by SKB [3, 12].

More general considerations regarding modelling of container lifetimes can be found in reference [59].

CONCLUDING REMARKS

In safety analyses, acts of future generations are usually not included[2, 3]. As one of the conclusions from a seminar on ethical aspects on nuclear waste, KASAM stated[60]: "A repository should be constructed so that it makes control and corrective measures unnecessary, while at the same time not making controls and corrective measures impossible. In other words, our generation should not put the entire responsibility for maintenance of repositories on coming generations; however, neither should we deny coming generations the possibility of control."

As a consequence of this, records are likely to be kept for a long period of time, including "deposition" of records in the repository itself. In this context, the author would suggest that this transfer of information should not be limited to records alone but should also include specimens of the material(s) used, and perhaps also a few containers that do not contain waste.

In this paper, certain features regarding the requirements and assessment regarding the future functioning of a container under uncertain conditions have been noted. Several of these features are specific to waste disposal and are not encountered in other areas of technology covered by corrosion R&D. Certain characteristics of corrosion pertinent for the assessment of behaviour during long periods of time have also been identified and discussed.

These features will be considered in the up-coming review of SKB's new R&D Program 92 and will also be utilized in a new safety analysis project, SITE 94, that the SKI has just started.

REFERENCES

[1] Final storage of spent nuclear fuel - KBS-3, Swedish Nuclear Fuel Supply Company (now SKB), May 1983.

[2] SKI Projekt-90, SKI Technical Report 91:23, August 1991.

[3] SKB-91. Slutlig förvaring av använt kärnbränsle. Berggrundens betydelse för säkerheten, [english translation of the title: "SKB-91. Final storage of spent nuclear fuel. The significance of the bedrock for the safety. "], Swedish Nuclear Fuel and Waste Management Company, May, 1992. To be published in English.

[4] Evaluation of SKB R&D Programme 89. The National Board for Spent Nuclear Fuel, March, 1990.

[5] Andersson, J., editor, The joint SKI/SKB scenario development project, SKI Technical Report 89:14, December 1989, also published as SKB Technical Report 89-35, December 1989.

[6] Fifth Report to the U.S. Congress and the U.S. Secretary of Energy, the United States Nuclear Waste Technical Review Board, June 1992.

[7] McEwen, T. and de Marsily, G., The potential significance of permafrost to the Behaviour of a Deep Radioactive Waste Repository, SKI TR 91:8, February 1991.

[8] Ahlbom, K., Äikäs, T. and Ericsson L., O., SKB/TVO ice age scenario, SKB Technical Report 91-32, June 1991.

[9] Manakatala, H., K. and Interrante, C., G., Technical considerations for evaluating substantially complete containment of high-level waste within the waste package, US NRC NURE.G./CR-5638, December 1990.

[10] Manaktala, H., Wu, Y., Nair, P., Interrante, C. and Bunting, J., Technical considerations for evaluating substantially complete containment of high.level waste, High level radioactive waste management, Vol 1, Las Vegas, USA, April 8-12, 1990.

[11] Rajainmäki, H., Nieminen, M. and Laakso, L., Production methods and costs of oxygen free copper canisters for nuclear waste disposal, SKB Technical Report 91-38, June 1991, also published as Nuclear Waste Commission of Finnish Power Companies, Report YJT-91-17, August 1991.

[12] Werme, L., Near-field performance of the advanced cold process canister, SKB Technical Report 90-31, September 1991, also published as Nuclear Waste Commission of Finnish Power Companies, Report YJT-90-20, December 1990.

[13] Auerkari, P., Leinonen, H. and Sandlin, S., Creep of OFHC and silver copper at simulated final repository canister-service conditions, SKB Technical Report 91-37, September 1991, also published as Nuclear Waste Commission of Finnish Power Companies, Report YJT-91-16, July 1991.

[14] Henderson, P., J., Österberg, J-O. and Ivarsson, B., Low temperature creep of copper intended for nuclear waste containers, SKB Technical Report 92-04, March 1992.

[15] Auerkari, P. and Sandlin, S., Long term creep strength of silver alloyed copper, Nuclear Waste Commission of Finnish Power Companies, Report YJT-88-14, December 1988.

[16] Pedersen, K., Potential effects of bacteria on radionuclide transport from a Swedish high level nuclear waste repository, SKB Technical Report 90-05, January 1990.

[17] Lagerblad, B., Chemistry and genesis of deep ground waters in crystalline bedrock, SKN/SKI to be published.

[18] Vallander, P. and Eurenius, J., Impact of a repository on permafrost development during glaciation advance, SKB Technical Report 91-53, December 1991.

[19] Lindbom, B. and Boghammar, A., Exploratory calculations concerning the influence of glacviation and permafrost on the groundwater flow system, and an initaial study of permafrost influences at the Finnsjön site - an SKB-91 study, SKB Technical Report 91-58, December 1991.

[20] Copper, Part one, Gmelins Handbuch der anorganischen Chemie, System-nummer 60, 1955 (in German).

[21] Copper, Part two, Gmelins Handbuch der anorganischen Chemie, System-nummer 60, 1958 (in German).

[22] Copper, Interscience Publishers, Encyklopedia of industrial chemical analysis, Vol 10, pp598-680.

[23] Gerhartz, W., editor, Copper, Ullman's Encyclopedia of Industrial Chemistry, fifth completely revised edition, volume A7, pp 471-593.

[24] Leidheiser, H., The corrosion of copper, tin, and their alloys, part one, Corrosion of Copper and its alloys, John Wiley and Sons, Inc, 1970.

[25] Pugh, E., N., Craig, J., V. and Sedriks, A., J., The stress corrosion cracking of copper, silver, and gold alloys, Proceedings of conference fundamental aspects of stress corrosion cracking, The Ohio State University, USA, September 11-15, 1967.

[26] Benjamin, L., A., Hardie, D. and Parkins, R., N., Investigation of the stress corrosion cracking of pure copper, SKB Technical Report 83-06, April 1983.

[27] The Swedish Corrosion Research Institute and its reference group, Corrosion resistance of a copper canister for spent nuclear fuel, SKB Technical Report 83-24, April 1993.

[28] Amcoff, Ö. and Holényi K., Stability of metallic copper in the near surface environment, SKN Report 57, March 1992.

[29] Marcos, N., Native copper as a natural analogue for copper canisters, Nuclear Waste Commission of Finnish Power Companies, Report YJT-89-18, December 1989.

[30] Aaltonen, P., Corrosion of pure OFHC-copper in simulated repository conditions, Part I, Nuclear Waste Commission of Finnish Power Companies, Report YJT-88-09, June 1988.

[31] Aaltonen, P., Corrosion of pure OFHC-copper in simulated repository conditions, Part II, Nuclear Waste Commission of Finnish Power Companies, Report YJT-90-07, April 1990.

[32] Beavers, J., A. and Thompson, N., G., Environmental effects on corrosion in the tuff repository, US NRC NUREG/CR--5435, February 1990.

[33] Farmer, J., C., McCright, R., D. and Kass, J., N., Survey of degradation modes of candidate materials for high-level radioactive waste disposal containers, overview, Lawrence Livermore National Laboratory on commission by US DOE, UCID--21362, June 1988.

[34] Gdowski, G., E. and Bullen, D., B., Survey of degradation modes of candidate materials for high-level radioactive waste disposal containers, volume 2 oxidation and corrosion, Lawrence Livermore National Laboratory on commission by US DOE, UCID--21362, August 1988.

[35] Farmer, J., C., Van Konynenburg, R., A. and McCright, R., D., Survey of degradation modes of candidate materials for high-level radioactive waste disposal containers, volume 4 stress corrosion of copper-based alloys, Lawrence Livermore National Laboratory on commission by US DOE, UCID--21362, May 1988.

[36] Farmer, J., C., Van Konynenburg, R., A., McCright, R., D. and Gdowski, G., E., Survey of degradation modes of candidate materials for high-level radioactive waste disposal containers, volume 5 localized corrosion of copper-based alloys, Lawrence Livermore National Laboratory on commission by US DOE, UCID--21362, May 1988.

[37] Maiya, P., S., A review of degradation behavior of container materials for disposal of high-level nuclear waste in tuff and alternative repository environments, Argonne National Laboratory, ANL--89/14, June 1989.

[38] Kass, J., Evaluation of copper, aluminium bronze, and copper-nickel for the Yucca mountain project, Lawrence Livermore National Laboratory, Corrosion of nuclear fuel waste containers, proceedings of a workshop in Winnipeg, Canada, February 9-10, 1988.

[39] Farmer, J., C., Gdowski, G., E., McCright, R., D. and Ahluwahlia, H., S., Corrosion models for performance assessment of high-level radioactive waste containers, Nuclear Engineering and Design, Vol 129, 1991, pp 57-88.

[40] McNeil, M., B. and Little, B., J., Corrosion products and mechanisms in long-term corrosion of copper, Scientific Basis for Nuclear Waste Management XIV, Boston, USA, November 26-29, 1990.

[41] Reed, D., T. and Van Konynenburg, R., A., Corrosion of copper-based materials in irradiated moist air systems, Scientific Basis for Nuclear Waste Management XIV, Boston, USA, November 26-29, 1990.

[42] Manaktala, H., K., Degradation modes in candidate copper-based materials for high-level radwaste canisters, Corrosion 90, Las Vegas, USA, April 23-27, 1990.

[43] Shoesmith, D., W., editor, Corrosion of nuclear fuel waste containers, Proceedings of a workshop held in Winnipeg, Canada, February 9-10, 1988.

[44] King, F. and LeNeveu, D., Prediction of the lifetimes of copper nuclear waste containers, Nuclear waste packaging, Focus'91, Las Vegas, USA, September 29 - October 2, 1991.

[45] King, F. and Litke, C., The corrosion of copper in NaCl solution and under simulated disposal conditions, Scientific basis for nuclear waste management XII, Pittsburgh, USA, 1989, pp 403-409.

[46] King, F. and Litke, C., Electrochemical behaviour of copper in aerated m mol per litre NaCl at room temperature, Parts 1-3, Atomic Energy of Canada Limited reports, AECL-9571, -9672, -9573, 1989.

[47] Hultqvist, G., Hydrogen evolution in corrosion of copper in pure water, Corrosion Science, Vol 26, No 2, pp173-177, 1986.

[48] Hultqvist, G., Chuah, G., K. and Tan, K., L., Comments on hydrogen evolution from the corrosion of pure copper, Corrosion Science, Vol 29, No 11/12, pp 1371-1377, 1989.

[49] Hultqvist, G., Chuah, G., K. and Tan, K., L., A SIMS study of reactions in the metal-oxygen-hydrogen-water system, Corrosion Science, Vol 31, pp 149-154, 1990.

[50] Simpson, J., P. and Schenk, R., Hydrogen evolution from corrosion of pure copper, Corrosion Science, Vol 27, No 12, pp 1365-1370, 1987.

[51] Eriksen, T., E., Ndalambada, P. and Grenthe, I., On the corrosion of pure copper in pure water, SKB Technical Report, TR 88-17, March, 1988.

[52] Handling and final disposal of nuclear waste, Programme for research, development and other measures, SKB R&D Programme 89, September, 1989.

[53] Berger, R., University of Uppsala, Institute of Chemistry, Uppsala, Sweden, private communication.

[54] Grauer, R., The reducibility of sulphuric acid and sulphate in aqueous solution (translated from German), SKB Technical Report 91-39, July 1990.

[55] Pedersen, K., Ekendal, S. and Arlinger, J., Microbes in crystalline bedrock. Assimilation of carbon dioxide and introduced organic compounds by bacterial populations in groundwater from deep crystalline bedrock at Laxemar and Stripa, SKB Technical Report 91-56, December 1991.

[56] Hwang, Y. and Pigford, T., H., Life of copper canister limited by mass transfer of sulfide, to be published by the SKI.

[57] Hermansson, H-P. and Rosborg, B., Studsvik Materials AB, Nyköping, Sweden, private communication.

[58] Pusch, R., Final report of the Buffer Mass Test - Volume II: test results, Stripa Project Technical Report 85-12, August 1985.

[59] Wu, Y. and Nair, P., Fast probabilistic performance assessment (FPPA) methodology evaluation, CNWRA 88-004, October 1988.

[60] Ethical aspects on nuclear waste, some salient points discussed at a seminar on ethical action in the face of uncertainty held by KASAM and SKN in Stockholm, Sweden, September 8-9, 1987, SKN report 29.

NOTE ON THE AVAILABILITY OF REPORTS:

Many of the references cited appear in the SKB, SKN and SKI series which are not available at all libraries (they do appear in the INIS system). SKB reports can be ordered from the Swedish Nuclear Fuel and Waste Management Company (SKB), Box 5864, S-102 48 Stockholm, Sweden. The tasks of the Swedish National Board for Spent Nuclear Fuel (SKN) were taken over by the Swedish Nuclear Power Inspectorate (SKI) in 1992. SKN and SKI reports can be ordered from The Swedish Nuclear Power Inspectorate, Box 27 106, S-102 52 Stockholm, Sweden.

Richard A. Perkins[1] and Shih-Hsiung Shann[1]

CORRELATION OF AUTOCLAVE TESTING OF ZIRCALOY-4 TO IN-REACTOR CORROSION PERFORMANCE

REFERENCE: Perkins, R. A. and Shann, S.-H., **"Correlation of Autoclave Testing of Zircaloy-4 to In-Reactor Corrosion Performance,"** Application of Accelerated Corrosion Tests to Service Life Prediction of Materials, ASTM STP 1194, Gustavo Cragnolino and Narasi Sridhar, Eds., American Society for Testing and Materials, Philadelphia, 1994.

ABSTRACT: As the service environment for core materials for Pressurized Water Reactors (PWRs) becomes more severe due to increases in water temperature, changes in water chemistry, longer cycle lengths and higher burnups; the corrosion performance of Zircaloy-4 which is used for fuel cladding and structural components becomes more important. The standard three day, 400°C steam autoclave test which has been used to evaluate the uniform corrosion resistance of Zircaloy-4 is no longer sufficient. Some cladding which has given acceptable results in this test has not exhibited satisfactory in-reactor corrosion performance. A series of cladding types for which in-reactor corrosion measurements are available has been subjected to extended autoclave testing in steam at 400 and 415°C. A better correlation to the in-reactor performance is obtained if the 400°C test is extended to at least 100 days or if the test temperature is raised to 415°C. At 415°C, a test duration of three days is best for evaluating Zircaloy-4 since extended time does not improve the correlation.

KEYWORDS: Zircaloy-4, uniform corrosion, autoclave testing, in-reactor performance

---

[1]Senior Engineers, Siemens Power Corporation, 2101 Horn Rapids Rd., Richland, Washington 99352, USA

INTRODUCTION

Zircaloy-4 is commonly used for nuclear fuel cladding and structural parts for the fuel assemblies to be used in Pressurized Water Reactors (PWR). The water temperature in the core of a PWR is in the range of about 310 to 330°C, and uniform corrosion with associated hydriding of the Zircaloy is the major concern. In order to operate the reactors more efficiently, increases in the reactor core water temperature, LiOH content in the water, cycle length, and end-of-life burnup are being utilized. All of these changes require better corrosion performance from the Zircaloy components in the reactor core. This emphasizes the need for a reasonably short, out-of-reactor corrosion test which can accurately predict the in-reactor corrosion performance of Zircaloy-4 in the PWR environment. The standard corrosion test in the nuclear industry for evaluation of uniform corrosion resistance is a steam autoclave test at 400°C for 3 or 14 days with 1500 psi steam. This corrosion test has given acceptable results for some cladding types which have performed poorly in-reactor. Therefore, due to the increased severity of the PWR environment, a better test for uniform corrosion resistance is necessary. Extension of the 400°C test to long times has been shown to give a better representation of the uniform corrosion resistance in-reactor [1,2]. Another option which has been evaluated is to increase the test temperature [2,3]. In this study both methods have been used for a set of Zircaloy-4 cladding samples representing eight different types of cladding for which in-reactor measurements are available. The cladding variants arise from the use of various Tube Reduced Extrusion (TREX) suppliers and cladding fabricators. The cladding types have been irradiated in four reactors, and autoclave tests have been performed at 400 and 415°C.

EXPERIMENTAL PROCEDURE

Poolside Oxide Thickness Measurements

Fuel rod oxide thickness measurements were made at poolside with an eddy current thickness probe which consists of a small, approximately 3 mm diameter, eddy current coil whose axis is held perpendicular to the fuel rod surface. A thin diamond disk protects the coil and keeps it from touching the fuel rod. The impedance of the eddy current coil changes depending upon the distance to the metal surface of the rod. Since a varying oxide layer thickness will move the coil closer to or farther away from the metal surface, the impedance change is a measure of the oxide thickness.

In practice, the probe impedance change is directly measured as oxide thickness (or more accurately, lift-off) by calibrating against standards of Zircaloy-4 cladding with well known oxide layer thicknesses. The standards are made by oxidizing Zircaloy-4 tubing of the same size as that used for the fuel rods in air or autoclaving in high pressure steam. The oxide thickness of the standards is measured on metallographic cross-sections of the oxidized cladding using an NBS traceable microscope length scale.

The oxide thickness measurements on fuel rods were made at poolside without removing the rods from the assemblies. A continuous oxide thickness trace was obtained for each span in between spacers. The maximum oxide thickness value measured on each fuel rod was used in evaluation of the corrosion data. The maximum thickness value usually occurred in the next to the highest span in the assembly where the fuel rod surface temperature is highest. Oxide thickness measurements on eight types of cladding irradiated in four reactors will be discussed.

Corrosion Model

For the analysis of the oxide thickness data, the MATPRO [4] corrosion model has been used. MATPRO uses a cubic rate equation in the pre-transition region, and a linear rate equation after transition:

Pre-Transition ($X < X_{tran}$):

$$X^3 = X_0{}^3 + 4.976E\text{-}9\ A(t - t_0)\exp(-15660/T) \qquad (1)$$

Post-Transition ($X > X_{tran}$):

$$X = X_0 + 82.88\ A(t - t_0)\exp(-14080/T) \qquad (2)$$

where

t_0 = time (days) at the start of a calculation step,
t^0 = time (days) since start of oxidation,
X_0 = oxide thickness (m) at time t_0 (days)
X^0 = oxide thickness (m) at time t^0 (days)
A = temperature dependent enhancement factor for in-reactor corrosion, and
T = temperature (K) at the metal-oxide interface.

The thickness of the oxide layer, X_{tran}, at the transition point is given by the equation

$$X_{tran} = 7.749E\text{-}6\ \exp(-790/T) \qquad (3)$$

The temperature dependent enhancement factor, A, for in-reactor corrosion is given by

$$A = 120.3\ \exp(-7.118E\text{-}3\ T_C) \qquad (4)$$

where T_c is the temperature of the oxide at the oxide-coolant surface. For increasing temperatures at the oxide-coolant surface of PWR fuel rods, the value for A decreases in magnitude.

For the evaluations described here, the MATPRO rate equations have been modified with a unitless multiplier, E, which is a fuel rod specific corrosion parameter. E is a constant for a given fuel rod and uniquely describes the corrosion susceptibility of the fuel rod based upon its power history if no significant changes in the reactor environment have been made. The corrosion parameter can also be used to normalize the oxide thicknesses measured at several different burnups to values for a single exposure level for comparing the behavior of differing cladding types. A more complete description of the application of the corrosion model has been given previously [5].

Autoclave Testing

The initial portion of each autoclave test was performed in a four liter, refreshed steam (1500 psi) autoclave system with a water flow rate of about five cc/min. The water was passed through two ion exchange columns and filtered before degassing with argon in a holding tank. The oxygen content of the water was maintained at 20 ppb or lower throughout the test. The temperature variation between six thermocouples located in the autoclave was about ±2C. When tests were extended beyond 14 days, the testing was performed either in the refreshed autoclave or a static autoclave (about two liters). Previous testing has verified that similar corrosion results are obtained in the two autoclaves. The static autoclave was started by filling it with water and boiling off sufficient water at ambient pressure to obtain approximately 1500 psi steam at the test temperature. Small amounts of water could be added to the autoclave to achieve the desired pressure. Also excess steam could be bled off.

Autoclave testing was performed at 400 and 415°C. The tests extended to about 100 days. Previous work has established that the oxidation mode is similar at these temperatures, but at higher temperatures a change in mode occurs such that the weight gain does not correlate with in-reactor performance.[3] The samples were tested in the as-received condition with a belt-polished outer surface and an etched inner surface. The samples were periodically removed from the autoclaves for weight gain determination. When duplicate samples were tested, the error bars for one standard deviation are given in the figures.

Samples which have shown high weight gains at the extended exposures have shown large local variations in oxide as has been discussed in Reference 3. They have a nodular appearance. This is observed on samples of Type G and H cladding at both test temperatures but no spalling occurred.

Cladding Types

All of the cladding variants which have been tested were fabricated using standard production processes. The cladding variants represent material from four TREX sources and fabrication procedures of four cladding vendors. In general, for a given type of cladding comparisons have been made between in-reactor measurements on tubing from a given lot and autoclave results on a piece of archived tubing from the same lot. However, the within-lot variability of type H cladding made this procedure ineffective, and average in-reactor and autoclave values were compared for this type.

Variability between lots was also observed for type G, both in-reactor and in the autoclaves. A low annealing parameter was common to both types. It appears that the low annealing parameter, together with the inherent variations in the annealing process, allowed the occurrence of values low enough to affect corrosion performance.

RESULTS

The autoclave results for four of the cladding types are plotted as a function of time in Figures 1 and 2 for the 400 and 415°C tests, respectively. For these four types, the weight gain for the sample representative of the poorest corrosion resistance is plotted. Types B and C possess good uniform corrosion resistance, and Types G and H have poor uniform corrosion resistance. At 400°C the weight gains for all four samples are essentially the same for about 50 days; then there is a separation among them. At 415°C a difference in corrosion resistance can be seen at three days.

In order to compare the autoclave performance to the in-reactor behavior, the normalized oxide thickness has been plotted versus the autoclave weight gain in Figures 3-6 for the four reactors. For all types except H, the maximum oxide thickness measurements for all the rods from a given cladding lot were normalized to a burnup of 30 GWd/MTU; then the average value and standard deviation were used to represent that lot of cladding. When duplicate autoclave samples were tested, the average and standard deviation of the weight gain are plotted. The in-reactor oxide thickness has been plotted versus the weight gain measured in the autoclave tests at three days and at an extended time. For Reactor 1 which had the most continuous spread in corrosion performance, the corrosion parameter has also been plotted versus the autoclave weight gain for the 415°C/3 day test in Figure 7. The correlation is the same as that which is seen in Figure 3c when the oxide thickness is plotted. When the reactor environment changes significantly between cycles,

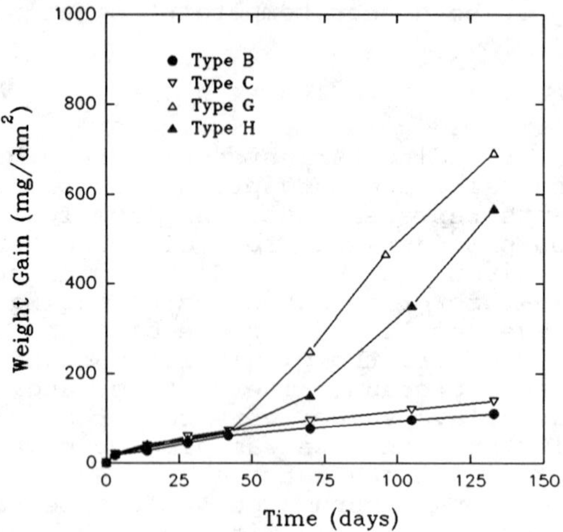

Figure 1 Weight gain as a function of time in the 400°C autoclave test for highest weight gain samples of Types B, C, G and H cladding (10 mg/dm^2 = 1 g/m^2).

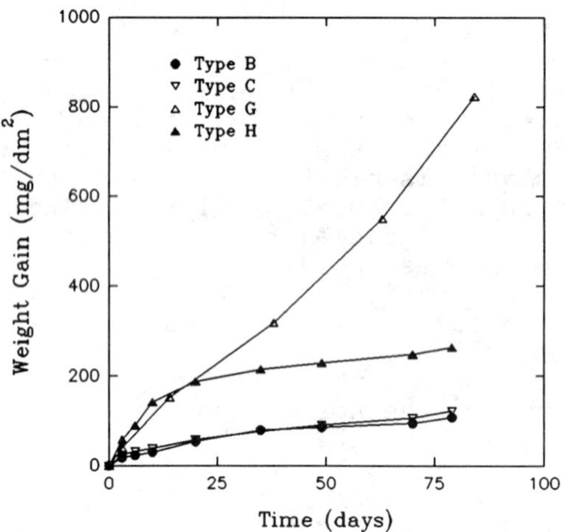

Figure 2 Weight gain as a function of time in the 415°C autoclave test for highest weight gain samples of Types B, C, G and H Cladding (10 mg/dm^2 = 1 g/m^2).

the corrosion parameter may also change. This was observed in Reactor 4 when the lithium concentration range used for one of the cycles was changed. An increase in corrosion rate was observed when the lithium content limit was increased from <2.2 ppm to <3.5 ppm.[6] Therefore, the oxide thickness is used to represent the in-reactor corrosion rather than the corrosion parameter.

DISCUSSION

Both the 400 and 415°C autoclave test have been used to try to predict the in-reactor uniform corrosion performance of Zircaloy-4. In Figures 3 to 6, the in-reactor oxide thickness is plotted versus the weight gain results obtained at 400°C for three and 133 days in parts a and b, respectively. Obviously, at three days there is no correlation between the two results. Similar autoclave results were obtained for cladding with high and low in-reactor oxide thickness values. After autoclave testing for 133 days, a much better correlation to the in-reactor behavior is observed. The observations for the 415°C autoclave test were the opposite. As seen in parts c and d of Figures 3 to 6, a correlation between the autoclave weight gain and the in-reactor performance is observed at three days, but the differentiation between the various cladding types decreases when the test is extended.

The transition in oxidation kinetics for the steam oxidation of Zircaloy from a cubic or parabolic time dependence to a linear time dependence occurs at an oxide thickness of about 2μm which is equivalent to a weight gain of about 3 g/m^2.[7] In the 415°C autoclave test, the weight gain for the cladding types with good corrosion resistance are somewhat less than the transition value after three days. Those cladding types with poorer corrosion resistance exhibit weight gains greater than about 3 g/m^2. Therefore, they are in the region of linear oxidation kinetics. This change in corrosion kinetics might account for the better differentiation between cladding types at three days exposure than at longer times for which all of the samples have exceeded a weight gain of 3 g/m^2, and thus are in the linear corrosion kinetics region.

At 400°C/3 days all of the samples are in the pretransition region, and the weight gains are not sufficient to distinguish between the different cladding types. As the exposure time is increased, and the samples enter the linear corrosion kinetics region, the differentiation between the various cladding types becomes more apparent.

The goal of having an autoclave test for Zircaloy-4 cladding is to be able to determine whether material is going to exhibit poor corrosion performance in the reactor environment. It is obviously desirable for this test to be

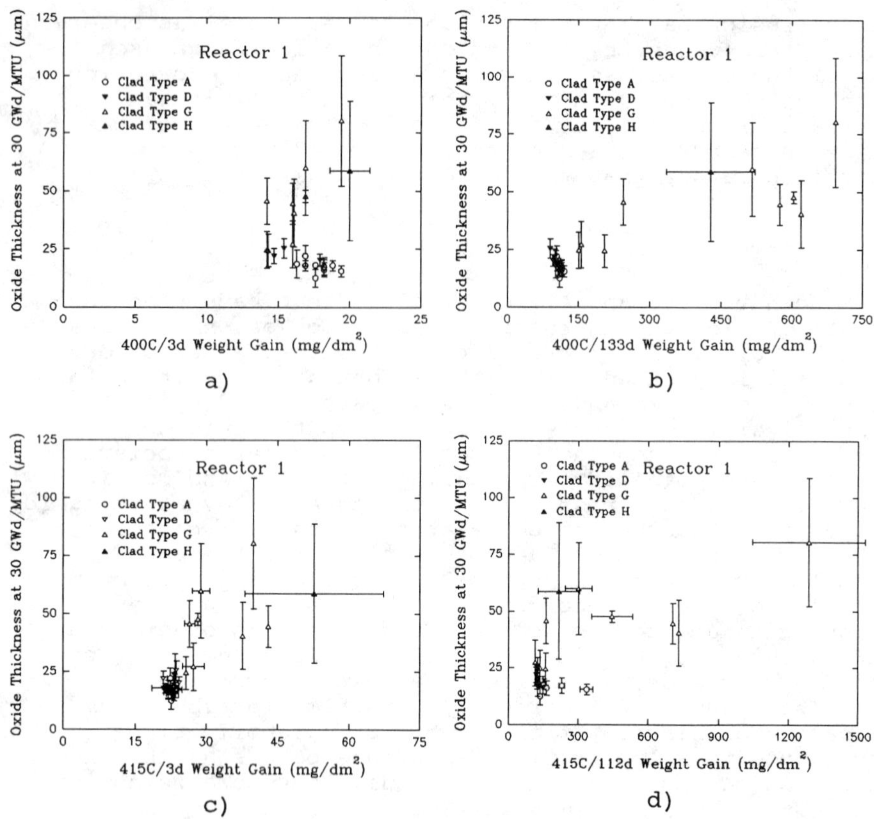

Figure 3 Correlation of in-reactor and autoclave performance of cladding irradiated in Reactor 1 (10 mg/dm^2 = 1 g/m^2).

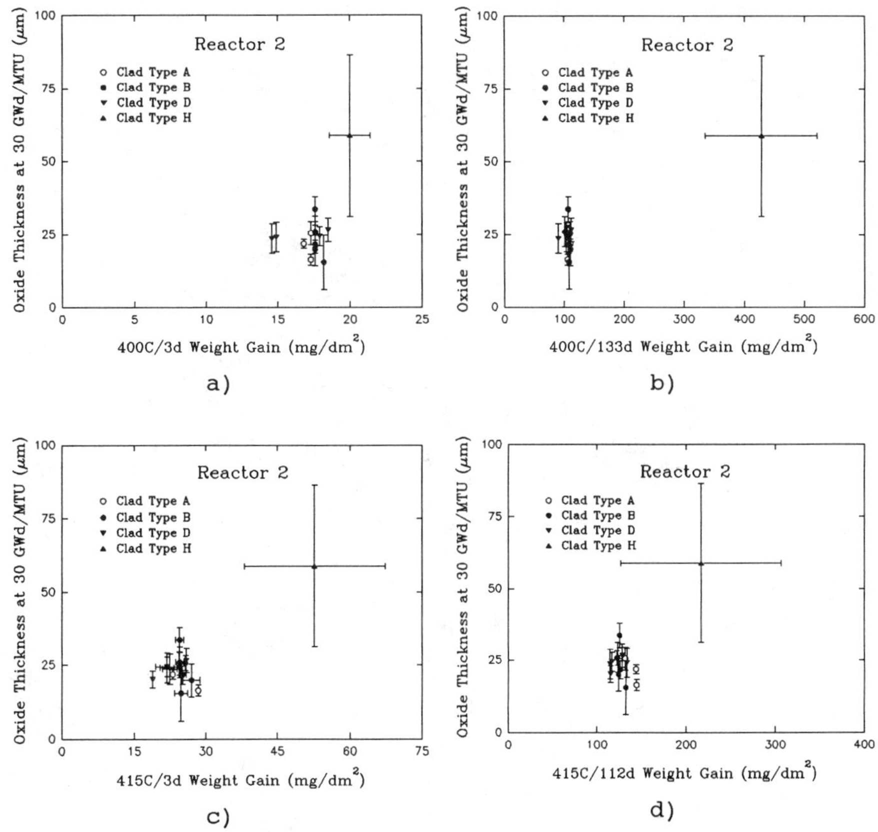

Figure 4 Correlation of in-reactor and autoclave performance of cladding irradiated in Reactor 2 (10 mg/dm^2 = 1 g/m^2).

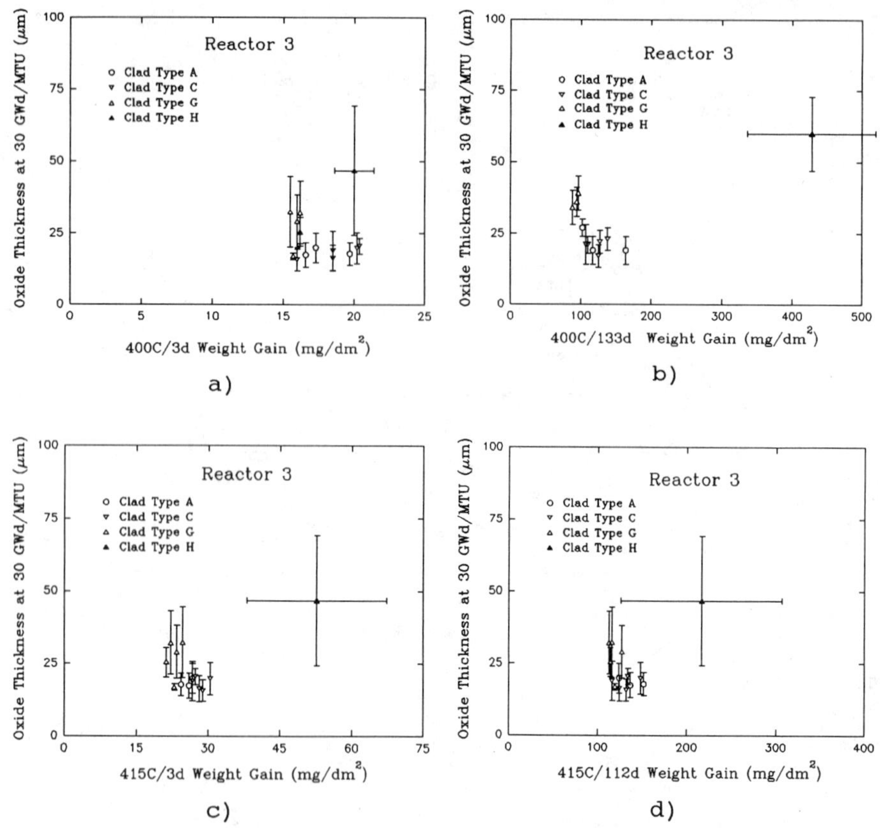

Figure 5 Correlation of in-reactor and autoclave performance of cladding irradiated in Reactor 3 (10 mg/dm^2 = 1 g/m^2).

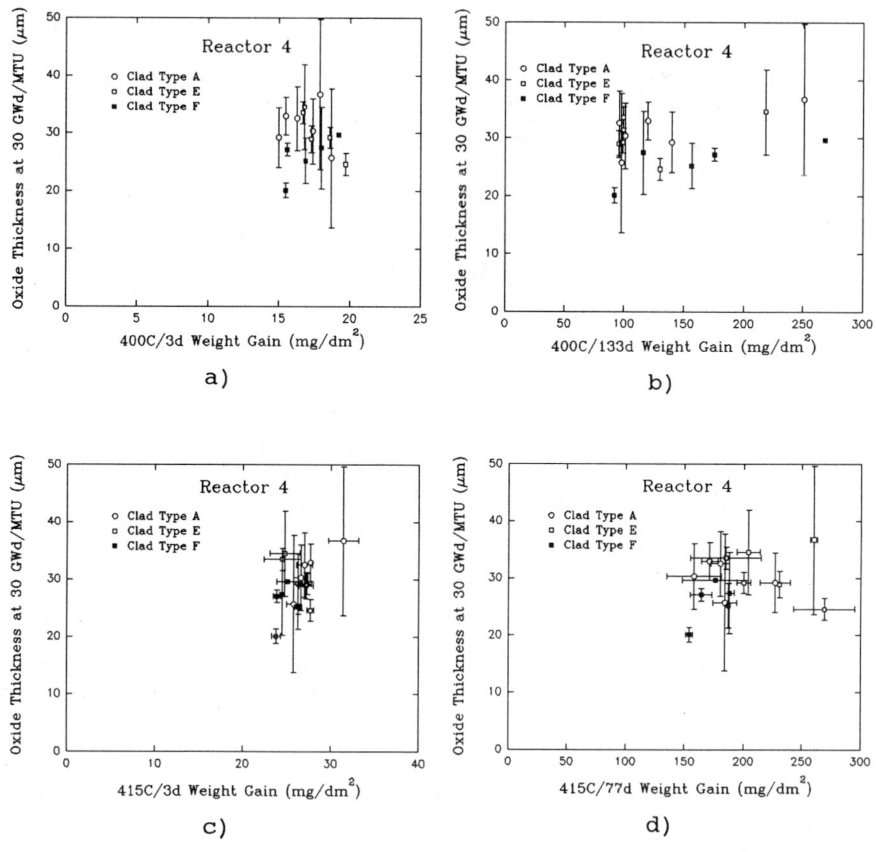

Figure 6 Correlation of in-reactor and autoclave performance of cladding irradiated in Reactor 4 (10 mg/dm^2 = 1 g/m^2).

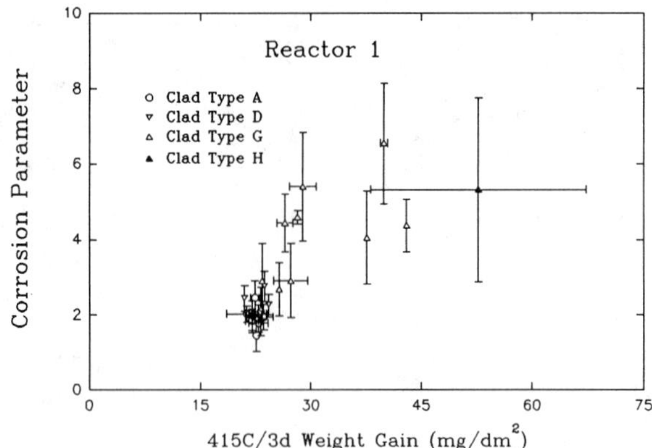

Figure 7 Correlation of the corrosion parameter to autoclave (415°C/3d) weight gain for cladding in Reactor 1 (10 mg/dm^2 = 1 g/m^2).

short in duration in order to be able to responsive to production demands. The use of the test for evaluating alloys outside the chemical composition range of Zircaloy-4 is desirable but more difficult to evaluate. The results from the performance of eight different types of production Zircaloy-4 cladding in four reactors have shown that the 400°C/3 day autoclave test is of no use in identifying cladding which will exhibit poor in-reactor corrosion performance. When extended to over 100 days, it is much more effective. However, if the test temperature is increased to 415°C, the cladding types which have exhibited poor in-reactor performance have had noticeably higher weight gains in three days. Therefore, increasing the test temperature to 415°C is the preferred method for achieving a suitable test in a time frame necessary for production conditions.

REFERENCES

[1] Schemel, J. H., Charquet, D., and Wadier, J.-F., Zirconium in the Nuclear Industry: Eighth International Symposium, ASTM STP 1023, American Society for Testing and Materials, 1989, pp 141-152.

[2] Rudling, P., Pettersson, H., Andersson, T., and Thorvoldsson, T., Zirconium in the Nuclear Industry: Eighth International Symposium, ASTM STP 1023, American Society for Testing and Materials, 1989, pp 213-226.

[3] Perkins, R. A. and Busch, R. A., Proceedings of the Fourth International Symposium on Environmental Degradation of Materials in Nuclear Power Systems-Water Reactors, 1990, pp 10-15 to 10-24.

[4] "MATPRO: Version 10, A Handbook of Materials Properties for Use in the Analysis of Light Water Reactor Fuel Rod Behavior," TREE-NUREG-1180, 1978.

[5] Van Swam, L. F. P., and Shann, S. H., Zirconium in the Nuclear Industry: Ninth International Symposium, ASTM STP 1132, American Society for Testing and Materials, 1991, pp 758-781.

[6] Shann, S. H., Van Swam, L. F. P., and Martin, L.A., Effects of Coolant Li Concentration on PWR Cladding Waterside Corrosion, ANS/ENS Topical Meeting on LWR Fuel Performance, Avignon, France, April 21-24, 1991.

[7] Garzarolli, F., Jorde, D., Manzel, R., Politano, J. R., and Smerd, P. G., Zirconium in the Nuclear Industry: Fifth Conference, ASTM STP 754, American Society for Testing and Materials, 1982, pp. 430-449.

Gino Palumbo[1], Alex M. Brennenstuhl[1], Francisco S. Gonzalez[2]

THE IMPORTANCE OF SUBTLE MATERIALS AND CHEMICAL CONSIDERATIONS IN THE DEVELOPMENT OF ACCELERATED TESTS FOR SERVICE PERFORMANCE PREDICTIONS

REFERENCE: Palumbo, G., Brennenstuhl, A. M., and Gonzalez, F. S., **"The Importance of Subtle Materials and Chemical Considerations in the Development of Accelerated Tests for Service Performance Predictions,"** Application of Accelerated Corrosion Tests to Service Life Prediction of Materials, ASTM STP 1194, Gustavo Cragnolino and Narasi Sridhar, Eds., American Society for Testing and Materials, Philadelphia, 1994.

ABSTRACT: The usefulness of accelerated laboratory tests for selecting materials for applications where corrosion is likely to be the main life limiting concern can be compromised by subtle, often ignored, microstructural and microchemical factors. Residual deformation and crystallographic orientation are metallurgical factors which can have a large effect on corrosion response and therefore must be considered when comparing materials. Microchemical differences are often addressed but there is still scope for improvement here. Less commonly used microprobing techniques such as secondary ion mass spectrometry (SIMS) can provide additional information on the segregation of trace impurities which can have an important influence on corrosion resistance.

Another difficulty with accelerated tests is that high levels of acceleration often mask differences in corrosion response. This may lead to the best candidate material for the service environment not being selected. To minimize this problem the accelerated test should simulate as closely as possible the service environment. This may mean that data will take longer to obtain, however, this is the inevitable price of a greater degree of confidence in the material that is finally selected.

KEYWORDS: microchemistry, microstructure, coincidence site lattice, residual deformation, texture and mesotexture, field simulations

---

[1]Research Scientist, Metallurgical Research Department, Ontario Hydro, 800 Kipling Avenue, Toronto, Ontario, M8Z 5S4

[2]Associate Research Engineer, Chemical Research Department, Ontario Hydro, 800 Kipling Avenue, Toronto, Ontario, M8Z 5S4

Introduction

The purpose of accelerated corrosion testing is to provide information which makes possible the selection of materials for applications where corrosion is likely to limit plant or equipment life. Numerous test procedures are available; some are based on electrochemical methods (closed circuit testing) such as anodic polarizations while others rely on open circuit chemical exposure. Accelerated corrosion tests have been developed for determining susceptibility to localized attack such as pitting, crevice corrosion, and stress corrosion cracking (SCC) in addition to general forms of corrosion. Generally, it is only possible to obtain order of merit information using accelerated tests. This is particularly the case with localized forms of corrosion. The rate of material degradation is difficult to establish because of the complications associated with determining the acceleration factor for the test.

Accelerated tests are only useful if they can reliably predict relative materials performance. When service failures are the result of misleading information from accelerated tests their utility is undermined. Reasons for misleading data are; i) subtle and difficult to characterize microstructural or microchemical variations between alloys or between batches of the same material may not be taken into account during the testing and ii) tests which result in a high level of acceleration may mask differences which become pronounced in the long-term in the service environment.

Careful microstructural and microchemical characterization of the materials of the materials under investigation before testing can in some cases eliminate this problem. In other cases it may be necessary to reduce the aggressiveness of the test environment before it is possible to discriminate between materials which display small but potentially important differences in corrosion resistance.

This paper presents examples of where small microstructural and microchemical effects can lead to differences in performance. In addition one case is given where accelerated tests were unable to discriminate between two materials with a similar nominal composition but which displayed widely different service performance.

Microstructural Considerations

The metallurgical condition of a specimen is usually assessed on the basis of bulk chemistry, final heat treatment, and in some cases grain size. The most frequently overlooked micro-structural parameters are those related to residual stress and crystallographic orientation. In the following section some examples are presented where these subtle micro-structural parameters have been shown to significantly influence bulk corrosion response in laboratory tests.

Residual Deformation

Although residual deformation is usually thought of as resulting from the bulk mechanical processing of material (e.g. cold rolling, extruding, and drawing operations), the most frequent cause of largely uncontrolled material deformation is in the surface preparation of corrosion test specimens. Indeed, any prior cold work applied to a test specimen becomes entirely irrelevant if it is subsequently tested in the mechanically polished/ground state.

Palumbo and Aust [1] investigated the effect of residual surface deformation on the passivation behaviour of high purity Ni (99.999%) in 2N H_2SO_4 at 30°C. Specimens were mechanically polished and subsequently electropolished for various times to effect variable extents of removal of the heavily deformed layer. Residual deformation following electropolishing was quantified by electron channeling pattern degradation in the scanning electron microscope (ECP/SEM) and correlated with dislocation density (ρ) [1]. Figure 1 summarizes the results of this study and shows the effect of dislocation density (infered from the average electron channelling pattern (ECP) line width) on both the passive charge (Figure 1a) and the minimum passive current density (Figure 1b) during potentiostatic polarization to +750 mV (SCE) from an initial cathodic potential of -450 mV (SCE).

Both passivation charge and passivation current density are observed to be significantly affected by residual surface deformation and show similar deformation dependence. At low levels of residual deformation, increases in dislocation density result in significant increases in the passivation parameters; however, maxima are noted at $\rho = 10^{10}$ cm^{-2}, beyond which further increases in dislocation density result in commensurate decreases in both the passivation parameters investigated. The observed dependencies were rationalized on the basis of the relative effects of surface-intersecting dislocations on the kinetics of ; (i) anodic dissolution, (ii) surface adsorption of passivating anionic species, and (iii) transformation of an initial chemisorbed film to a stable oxide [1]. Moreover, the results underscore the potential variability in both laboratory test performance and in long term service performance which can arise from subtle variations in surface preparation methodology.

Crystallographic Orientation

Crystallographic orientation and misorientations (i.e. texture and mesotexture) are usually governed by the entire thermomechanical processing history of material, not just the final treatment. The influence of texture on corrosion response can be categorized on the basis of; (i) grain interior (i.e. surface orientation), and (ii) grain boundary (i.e., mesotexture effects.

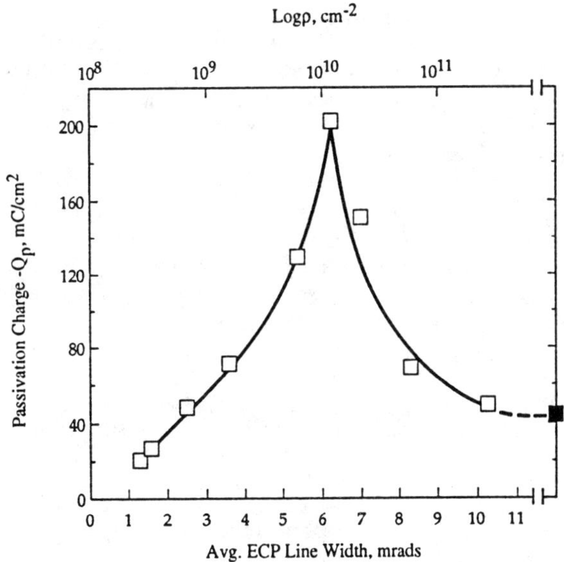

FIG. 1a--Effect of dislocation density on passive charge (Qp).

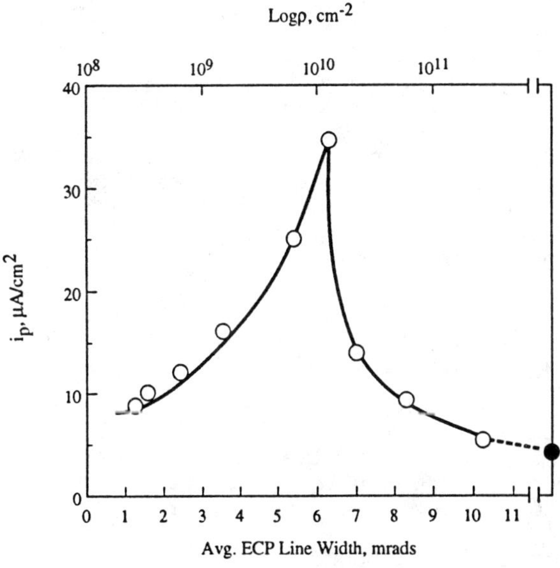

FIG 1b--Effect of dislocation density on (ip) the minimum passive current density.

Figure 2 shows the effect of surface orientation on the pitting susceptibility of fully electropolished 99.999% Ni electrodes following potentiostatic polarization at a potential of +300 mV (SCE) for three hours in 0.3 M NaCl +0.3 M Na_2SO_4 at 30°C. Significant variations in the pit density were noted, amongst the different grain surfaces.

Crystallographic orientations within approximately 5° of the low index (001), (011), and (111) typically displayed a high resistance to pitting (see Figure 2c) as evidenced by reduced pit densities. Minimal pitting attack was also observed with a grain having a surface orientation within 0.3° of (113) (i.e. grain 'V' in Figure 2). It should be noted that the (113) plane is the only plane other than (001), (011), and (111) in FCC materials to have atomic packing density greater than unity. These observed crystallographic effects on pitting demonstrate the potential variability in both laboratory tests and field performance which can arise in the absence of texture characterization/control. Furthermore, potential benefits from attaining preferred orientations in service components is clearly demonstrated. It should be noted that surface deformation (e.g. from mechanical polishing of surfaces) can completely eliminate crystallographic effects in pit initiation [2]; however, crystallographic influences are likely to be present and persist throughout most stages of pit propagation.

The crystallographic misorientation of adjoining crystals has been shown, through the coincidence site lattice (CSL) model of interface structure [3] to govern the defect character of grain boundaries. The CSL model and its associated Sigman-criteria (sigma is equal to the volume of the CSL unit cell/volume of the lattice unit cell) has been shown in numerous studies to be applicable to a wide range of interfacial properties [4], whereby 'special' properties are expected of grain boundaries having low sigma CSL character. The applicability of the CSL sigma-criteria to intergranular SCC susceptibility for alloy N06600 in high temperature aqueous media has recently been established [5,6]. Figure 3 shows a stressed (to 150% of nominal 0.2% yield strength) C-ring specimen of mill annealed N06600 following 4000 hour autoclave exposure to 10% NaOH at 350°C [5].

The CSL classification $(\Sigma, \Delta\theta)$ was up to $\Sigma = 49$ using Brandon's criterion [7] for the maximum allowable angular deviation. Figure 4 summarizes the results of this analysis whereby several of the grain boundaries having Σ < or equal to 25 were found to display selective immunity.

The coherent twin boundary (i.e. annealing twin - Σ 3 with (111) boundary plane) was found to display complete immunity to cracking. It should be noted that the relatively high susceptibility of LAGB's was due to many of these boundaries extending up to 15°.

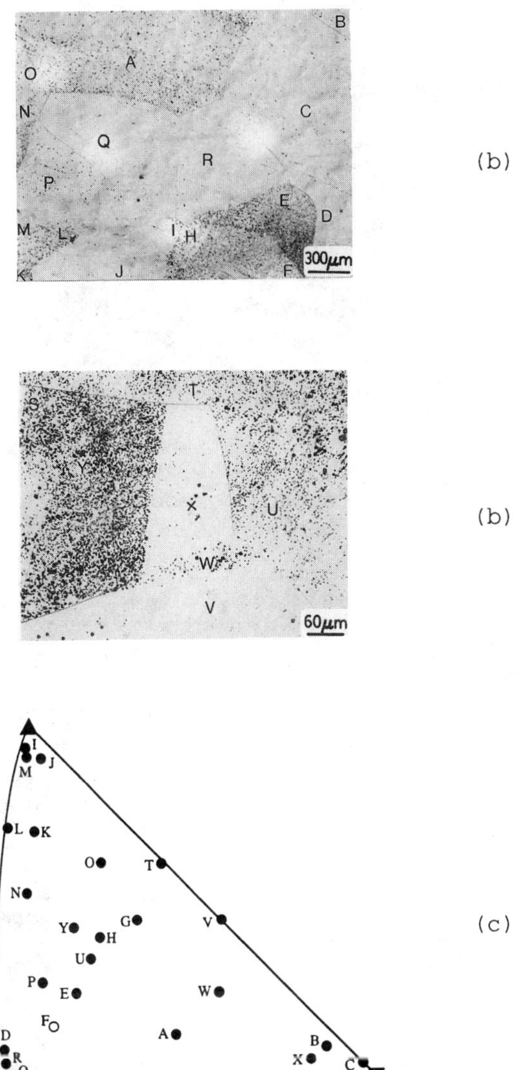

FIG. 2--The effect of surface orientation on pit distributions at E_b = 300 mV. (a,b) optical micrographs of pitted specimens. (c) Standard sterographic triangle (SST) showing the determined surface orientations (±2°) of grains labelled in a and b.

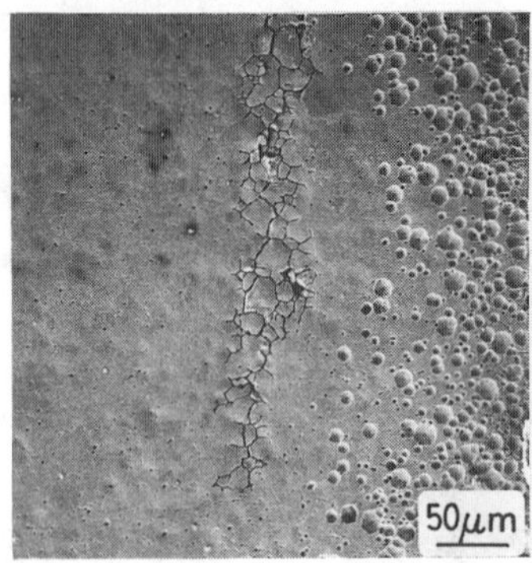

FIG. 3--A mill annealed N06600 'C'-ring specimen stressed at 150% of yield following a 4000 hour exposure to 10% NaOH at 350°C. The crack path was analyzed using the electron beam channeling pattern technique.

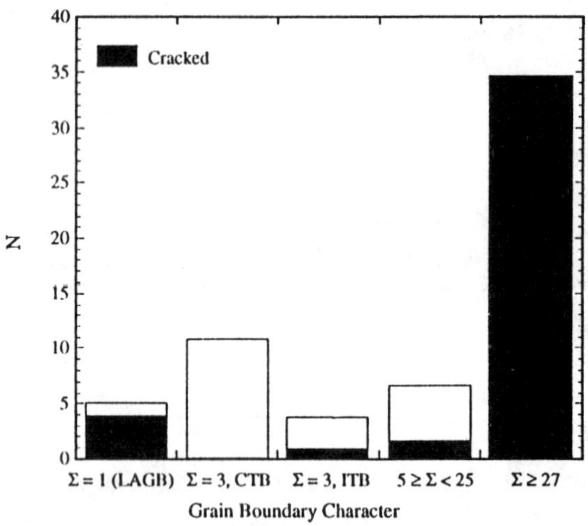

FIG. 4--CSL sigma character versus susceptibility to intergranular cracking for the specimen shown in FIG.3. Data for low angle boundaries (LAGB), coherent twin boundaries (CTB) and incoherent twin boundaries (ITB) are also shown.

The implications of such boundary variability in cracking susceptibility on bulk material response has recently been demonstrated [8] in a geometric model in polycrystalline aggregates. By considering that a propagating crack must terminate and reinitiate at triple junctions (i.e. line of intersection of three crystals) in the microstructure, the influence of the fraction of crack resistant interfaces in the distribution was established. Figure 5 summarizes the results of this analysis and gives some of our experimental data.

The results clearly demonstrate the potential impact of mesotexture on cracking susceptibility, particularly the performance uncertainty associated with low 'special' grain boundary fractions (i.e., < or = 15%) typical of conventional materials.

Microstructural Considerations

Microchemical considerations are often given more attention than microstructure when selecting materials of the same nominal composition. However, subtle differences can be missed if the analytic techniques used are not sufficiently sensitive. Often the less common techniques such as secondary ion mass spectrometry (SIMS) have to be employed for trace element determination. A SIMS study of grain boundary chemistry of two batches of alloy N04400 (batches 'A' and 'B') is used to illustrate grain boundary segregation which might go unnoticed when the usual microchemical characterization techniques (AES, SEM/EDS, and ATEM) are employed.

SIMS was selected for probing the 'A' and 'B' batches because of its large dynamic range; this makes it possible to detect elemental concentrations down to the part per billion range. A powerful feature of this technique is that spatial distribution of such trace concentrations can often be mapped if the proper SIMS instrumentation is available. The analysis was performed to identify grain segregation at the grain boundaries and grain interiors for several elements.

SIMS images taken within the bulk of the 'B' alloy are shown in Figure 6. Boron is strongly segregated at the grain boundaries. Bulk chemical analysis indicated that both the 'A' and 'B' materials contained approximately 10 ppm of boron.

Almost the entire boundary region contains boron. Only in a few regions can boron be observed as an intergranular segregate. No correlation could be established between the distribution of boron at grain boundaries and any of the other elements, suggesting that boron may be a non-equilibrium segregant. Boron images for the 'A' material revealed a distribution within the material very different from that in the 'B' batch. Boron is randomly distributed throughout the material with no preference for the grain boundaries. The areas of high boron intensity are associated with non-metallic inclusions. Segregation like that seen in the 'B' material can have a dramatic impact on corrosion response because of the associated changes in grain boundary free enthalpy.

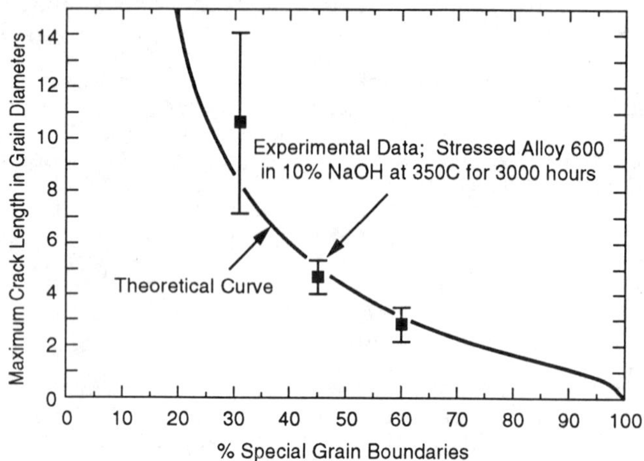

FIG. 5--Maximum crack length in grain diameters versus % special grain boundaries. The theoretical curve and experimental data are shown.

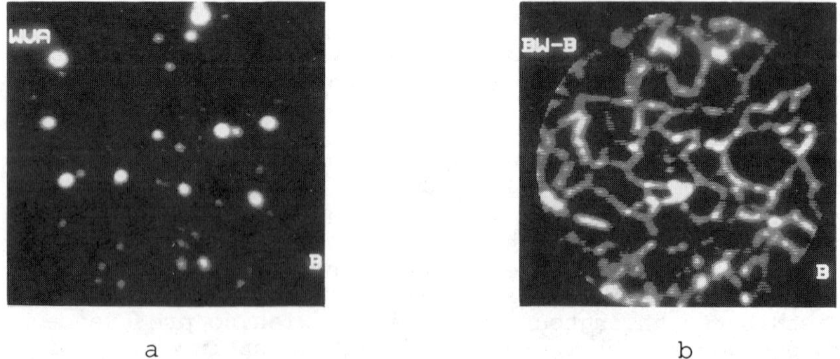

FIG.6--Boron distribution in a) 'A' material and b) in 'B' material. These images give a 150 μm field of view.

Corrosion data for the 'A' and 'B' material are given in the next section of this paper.

The Failure of Accelerated Tests for Predictions Alloy Performance

The example that we present here relates to steam generator performance at Ontario Hydro's Pickering Nuclear Generating Plant. The Pickering 'A' station consists of four reactors which have been running for more than twenty years. The steam generators (SG's) are tubed with alloy N04400 which is not commonly used in the nuclear industry; as a consequence qualifying laboratory test data are quite limited. The operating performance of this material has been outstanding; no corrosion failures have occurred to date.

At the same time the station records show that the operating boiler water chemistry has been less than optimal. High concentrations of ionic impurities, which have mostly originated from condenser in-leakage have been present in the steam generator for long periods of time. Chloride concentrations in the bulk water have reached values of up to 30 ppm which is approximately one thousand times the concentration level recommended by the water chemistry specifications.

Station 'B' which is a duplicate of 'A' is situated adjacent to the 'A' station. The oldest unit in 'B' has operated for seven years and the newest unit for five years. Both stations use the same cooling water in the condensers. The operating chemistry history of the 'B' station has been similar to that of the 'A' station. The steam generators have also operated with high concentration of contaminants from condenser in-leakage for long periods of time.

Tubes failed during operation at the oldest of the 'B' station. Underdeposit corrosion was the cause of the failures. Non-destructive evaluation of the boilers indicated numerous defects. The tube deposits are comprised of feed train corrosion products and lake water contaminants. Corrosion of the copper bearing feed train has been significant as is indicated by the sludge loadings in the low flow areas of the steam generators; however, the sludge loading at the 'A' stations SG's is quite similar, although larger due to the longer operating times.

A similar difference in corrosion resistance was also observed with the low temperature shutdown heat exchangers (SHX's). The tubes employed for these HX's are identical to those used for the SG's. Underdeposit corrosion occurred in untreated Lake Ontario water, on environment vastly different from that which exists in the sludge pile region of SG's had resulted in tube perforation.

The widely different corrosion resistance of N04400 SG's and SHX's tubed with material of the same nominal composition has led to a detailed investigation into the cause of the difference. This investigation included subjecting the different batches of material to accelerated corrosion tests. Anodic polarizations, scratch testing, exposure to a sodium chloride/hydrochloric acid mixture and to ferric chloride were used to determine if it was possible to discriminate between the two materials. In addition to the accelerated tests a plant simulation was also carried out.

Anodic Polarizations

Table 1, gives the results of anodic polarizations performed at 70°C in untreated Lake Ontario water. Pickering 'B' material consistently yielded the highest pitting potential (E_{pit}). Similar corrosion potentials, and protection potentials were obtained for the two materials.

Scratch tests were carried out to establish repassivation time constants. Testing was comprised of polarizing the samples at two potentials in the passive range: -200 mV (SCE) and -100 MV (SCE). At these potentials, a scratch caused the current to increase as the bare metal surface was exposed to the solution; as the surface repassivated, the current decreased until near steady state conditions were again attained. Of prime interest was the time required for the current to return to its previous passive values. Typical time constants for the two batches are given in Table 2. The current response due to the scratch were in the range of 4-10 μm cm^{-2}. It can be seen from these results that there was not a significant difference in the repassivation behaviour of the two alloys, regardless of the initial polarization potential.

Table 3 shows the breakdown time for the 'A' and 'B' alloys. This time is an indication of the transition from the passive to the active state. In the solution investigated the passive films on these alloys were not completely stable. However, as with the scratch tests there was not a significant difference between the two alloys.

Immersion Tests

Results from immersion tests are shown in Table 4; again these tests were unable to discriminate between the Pickering 'A' and 'B' materials. The immersion tests were conducted on material which was heat treated to simulate the oxide on the surface of the tubes in the SG's and SHX's.

Field Simulations

Samples of the 'A' and 'B' tubing were exposed to a simulated SG environment. Conditions of the test closely followed the chemistry and design in an operating unit. The testing focused on the simulation of the chemical and thermohydraulic conditions in the most vulnerable part of the SG's, i.e. the area on top of the tube sheet where corrosion products (sludge) deposit. Sludge accumulates and forms a porous body. The heat transfer process on the tube surface and the non-nucleated type of boiling under the

TABLE 1--Results of Anodic Polarizations Performed on N04400 Tube Material from the PNGS-A and PNGS-B Nuclear Generating Station

Material	$E_{corr.}$	E_{pit}	$E_{protect.}$
PNGS-B	0.043	0.598	0.0478
PNGS-A	0.028	0.298	0.0508

Note: Potentials are versus S.H.E. tests conducted at 70°C using a scan rate of 1 mV s^{-1}.

TABLE 2--Repassivation Times N04400 in 3 wt% deaerated NaCl 23°C

Material	E_{app}/mV(S.H.E.)	Repassivation Time (s)
PNGS-A	0.080	1.41 ± 0.1
	0.180	0.39 ± 0.05
PNGS-B	0.080	1.71 ± 0.35
	0.180	0.49 ± 0.15

TABLE 3--Open Circuit Potential Decay Breakdown Time for N04400 Alloy in 3% NaCl at 23°C

Material	$E_{app.}$ (S.H.E.)	Breakdown Time, t_b (s)
PNGS-A	0.180	1181 ± 116
PNGS-B	0.180	1029 ± 7

TABLE 4--Results of Accelerated Immersion Tests

Material	Test Environment	%Weight Loss
PNGS-A	Aqueous solution containing 3.5% NaCl, 1.0 vol. % HCl at 20° pH 1.1 (100 hr exposure)	0.26
PNGS-B	As above	0.27
PNGS-A	Aqueous solution containing 10 wt% $FeCl_3$ at 20°C (100 hr exposure)	15.3
PNGS-B	As above	13.6

deposit are conducive to the accumulation of soluble non-volatile impurities some of which, (chloride and sulphate) are aggressive to the tube material).

The chemistry employed during this test simulated a condenser in-leakage situation, a condition which leads to high dissolved oxygen and lake water contaminants. Also, the test samples were designed to include a sludge pile. The inclusion of these factors in the test leads to a measure of acceleration. In an operating system sludge usually takes several years to buildup to a depth which would lead to corrosion problems and lake water contaminants are not a concern until condenser leaks occur, again this usually takes several years. The test was carried out for a period of four weeks.

Destructive examination of the test probes after testing showed intergranular attack (IGA) which only progressed up to two grains in depth. However, the type of IGA was different in the 'A' and 'B' material. In the 'A' material the whole grain interior seemed to be attacked in a general manner, in 'B' samples dissolution had preferentially taken place along grain boundaries (see Figure 7).

These different modes of attack appear related to station performance which illustrates the important of careful testing when trying to assess the performance of an alloy for field use.

Clearly, unlike the tests with high acceleration the more elaborate simulation revealed slight differences in corrosion resistance. This result underlines the importance of a test which simulates the operating environment.

Conclusions

The usefulness of accelerated tests in providing the basis of materials selection can be compromised by testing materials which are subtly different form than those used in the field. Slight microstructural and microchemical differences can lead to large differences in performance in the service environment. Microstructural factors which are commonly neglected are:

i) Residual deformation introduced during sample preparation.

ii) Crystallographic orientation and misorientation resulting from the thermomechanical processing history.

iii) Crystallographic misorientation of adjoining crystals, i.e. the CSL character of the interface structure.

Subtle microchemical differences resulting from trace element segregation which often cannot be detected with the more common microprobe techniques, such as SEM/EDS, AES, and ATEM, may also influence corrosion resistance. If these differences are not identified and taken into account misleading data may result.

In addition to recognizing the importance of subtle microstructural and microchemical factors on corrosion tests with high acceleration can mask differences in corrosion response of

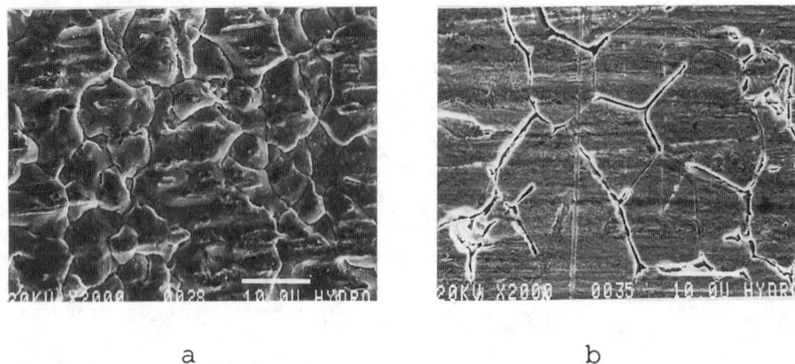

a b

FIG. 7--A topographical view of the 'A' and 'B' alloys showing a) general attack of the 'A' material and b) intergranular attack of the 'B' N04400.

materials. This is likely to be especially the case when attempting to assess long-term behaviour in the actual operating environment. This problem can be reduced by developing a balanced test which better simulates the service environment but at the same time introduces a measure of acceleration.

References

[1] Palumbo, G. and K.T. Aust, (1989) Microstructural Science, Int. Metall. Soc. Proc., (ed. P.J. Kenny et al), Vol. 17,275.

[2] Palumbo, G. and K.T. Aust (unpublished).

[3] Kronberg, M.L. and F.H. Wilson (1949), Trans. TMS-AIME, 185, 501.

[4] Palumbo, G. and K.T. Aust, in Materials Interfaces: Atomic Level Structure and Properties (eds. D. Wolf and S. Yip) Chapman and Hall, London, P. 190 (1992).

[5] Palumbo, G., P.J. King, K.T. Aust, U. Erb and P.C. Lichtenberger, (1992) in Structure Properties of Interfaces in Materials (eds. W.A.T. Clark, U. Dahmen, C.L. Briant) MRS, Pittsburgh, PA, 311.

[6] Crawford, D.C. and G.S. Was (1992), Metall. Trans. A, 23, 1195.

[7] Brandon, D.G. (1966), Acta Metall., 14, 1479.

[8] Palumbo, G., P.J. King, K.T. Aust, U. Erb and P.C. Lichtenberger, (1991). Scripta Metallurgica et Materialia 25, 1775.

Juri Kolts[1] and Erwin Buck[2]

CORROSION LIFE PREDICTION OF OIL AND GAS PRODUCTION PROCESSING EQUIPMENT

REFERENCE: Kolts, J. and Buck, E., "Corrosion Life Prediction of Oil and Gas Production Processing Equipment," Application of Accelerated Corrosion Tests to Service Life Prediction of Materials, ASTM STP 1194, Gustavo Cragnolino and Narasi Sridhar, Eds., American Society for Testing and Materials, Philadelphia, 1994.

ABSTRACT: A large number of tools are used in corrosion life prediction of new oil and gas facilities during engineering of these projects. The corrosion prediction tools are generally based on laboratory tests and are occasionally verified by field performance. These tools are described in this paper.

KEYWORDS: corrosion, prediction, oil and gas, carbon dioxide, erosion, inhibition, velocity

INTRODUCTION

Engineering of new projects requires that the service life of the equipment exceeds or is equal to the design life. Thus, materials selection, maintenance options, and process design are considered in predicting the life of each option. The various phases of design require that each option be evaluated for its potential service life, thus, design is considered as an exercise in life prediction.

These predictions have always been an important phase of engineering projects because of the large capital investment in these projects. It is becoming even more important in offshore developments. The costs of repair or retrofitting corrosion control on offshore production equipment is extremely expensive. Thus, the initial life assessment must be accurate. Also, there are an increasing number of reliable material and corrosion control alternatives. The selection from among these alternatives may have a substantial economic impact on projects.

Corrosion service life prediction during engineering of new oil and gas developments is done early in conceptual engineering of pipelines, flowlines, and process facilities for estimating the cost of these projects. Because the selection of materials has a large economic impact on the capital expenditure, corrosion and materials experts become involved early in the engineering phase.

A "design life" is specified, not simply from the materials standpoint. Maintenance, repair, and failure prevention are continuing operating functions that are essential components of the design life. The relative cost and implications of maintenance and repair are considered in determining the life. For example, a corrosion

[1]Research Fellow, Production Technology, Conoco Inc., P.O. Box 1267, Ponca City, OK 74602-1267.

[2]Staff Scientist, Production Technology, Conoco Inc., P.O. Box 1267, Ponca City, OK 74602-1267.

perforation in a land-based, unburied water injection pipeline may require only incidental repair costing a few hundred dollars while a perforation in an offshore production riser may cost many millions of dollars. A perforation in one may constitute failure while not in the other.

Often a "design life" is more conservative than real life. For example, a pipeline with pits penetrating 30 percent of the pipe wall may be fit for its present purpose, yet such corrosion would be considered excessive if predicted in the design phase of a project. A simple example is a system pressure that has decreased from that of the design basis.

The prediction of the life of a project must consider the technological capability of a company or division. A given engineering option for one organization may not have the same life as for another organization. One option could be chosen over another because of the technical resources available. An obvious organizational difference is differing maintenance philosophies. Thus, life prediction must consider the capabilities of its captive technical community.

With these considerations in mind, the methodology of corrosion life prediction for developmental projects will be described. The focus of this paper will be on systems containing no H_2S. For sour environments, cracking- and hydrogen-induced damage must also be considered. An excellent review is available [1] to assist in this.

PRODUCTION PROFILE

In most projects, the process conditions are predicted for the life of a development. The trends that are predicted include temperature and pressure profiles, water and hydrocarbon production rates, changes in water phase composition, fluid velocities, and gas compositions. These are bases for estimating corrosion rates or cracking likelihood. The changing trends in process conditions can vary in a predictable way, thus, life prediction must account for the changing environment.

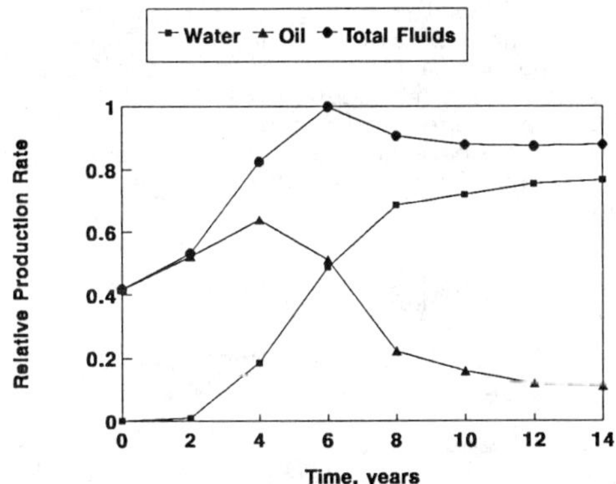

Figure 1, Example of Relative Production Profile for a field, Ratio to Maximum Total Fluids Produced

For example, Figure 1 displays the estimated production profile and water production rate for a given field. For **oil** fields, early production is generally associated with little or no water, and in early life, the environment may not be corrosive. Thus, short-term experience in oil fields may not aid in proper materials selection.

As an **oil** field ages, water breakthrough occurs and the environment becomes corrosive. As the field ages further, the water cut (fraction) increases, temperature often increases, gas to oil ratio (GOR) decreases, pressure usually decreases, and produced water composition changes. Often the H_2S concentration changes drastically. Corrosion life prediction must take into consideration the changing environment. Each of these changing environmental conditions has a specific effect on the corrosivity of the environment. Thus, corrosion life prediction must account for all factors throughout the system lifetime.

As a **gas** field ages, the temperature and production rate decrease. However, the number of wells may increase, thus, the flow through a pipeline may peak during life of a given field. Other developments in the region may provide additional fluctuations in the flow rate and fluid composition in a pipeline.

As shown above, the environmental conditions will be variable. There has been much research to predict the effect of the various environmental parameters so that a cumulative life prediction can be made. The following discussion addresses each of the parameters of importance.

Significant Parameters Affecting Corrosion

Temperature, pressure, and particularly partial pressure of CO_2 and H_2S, liquid composition, fluid velocity both (maximum and minimum), concentration of hydrate inhibitors (glycol or methanol), effectiveness of corrosion inhibitors, water fraction (water cut), oxygen content of injection waters, biocide type and content, flow regime, and corrosion accelerating/inhibiting properties of liquid hydrocarbon phase all affect corrosion rates. Oil and gas facilities are designed, built, and operated safely and profitably in an environmentally sound way using the following prediction techniques.

Water Composition

Research has shown the effects of water composition on corrosivity [2],[3],[4],[5]. The in-situ pH [6],[7],[8],[9] is probably the most important variable. The pH is strongly affected by alkalinity [10]. A program that predicts in-situ pH in CO_2 containing environments has been developed for the industry [11] to estimate the pH for a broad variety of environmental conditions. A regression equation is commonly [12] used to estimate pH, although it is less accurate and is not valid for solutions containing little or no alkalinity.

$$pH = \log\left(\frac{Alk}{P_{CO_2}}\right) + 8.68 + 4.05 \times 10^{-3}T + 4.58 \times 10^{-7}T^2 - 3.07 \times 10^{-5}P - 0.477IS^{0.5} + 0.193IS \quad (1)$$

Where

$$Alk = \frac{mg/l\ HCO_3^-}{61,000}$$

PCO_2 = *partial pressure of* CO_2, *psia*

T = *Temperature, F*

$$IS = \frac{mg/l\ TDS}{58,500}$$

P = *Total pressure, psia*

Water analyses are often obtained for well tests. In short duration well tests, the fluid may be contaminated with mud, which results in underestimating the H_2S content and overestimating the alkalinity. Also, water composition can change with time. As injection water (often seawater) breakthrough occurs, the alkalinity and scaling tendency can change, correspondingly affecting the corrosivity.

Water composition is included in corrosion prediction of most developments. However, since the composition from well to well or from day to day may be highly variable, it may be difficult to apply this information, and often the most severe condition is taken as representative, especially for gas fields.

Water Production Rate

Another important parameter is the water production rate. For oil wells, the parametric relationship between oil fraction and corrosion has not been quantified, although there are estimates of critical oil fraction [13],[14],[15] to reduce corrosion to nominal levels.

For gas wells, a correlation to relate water production rates with corrosivity has been formulated [16]. While the trends appear valid, the water production rate in gas fields is not easily predictable and is subject to operating practice. Thus, this method is not frequently applied to gas developments.

Gas Composition

The partial pressures of CO_2 and H_2S are dominant variables affecting corrosivity. While composition of the gas phase may not vary significantly with time, the pressure does change. Thus, the partial pressure of CO_2 or H_2S and, correspondingly, the corrosivity may vary significantly.

CO_2 Corrosion--Prediction equations relating the corrosivity to CO_2 partial pressure have been extensively developed [17],[18],[19],[20].

This technique is one of the most widely used methods for predicting corrosion in CO_2 containing environments. This method is excellent but must be used within its limitations. The technique does not fully consider the effects of naturally occurring alkalinity in waters, nor does it treat corrosion inhibition nor velocity effects comprehensively. Nevertheless, even with these small shortcomings, corrosion prediction in CO_2 containing environments is the most well developed corrosion prediction technique in oil and gas production.

One commonly misunderstood factor is the effect of total pressure on partial pressure in a liquid phase. Once the bubble point pressure has been reached, an increase in the pressure has only a small effect on the partial pressure in a system.

The occurrence of pitting corrosion is of importance under some conditions. The present statistical[21] or mechanistic[22] methods require more information than is usually available to make corrosion predictions in the design phase. These tools are useful when detailed information is available.

H_2S Corrosion--Regression equations relating H_2S partial pressure to corrosion rate are lacking. In oil wells, souring of a reservoir is inevitable. Good biological control can maintain the H_2S content to a few ppm; however, H_2S contents in the many thousands of ppm can result with improper biological control of the injected waters. Other reservoirs are naturally sour. The only present guidance is that offered by NACE MR 01-75 for sulfide cracking. More work is needed to further understand and predict corrosion in H_2S containing environments.

Temperature of Production

Again, for corrosion controlled by CO_2 partial pressure, the effects of temperature have been developed and are available as corrosion prediction tools. These are included in the CO_2 corrosion prediction techniques. In water injection systems, the influence of temperature is included in oxygen controlled corrosion. For microbially influenced corrosion or in H_2S related corrosion, temperature effects are not yet predictable.

Oxygen Content

Oxygen is generally not present in produced fluids; it is introduced unintentionally as leaks or as a result of introduction of oxygen containing fluids in the process. An equation relating oxygen content to corrosivity, especially for seawater injection systems, has been developed [23]. This equation assumes that the corrosion rate is governed by oxygen transport and thus the predicted rate is, as is the deWaard and Milliams equation for CO_2, a maximum rate.

$$R\,(mm/y) = \frac{5.65 C_{O_2} v}{Re^{0.125} Pr^{3/4}} \qquad (2)$$

Where

R = corrosion rate, mm/y

C_{O_2} = oxygen concentration, ppb

v = velocity, m/sec

Re = Reynold's number

Pr = Prandtl number

For seawater, taking into account the temperature dependence of the diffusivity [24], the equation becomes:

$$R\,(mm/y) = \frac{c_{O_2}}{189}\left(\frac{v^7}{d}\right)^{\frac{1}{8}} \exp\,(T/42.6) \qquad (3)$$

Where

R = *corrosion rate, mm/y*

c_{O_2} = *oxygen concentration, ppb*

v = *velocity, m/*sec

d = *inside pipe diameter, meters*

T = *temperature, C*

Using Equation (3), a corrosion rate of 0.91 mm/y is obtained for an O_2 concentration of 20 ppb, pipe diameter of 0.15 m, a velocity of 4 m/sec, and a temperature of 30°C. This exceptionally high rate can not be sustained and is only an instantaneous rate on bare metal. The maximum sustainable O_2 corrosion rate is controlled by the stoichiometry of the corrosion process. The stoichiometric equation assumes that all O_2 oxidizes Fe to Fe^{+2} and that it does so uniformly throughout the length of pipe. In the rate equation following, the symbols have the same meaning as in Equation (3).

$$R\,(mm/y) = 3.5\times\frac{c\times v\times d}{L} \qquad (4)$$

Where

L = *pipe Length, meters*

A corrosion rate of 0.04 mm/y is obtained from Equation (4) using the same values as in the example and a pipe length of 1000 meters. Interestingly, by equating Equations (3) and (4) the length of pipe required to consume all O_2 is calculable and is 46 meters for the present example.

These equations provide a useful tool in corrosion prediction. More work is needed to verify these predictions with experiences in field performance. In the absence of such information, direct experience with operating seawater injection systems [25],[26],[27],[28],[29],[30],[31],[32],[33],[34],[35], is often relied upon to establish corrosion allowances.

Effect of Oil Composition

The composition of the oil phase may have a significant impact on corrosivity [36],[37]. Some oils are inhibiting while others are corrosive. Organic acids may play a role in determining the corrosivity. The cited references suggest that some chemical species in naturally occurring oils affect corrosion rates. Thus, it is sound practice to perform laboratory tests with actual field retrieved oil samples to establish corrosivity.

Glycol Content

Monoethylene glycol, or possibly diethylene glycol, is added to wet gas lines to inhibit the formation of gas hydrates. The gas hydrates can form above the freezing point of water and pose potential plugging and other operating problems. Glycol provides corrosion inhibition at relatively high concentrations. Shell [38], [39] has been the leader in the use of glycol for corrosion control, but others are planning its use [40]. Correlations for assessing the inhibition by glycol have been developed but have not yet been released to the public literature [41]. An example of inhibition at a high temperature is presented in Figure 2. As temperature is decreased, corrosion rates are reduced significantly.

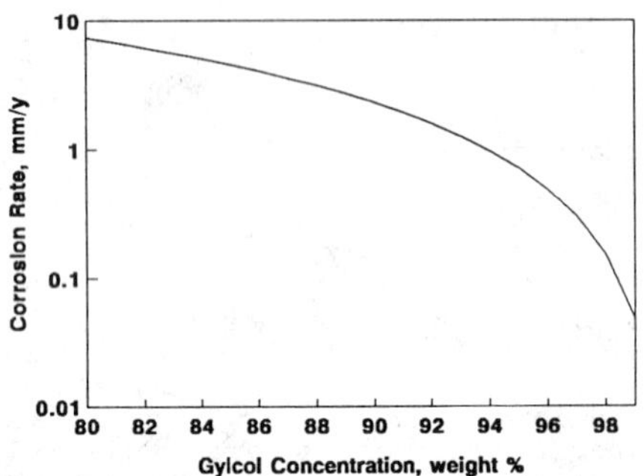

Figure 2, Example of Corrosion Rates of Steel in a Range of Glycol (MEG in Water) Concentrations, at 70C

Methanol Content

Methanol is used as an alternative gas hydrate inhibitor. In gas pipelines, it may reach volume fractions approaching 0.5 of the condensed water phase. Increasing methanol concentration also decreases corrosion rate, although a regression equation on this effect or an understanding of the mechanism has yet to be developed.

Erosion Velocity Maximum--API RP 14E [42]

A maximum velocity above which erosion may be a problem can be estimated by API RP 14E.
The equation,

$$V = \frac{C}{\rho_m^{0.5}} \tag{4}$$

where
V = velocity, ft/sec
C = empirical constant
ρ_m = gas/liquid mixture density at pressure and temperature, lb/ft^3

is often utilized when abrasive particles are not in the fluid. There is controversy in assigning the empirical constant [43]. Often used values are 160 for stainless steel, 140 for coated low alloy steel, and 125 for bare low alloy steel. These values are justified if no sand is present.

Presently, there is insufficient knowledge for the prediction of maximum flow velocities that establish limitations on inhibitor effectiveness. More work is being done in this area.

Pipeline Inhibition

Pipeline modeling [44],[45] used for establishing conditions when inhibitors are dispersed in the line and laboratory testing [46] of their efficacy are the bases for corrosion life prediction.

Experience with multiphase pipelines [47],[48],[49],[50] validates the model's life prediction.

Laboratory simulations are frequently used to study pipeline corrosion behavior. The experimental difficulty is that it is not always clear what parameter(s) to scale or how they scale. The situation with inhibitor efficacy is much better; the experimental techniques are sufficient to answer the simple will or won't it question.

Flow loops are often used to model full size pipelines. However, bench top modelling of real world systems is seldom obvious; velocity/flow modelling is no exception. A method of scaling laboratory flow loops to full sized pipes equates wall shear stresses. An equation [51] for wall shear stress in turbulent fluid flow is:

$$\tau_w = 0.0395 \times Re^{-0.25} \rho u^2, \tag{5}$$

where the Reynold's number, Re is

$$Re = \frac{\rho u d}{\mu}, \tag{6}$$

and ρ, μ, u, and d are the fluid density, viscosity, velocity, and pipe diameter, respectively. Equating the wall shear stress for the same fluid flowing in different diameter pipes gives

$$u_2 = u_1 \times \left(\frac{d_2}{d_1}\right)^{\frac{1}{7}}. \tag{7}$$

For models scaled with equal wall shear, the model velocity in a 3/8" pipe should be 60 percent of that in an 8" pipe.

The test model could also be scaled on the mass transport of corrodent to the pipe wall. Silverman [52] relates the mass transport coefficient to the wall shear as:

$$k = 0.0889 \times \left(\frac{\tau_w}{\rho}\right)^{0.5} Sc^{-0.704}, \tag{8}$$

where the Schmidt number, Sc is

$$Sc = \frac{\mu}{\rho D}, \tag{9}$$

and D is the diffusion coefficient of the corrodent. Again, equating two systems differing only in pipe size and liquid velocity provides the same seventh order dependence as in the wall shear case.

Rotating Cylinder

Most laboratories use some method of assessing the effects of velocity on corrosion. One common method is the rotating cylinder electrode. This method is useful in evaluating the effects of velocity on corrosion inhibition. It is used in the selection of inhibitors. Scaling the data from the rotating cylinder electrode to pipelines on the basis of shear stress or mass transport is shown in Figure 3. Because of lack of confirmation with field experience, these relationships should still be considered only semi-quantitative. However, as shown in the figure, scaling by either shear or mass transport provides roughly the same dependence. Thus, the mechanism of corrosion between these two choices need not be made when using this laboratory technique.

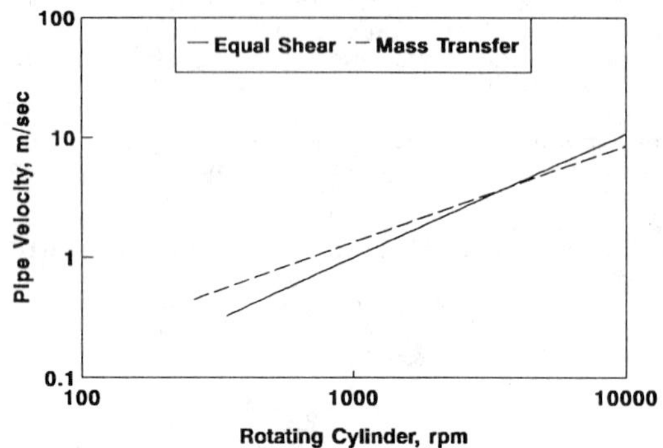

Figure 3, Relationship between Rotating Cylinder Rotation Speed and Mean Liquid Velocity in a Pipe for Equal Shear and Equal Mass Transfer, cylinder diameter = 7.5 mm

Sand Deposition

Sand deposition is important in inhibited pipelines. Inhibitors work only when they can reach the steel surface. Sand deposits prevent such contact. A prediction of sand deposition [53] for pipelines has been developed, which permits prediction of the flow regimes in pipes when corrosion inhibition is effective.

Water Entrainment

The equations of Wicks and Frazer [54] were derived to predict the minimum velocity in uninhibited oil pipelines at which water will drop from a suspension, thus providing a potential for corrosion. This establishes a minimum velocity in design of pipelines to ensure that H_2O is entrained in the oil phase. The calculated velocities are usually in the 0.3-0.6 meters per second range.

A minimum velocity is also important during corrosion inhibition of flowlines. At sufficiently high velocities, oil-water mixing occurs both to transport inhibitors to water wetted surfaces and to improve the efficiency of inhibition. While the Wicks and Frazer treatment has been used to predict the velocity at which mixing occurs, preference is presently placed on laboratory experimental methods [55] coupled with flow modeling.

Biological Control

There are presently no prediction models that deal with microbially influenced corrosion (MIC). It is industry practice to rely on maintenance and, preferably, preventative procedures to assure that MIC does not occur. Thus, the degree of MIC is usually not predicted in the engineering of projects, although equipment and procedures necessary to deliver biocides are necessary engineering considerations for most projects.

Corrosion Resistant Alloys

The life prediction methodology for corrosion resistant alloys is presently evolving and field experience, or linear laboratory test extrapolation, is used as a tool [56],[57],[58],[59],[60],[61],[62],[63],[64],[65],[66]. Although the volume of laboratory test data is expansive, the field experience database to verify the laboratory information is only now being accumulated.

DISCUSSION

The metallurgists and corrosion scientists will continue to use and improve the corrosion life prediction techniques. The increasing cost of developing new discoveries and the mounting pressures to remain competitive will require use of the lowest cost but reliable and safe developments, which will require an increasing need for accurate predictive tools.

Continuing improvement and updating of some of the above prediction tools is occurring; for example, the de Waard and Milliams nomogram has had a number of substantiative changes in recent years, greatly increasing the range of environments covered. Also, the Crolet prediction methodology is evolving. A combination of the two methods may yet occur.

There has been recent work in reexamining the correlation of Fe counts and production rates with corrosion [67]. There is continuing examination of the API RP 14E criteria, and further evolution of this method should occur in the near future [68].

REFERENCES

[1] Dewsnap, R. F., Jones, C. L., Lessells, J., Morrison, W. B., Rudd, W. J., Walker, E. F., Wilkins, R., A Review of Information on Hydrogen Induced Cracking and Sulfide Stress Corrosion Cracking in Pipeline Steels, Offshore Technology Report OTH 86 256, Department of Energy, Her Majesty's Stationery Office, 1987.

[2] Davies, D. H., Burstein, G. T., The Effects of Bicarbonate on the Corrosion and Passivation of Iron, Corrosion, Vol. 36, No. 8, 1980.

[3] Burstein, G. T., Davies, D. H., The Effect of Anions on the Behavior of Scratched Iron Electrodes in Aqueous Solutions, Corrosion Science, Vol. 20, p. 1143, 1980.

[4] Videm, K. and Dugstad A., "Effect of Flow Rate, pH, Fe+2 Concentration, and Steel Quality on the CO_2 Corrosion of Carbon Steels," Corrosion/87, Paper No. 42, March 9-13, 1987, San Francisco, CA.

[5] Lotz U., Sydberger T., CO2 Corrosion of C-Steel and 13% Cr Steel in a Particle Laden Fluid, Paper No. 440, Corrosion/87, 1987, NACE.

[6] Bonis, M. R., Crolet, J.-L., Basics of the Prediction of the Risks of CO_2 Corrosion in Oil and Gas Wells, Paper No. 466, Corrosion/89, April 17-21, 1989, NACE.

[7] Crolet, J.-L., Bonis, M. R., "The Very Nature of the CO_2 Corrosion of Steels in Oil and Gas Wells, and the Corresponding Mechanisms," Oil Gas European Magazine, Special Print, Vol. 10, February 1984, pp. 68-76.

[8] Crolet, J.-L., Bonis, M. R., "pH Measurements Under High Pressures of CO_2 and H_2S," National Association of Corrosion Engineers, Materials Performance, Vol. 23, No. 5, pp. 35-42, May 1984.

[9] Crolet, J.-L., Bonis, M. R., "Prediction of the Risks of CO_2 Corrosion in Oil and Gas Wells", Offshore Technology Conference, Houston, Texas, May 7-10, 1990, OTC, pp. 99-106.

[10] A. Miyasaka, Thermodynamic Estimation of pH of Sour and Sweet Environments as Influenced by the Effect of Anions and Cations, Paper No. 5, Corrosion/92, NACE, 1992.

[11] CORMED Program, Computer program available from Elf Aquitaine, J. L. Crolet.

[12] L. A. Rogers, M. B. Tomson, J. M. Matty, L. R. Durrett, Saturation Index Predicts Brine's Scale-Forming Tendency, Oil & Gas Journal, April 1, 1985, p. 97.

[13] U. Lotz, Velocity Effects in Flow Induced Corrosion, Paper No. 27, Corrosion/90, 1990, NACE.

[14] R. A. Horner, Case Histories - Oil Industry, Paper No. 11, Seminar on Internal Corrosion Control and Monitoring in the Oil, Gas, and Chemical Industries, 3/4 March 1987.

[15] J. Jelinek, Corrosion Problems and Control of North Sea Pipelines, UK Corrosion/86, 1986.

[16] P. A. Burke, R. H. Hausler, Assessment of CO_2 Corrosion in the Cotton Valley Limestone Trend, Materials Performance, August 1985, p 26.

[17] C. de Waard and D. E. Millians, Corrosion, Vol. 31, No. 5, 1975, p. 177.

[18] C. DeWaard, U. Lotz, D. E. Milliams, Predictive Model for CO_2 Corrosion Engineering in Wet Natural Gas Pipelines, Paper No. 577, Corrosion/91, NACE, 1991.

[19] M.J.J. Simon Thomas, C. de Waard, L. M. Smith, Controlling Factors in the Rate of C Corrosion, U.K. Corrosion'87, Brighton, October 1987.

[20] L. M. Smith, C. de Waard, Selection Criteria for Materials in Oil and Gas Processing Plant, U.K. Corrosion'87, Brighton, October, 1987.

[21] M. H. Achour, Juri Kolts, A. H. Johannes, "Statistical Modeling of CO_2 Pitting Corrosion for Downhole Applications," Paper No 141 CORROSION/93, NACE, 1993

[22] M. H. Achour, Juri Kolts, A. H. Johannes, Guahai Liu, "Mechanistic Modeling of Pit Propagation in CO_2 Environment Under High Turbulence Effects", Paper No. 87, CORROSION/93, NACE, 1993

[23] J. W. Olfield, G. L. Swales, B. Todd, Corrosion of Metals in Deaerated Seawater, Proceedings of the 2nd Corrosion Conference, BSE and NACE, Jan, 1981.

[24] Private communication, John Brown Engineering, May, 1990.

[25] J. P. Smith, Commissioning and Operational Experiences of Iiapco's Java Seawater Treating Facility, 5th Offshore Southeast Asia, February 21-24, 1984, Singapore.

[26] F. G. Blakburn, Detection, Monitoring, and Control of Bacterial Corrosion in a Large Middle-East Oil Field Waterflood, Paper No. 289, Corrosion/85, NACE, 1985.

[27] J. S. Brown, L. R. Dubreuil, R. D. Schneider, Seawater Project in Saudi Arabia--Early Experience of Plant Operation, Water Quality, and Effect on Injection Well Performance, SPE Paper No 7763, 1979, Society of Petroleum Engineers.

[28] L. A. Adatoba, L. A. Neyin, Meren Field Seawater Injection System, SPE LS112, 8th Annual International Conference of SPE, Lagos, Nigeria, August 15, 1984.

[29] R. W. Mitchell, The Forties Field Seawater Injection System, J. of Petroleum Technology, June 1978, p. 827.

[30] R. F. Flammang, T. M. Medcalfe, P. T. Gulligan, Problems Associated With the Operation of Offshore Seawater Injection Systems, SPE Paper No. 5770, Society of Petroleum Engineers of AIME, 1976.

[31] Mohamad, Ibrahim El-Hattab, Gupco's Experience in Treating Gulf of Suez Seawater for Waterflooding El-Morgan Oil Field, SPE 10090, 1981.

[32] C. K. Chang, Water Quality Considerations in Malasia's First Waterflood, 5th Offshore Southeast Asia, February 21-24, 1984, Singapore.

[33] M. W. Joosten, J, Kolts, L. H. Wolfe, L. S. Taylor, Materials Considerations for Critical Service Subsea Xmas-Trees and Tubing Hangers, International Conference on Subsea Production Systems: Prevention of Corrosion Problems, Trondheim, Norway, Jan. 1988.

[34] William. I. Cole, Design and Operation of Seawater Injection Systems, SPE 2200, 1968.

[35] J. O. Ogletree, R. J. Overly, Seawater and Subsurface Water Injection in West Delta Block 73 Waterflood Operations, Journal of Petroleum Technology, June, 1977, p623

[36] K. D. Efird, R.J. Jasinski, Effect of the Crude Oil on Corrosion of Steel in Crude Oil/Brine Production, Corrosion, 45, 2 (1989), p 165.

[37] U. Lotz, L. van Bodegom, C. Ouwehand, The Effect of Type of Oil or Gas Condensate on Carbonic Acid Corrosion, Paper No. 41, Corrosion/90, 1990, NACE.

[38] L. van Bogedom, K. van Gelder, J.A.M. Spaninks, M.J.J. Simon Thomas, Control of CO_2 Corrosion in Wet Gas Lines by Injection of Glycol, Paper No. 187, Corrosion/88, NACE, 1988.

[39] L. van Bogedom, K. van Gelder, M.K.F. Paksa, L. van Raam, Effect of Glycol and Methanol on CO_2 Corrosion of Carbon Steel, Paper No. 87, Corrosion/87, NACE, 1987.

[40] J. L. Crolet, J. P. Samaran, The use of the anti-hydrate treatment for the prevention of CO_2 corrosion in long crude gas pipelines, 3rd International Oil Field Chemicals Symposium, Geilo, Norway, March 1992.

[41] E. Skramstad, Investigation of Corrosion in Wet Trunklines - Glycol Inhibition, Veritec Offshore Technology and Services, Report #89-3442, 15 December 1989.

[42] API RP 14E, "Recommended Practice for Design and Installation of Offshore Production Platform Piping Systems," 4th Ed. April 15, 1984.

[43] Salama, M. M., Venkatesh, E. S., Evaluation of the API RP 14E Erosional Velocity Limitation for Offshore Wells, OTC 4485, Offshore Technology Conference, 1983.

[44] Achour, M., Stuever, W., Buck, Erwin, Kolts, Juri, Environmental Factors Affecting Corrosion Inhibition of Wet Gas Pipelines with an Oil Soluble Inhibitor, OTC 6600, Offshore Technology Conference, 1991.

[45] Kolts, Juri, Buck, Erwin, Erickson, D. D., Achour, M., Stuever, W., Corrosion Prediction and Design Considerations for internal Corrosion in Continuously Inhibited Wet Gas Pipelines, UK Corrosion 90, Vol. 3, p 217, 1990.27.

[46] Erwin Buck and Juri Kolts, "Philosophy of Corrosion Inhibitor Selection for Major Projects," Corrosion/UK, Manchester, 1992.

[47] E. V. Seymour, Design detailed for Australia's North West Shelf Pipeline, Oil and Gas Journal, August 31, 1981, p 51.

[48] Case History - Application of a Gas Flow Line Corrosion Inhibitor in an Australian Gas Field, unpublished information from Catoleum Corp.

[49] Ho-Chung-Qui, D. F. and Williamson, A.I., "Corrosion Experiences and Inhibition Practices in Wet Sour Gas Gathering Systems," Corrosion/87, Paper No. 46, March 9-13, 1987, San Francisco, CA.

[50] Frazer, I., Joosten, M. W., Buck, Erwin, Kolts, Juri, Jones, Glyn, Evaluations, inspections reveal North Sea pipeline free of corrosion, Oil and Gas Journal, April 6, 1992, p51

[51] D. C. Silverman, "Rotating Cylinder Electrode - An Approach for Predicting Velocity Sensitive Corrosion," Paper No. 13, Corrosion/90, Las Vegas, Nevada, 1990.

[52] D. C. Silverman, "Rotating Cylinder Electrode - An Approach for Predicting Velocity Sensitive Corrosion," Paper No. 13, Corrosion/90, Las Vegas, Nevada, 1990.

[53] S. Angelsen, O. Kvernvold, M. Linglem, S. Olsen, Paper D2, Long Distance Transport of Unprocessed HC, Sand Settling in Multi-Phase Flowlines, 4th International Conference on Multi-phase Flow, Nice, France, June, 1989.

[54] M. Wicks and J. P. Fraser, "Entrainment of Water by Flowing Oil," Materials Performance, May 1975, pp. 9-12.

[55] J. Paul Jepson, Unpublished results from a joint industry program, Multiphase Flow Research Center, Ohio University, 1992.

[56] Crolet J. L., Bonis M. R., Experience in the Use of 13% Cr Tubing in Corrosive CO2 Fields, SPE Production Engineering, September 1986, p. 344.

[57] Alkire, J. D., Ciaraldi, S. W., "Failures of Martensitic Stainless Steels in Sweet and Sour Gas Service," Corrosion 88, Paper No. 210, March 21-25, 1988, p. 12.

[58] Duncan, R. N., "Materials Performance in Khuff Gas Service," Materials Performance, July 1980, pp. 45-53.

[59] Ciaraldi, S. W., Some Limitations on the Use of 13Cr Alloys for Corrosive Gas and Oil Production, Paper No. 71, Corrosion/90, 1990, NACE.

[60] Klein, L. J. , Thompson, R. M., Moore, D. E., 9Cr-1Mo and 13Cr Tubulars for Sour Service, Paper No. 52, Corrosion/90, 1990, NACE.

[61] Sisak, W. J. and Gordon, J. R., Exxon Co., Laboratory and Field Evaluations of Clad Christmas Tree Equipment," 1989 Offshore Technology Conference, pp. 275-280.

[62] Bunch, P. D., Hartmann, M. P., and Bednarowicz, T. A., "Corrosion/Galling Resistant Hardfacing Materials for Offshore Production Valves," Offshore Technology Conference, May 1-4, 1989, pp. 263-274.

[63] Chen, W. C., Corrosion Performance of 13 Cr Stainless Tubulars in Sour Service, Paper No. 61, Corrosion 90, 1990, NACE.

[64] Combes, J. D., Kerr, J. G., and Klein, L. J., "13Cr Tubulars Solve Corrosion Problems in the Tuscaloosa Trend," Petroleum Engineer International, March 1983, pp. 50-70.

[65] Milliams, D. E., "The Combat of Downhole Corrosion Caused by Sweet Gas," First Cabval Symposium 1985, pp. 1-12.

[66] Corwith, J. R., "The Downhole Corrosion Control Program at Ekofisk," Noroil, February 1984, pp. 47-49.

[67] Hausler, R. H., Garber, J. D., "The Copra Correlation Revisited," Corrosion/90. Paper No 24, 1990 NACE.

[68] Erosion/Corrosion Research Center, University of Tulsa, unpublished results.

Carmen Andrade[1] and M. Cruz Alonso[2]

VALUES OF CORROSION RATE OF STEEL IN CONCRETE TO PREDICT SERVICE LIFE OF CONCRETE STRUCTURES

REFERENCE: Andrade, C. and Alonso, M. C., **"Values of Corrosion Rate of Steel in Concrete to Predict Service Life of Concrete Structures,"** Application of Accelerated Corrosion Tests to Service Life Prediction of Materials, ASTM STP 1194, Gustavo Cragnolino and Narasi Sridhar, Eds., American Society for Testing and Materials, Philadelphia, 1994.

ABSTRACT: Concrete is increasingly being used for encapsulating dangerous wastes although the experience with modern reinforced concrete is not longer than 100 years. In order to predict service lifes of 300-500 years for LLW (low level waste) facilities, some kind of extrapolation is therefore necessary. Due to values of corrosion intensity recorded when steel actively corrodes in concrete, it is not recommended to consider a propagation period in the design of LLW disposals. The design hypothesis should be to keep the steel passive unless unexpected changes in service life, occur. If the steel remains around values of 0.01 $\mu A/cm^2$ (normal in a passive steel embedded in concrete) the stability of the structure could be provided along the foreseen expected life. A final comment is presented on the erroneous assumption usually made that relates corrosion rates to the oxygen diffusion rate through concrete cover.

KEYWORDS: rebar, corrosion intensity values, long term performance, concrete service life.

Concrete is being increasingly used for encapsulating dangerous wastes because, in comparison with other materials of similar or higher impermeability, it is easier to manufacture and lower in cost.

The challenge of concrete is, in this case, its long term stability, because the longest use of modern reinforced concrete is about one hundred years, although it is well known, that Roman non reinforced concretes have survived in a very good state of conservation. These old concretes were mainly based in mixes of pozzolanic materials with lime and

[1]Research Professor, Institute of Construction Sciences "Eduardo Torroja". CSIC. Spain.

[2]Researcher, Institute of Construction Sciences "Eduardo Torroja". CSIC. Spain.

aggregates. Therefore concrete may have an adequate longevity providing it is prepared with appropiate raw materials and has no reinforcement embedded.

In fact the weakest part of modern concrete is the reinforcing steel, because it is a unstable material which tends to be oxidized by oxygen or other substances. The optimum choice for a disposal unit should be indeed, to use non reinforced concrete, which in many cases could be feasible if vaults or any other structure, are designed to work only in compression. Engineers and project Managers in general, should consider this approach based on the principle that real trials of 2000 year concretes have been conducted before us.

When reinforced concrete is the only choice for a disposal unit numerous precautions must be taken in order to be sure that steel will last the predicted service life.

In the present communication, some comments are presented on what corrosion rates of the steel embedded in concrete should be used to predict its long term (300-500 years) behaviour. The predictions are mainly related to low level radioactive waste management and disposal (LLW) facilities, which are at present being developed.

SERVICE LIFE OF CONCRETE STRUCTURES

Long term concrete durability or service life prediction of concrete structures in general is an area in rapid progress, both in relation to normal buildings and public works (needed lifes about 100 years) [1-3] and of disposal units for LLW (300-500 years) [4-5].

The majority of literature references dealing with LLW [4-7] follow the approach of describing first the different degradation processes which the concrete can undergo when used as buried structure, since most of the LLW facilities are designed to be kept underground, although above the water table. Figure 1 shows an example of a facility where the concrete units are encapsulated in mortar filling the gaps between them and all finally is covered with a thick ground layer.

The degradation mechanisms are the common ones such as sulphate attack, frost, leaching, alkali-aggregate reaction, etc. In addition to simply describing the processes, some authors summarize the degradation rates and therefore, offer ways of calculating the expected life of these disposal units before an unacceptable limit of deterioration is reached [1-5].

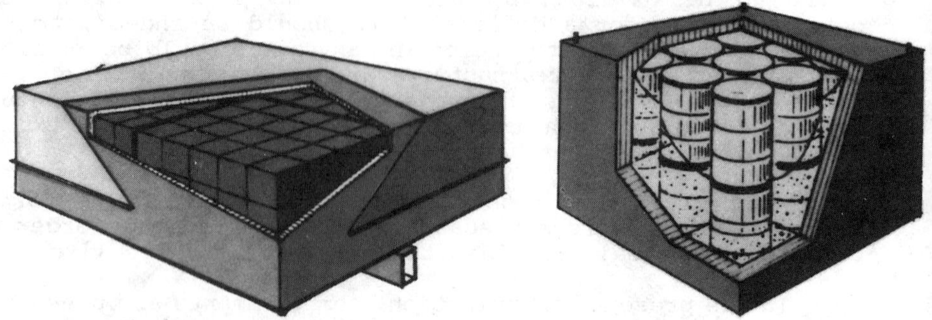

FIG. 1 - Illustration of earth-mounded concrete bunker disposal.

In the particular case of the corrosion of reinforcements the most widely used model is by Tuutti [1] and is shown in Fig. 2.

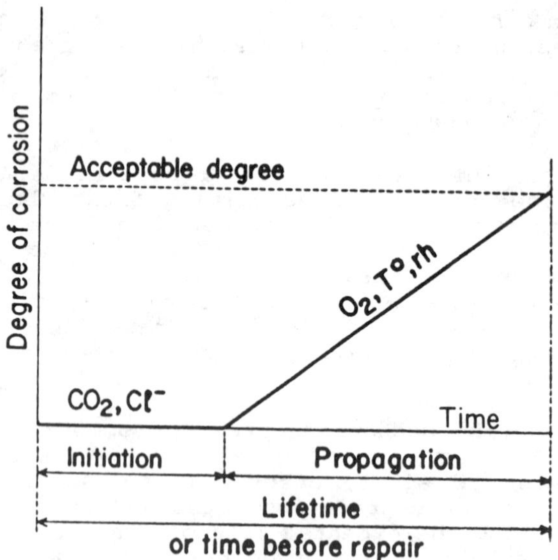

FIG. 2 - Service life schematic model of Corrosion of Steel Reinforcement in Concrete.

The <u>initiation period</u> t_o consists of the time from the erection of the structure until the steel depassivates due to an external aggressive agent reaching the reinforcement surface. The most common aggressive agents are the chloride ions, and the carbon dioxide which induces carbonation of the concrete cover.

The time taken by either chlorides or the carbonation front to reach the rebar surface can be calculated by some mathematical formulae [1,8-10], which in general are based on "the square root of the time":

$$x=k\sqrt{t}$$

x= cover thickness penetrated (mm)
t= time (years)
k= coefficient (mm/year$^{0.5}$) which depends on concrete and ambient characteristics.

These calculation procedures are increasingly being used and in the near future they will be likely incorporated to National Codes and Standars [9] in order to predict the minimum period in which the rebar will be free of corrosion.

The <u>propagation period</u>, t_p illustrated in Fig. 2 comprises the time from the steel depassivation until a certain "unacceptable" degree of damage is developped in the structure. This maximum amount of damage which is tolerated, is still a subject of controversy, therefore it will not be discussed here as being out of the scope of the present communication.

This period t_p, is clearly a function of the rate at which the steel corrodes. The steel corrosion rate is well known to depend on the ambient humidity (concrete resistivity and oxygen availability), temperature and Cl^-/OH^- ratio [1,11]. However there are very few researchers [12-16] who have quantified this parameter that, on the other hand, are crucial in some circunstances.

Thus for instance, Fig. 3 depicts the changes recorded in a carbonated specimen when the ambient relative humidity is changed. Fig. 4 illustrates the corrosion intensity evolution of steel bars embedded in concrete immersed in natural sea water. Before the chloride reaches the rebar the corrosion intensity values remain below a threshold of 0.1 - 0.2 $\mu A/cm^2$. When steel depassivates a dramatic increase of one to two orders of magnitude is recorded. The Icorr then progressively increases with time.

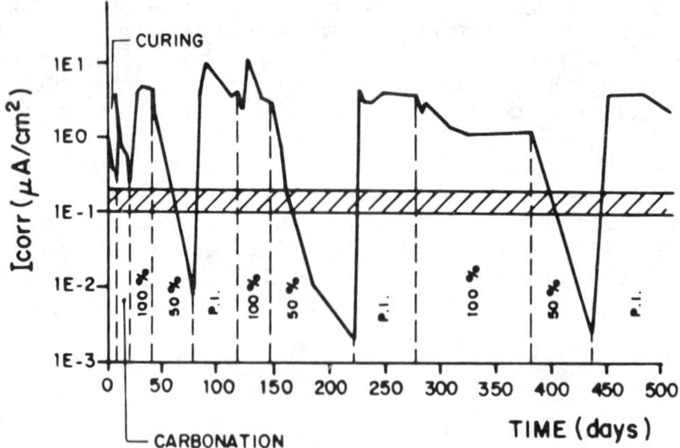

FIG. 3 - Corrosion Intensity values recorded along time in rebars embedded in carbonated mortar in function of changes in the relative humidity of the ambient [16].

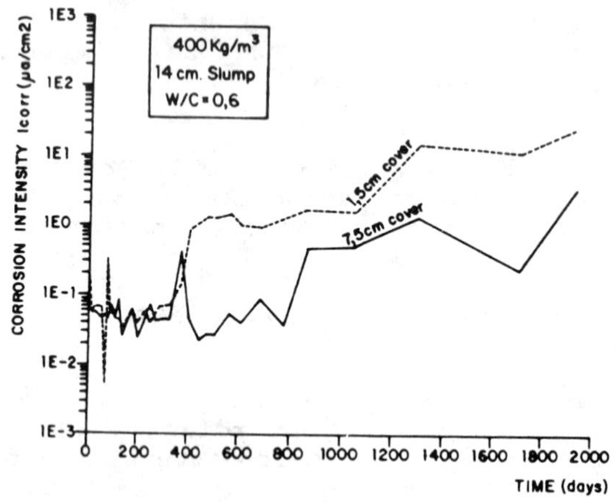

FIG.4 - Corrosion intensity evolution of steel bars embedded in concrete submerged in natural sea water [11].

In order to obtain the corrosion intensity, Icorr, values shown in Figs. 3 and 4, the Polarization Resistance, Rp, technique has been employed, as described by Stern and Geary [17]. This non destructive technique represents a fast and simple measurement method. Through the measurement of hundreds of values in laboratory experiments it was possible to map the expected values of Icorr as a function of the ambient aggressivity and relative humidity. In Fig.5, which has been previously published [11], are presented the different Icorr levels recorded in the experimentation.

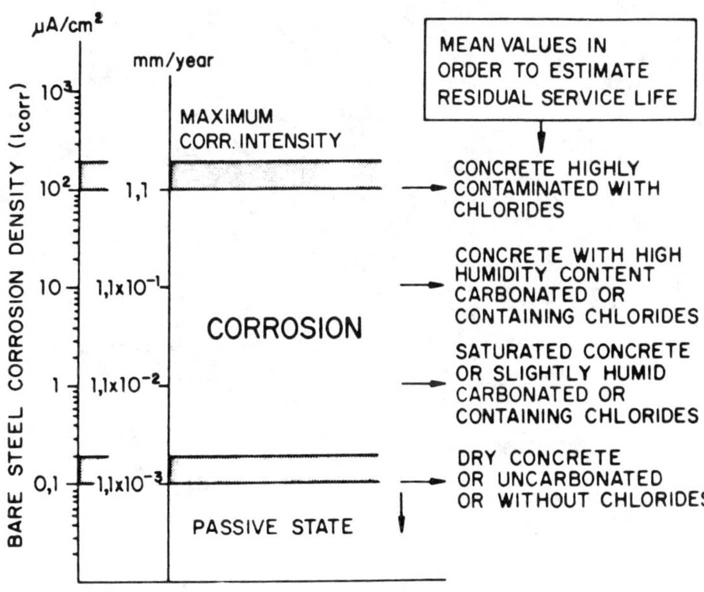

FIG. 5 - Ranges of Icorr values for active corrosion of steel embedded in concrete [11].

During these last years, Rp measurements have also been taken in real structures, and the recorded values agree very well with those measured in laboratory experiments [18]. Therefore the ranges shown in Fig. 5 can be considered as valid in an actual practice situation.

Finally, these Icorr levels have been implemented in Tuutti's model as is shown in Fig. 6 for bars of diameters of 10 and 20 mm. From the figure is possible to deduce that a loss of 5% in cross section can be reached in about 10 years in bars of 10mm diameter if the corrosion rate is 1μA/cm^2 and in approximately the double of years, if the bar is of 20mm diameter.

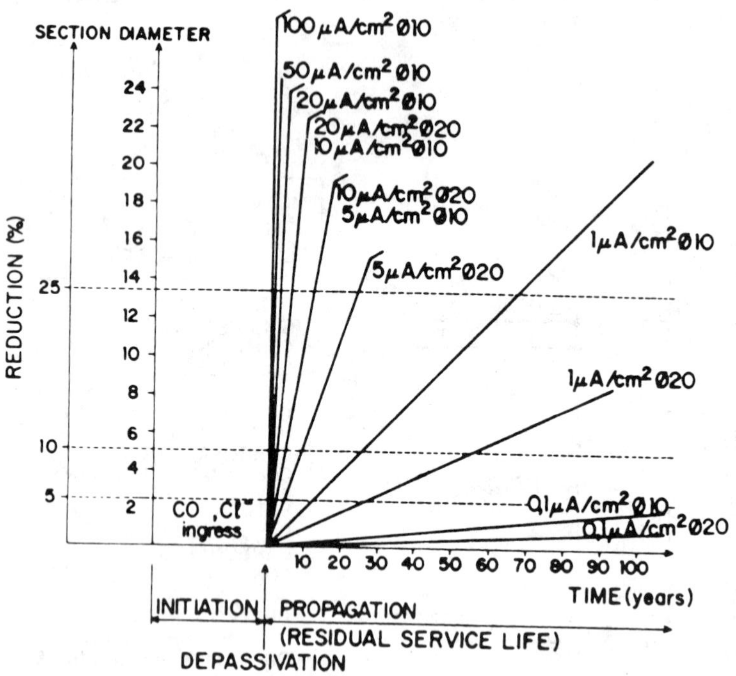

FIG. 6 - Propagation period in function of Icorr values.

STEEL CORROSION RATES EXPETED IN LLW DISPOSALS

In the prediction of service life of reinforced concrete to be used in LLW disposals, among others, a particular point must be emphasized related to steel corrosion: this point is that no corrosion, and therefore, no propagation period must be considered in the design phase. That is, the concrete has to be designed in order to completely avoid corrosion during the period of 300-500 years taken as reference.

If corrosion (propagation period) is likely to occur during the service life, the theoretical safety is very much reduced and a high risk of loosing the barrier characteristics exists before accomplishing the expected life.

Two consequences of this statement are that LLW disposals, at least, must not be located in places where chlorides could reach the rebar and that the carbonation front must not reach the rebar during the life of the structure. In addition, all other degradation processes, which might affect cover integrity and impermeability (leaching, sulfate attack etc), have to be avoided as part of the design phase.

This statement comes from the fact, deduced from Fig. 6, that once steel is depassivated, the time needed to reach an unacceptable degree of deterioration is too short compared with the expected service lifes (300-500 years). Therefore active corrosion must not be taken into consideration in the design.

Even corrosion rates considered in the boundary between active/passive state could be too high to insure such long term stability. Thus, Icorr values of $0.1 \mu A/cm^2$, for instance, are equivalent to penetration rates (assuming generalized corrosion) of 1.1 μm/year that in 500 years will give a loss in bar radius of 550 μm. This loss is several times higher than the steel loss able to induce cover cracking (about 10-50 μm is enough to induce cracking with cover thickness of 2-3 cm). Therefore, in order to avoid damage during 500 years life of a reinforced concrete the maximum corrosion rates values, must not be higher than 0.01 $\mu A/cm^2$ (equivalent to penetration rates of 0.1 μm/year).

What is the normal Icorr value of passive steel? As has been previously mentioned in Fig. 5 it has been empirically considered to be in the passive state when the recorded Icorr values are below $0.1 \mu A/cm^2$ (1.1 μm/year). However, much lower values are measured when the steel remains passive and the concrete hardens and increases its resistivity. Then, values of $0.01 \mu A/cm^2$ or lower are normal both in laboratory experiments [15] and in real structures [18].

DEPENDENCE OF REBAR ICORR VALUES ON OXYGEN AVAILABILITY

It has been a general assumption that rebar corrosion rate is a function of oxygen availability, in such a way that corrosion rates are predicted from oxygen diffusion rate calculations through concrete cover. Hence, most of the literature reviewed dealing with prediction of service life of LLW disposals, are based in this erroneous assumption.

Actually oxygen is needed to maintain active corrosion, but to derive from this fact that the rate is linearly a function of the amount of oxygen at the rebar surface, is an erroneous deduction which forgets the dependence of the corrosion rate on concrete resistivity and Cl^-/OH^- ratio, for instance.

A logical argument can illustrate the mistake: if the corosion rate is dependent on the oxygen flux, the corrosion rate would increase when concrete dries (the pores are more open and the pathway for oxygen is higher) and viceversa, the corrosion would decrease as concrete becomes wetter, which is clearly erroneous.

Therefore all predictions based on the diffusion rate of oxygen through concrete cover must be regarded as not reliable, although, what is true is that when actively corroding in completely water saturated concrete, the maximum corrosion rate is fixed by the oxygen availability and maximum values of 1μA/cm^2 are noticed [19].

The behaviour is completely similar to that of steel suffering atmospheric corrosion and is illustrated in Figs. 7 and 8. There, the corrosion rate is plotted as a function of the water layer thickness [20].

- When this layer is very small (Fig. 8, very dry concrete) the rate is low due to the high concrete resistivity.
- As the relative humidity increases, the layer thickness does but resistivity decreases, and therefore, Icorr will increase as Fig. 7 depicts, until the water layer thickness is high enough to control by diffusion the oxygen path to steel surface.
- At that moment (usually when concrete is completely water saturated as Fig. 8 illustrates) Icorr is controlled by oxygen availability which, although low, is not low enough to stop corrosion completely. Then cover thickness appears as a controlling rate factor because oxygen needs to run through the water from outside to the steel surface. In such conditions, resistivity is low enough for not being a controlling parameter.

All these comments apply to actively corroding steel. When steel remains passive they do not apply because, the Icorr is always low irrespective of concrete humidity content.

Therefore, in order to predict service life as a function of corrosion rates when steel is expected to remain passive, values of about 0.01 μA/cm^2 can be used. The value of 0.01 μA/cm^2 represents a theoretical homogeneous penetration (loss in bar radius) of 55μm in 500 years or 33μm in 300 years.

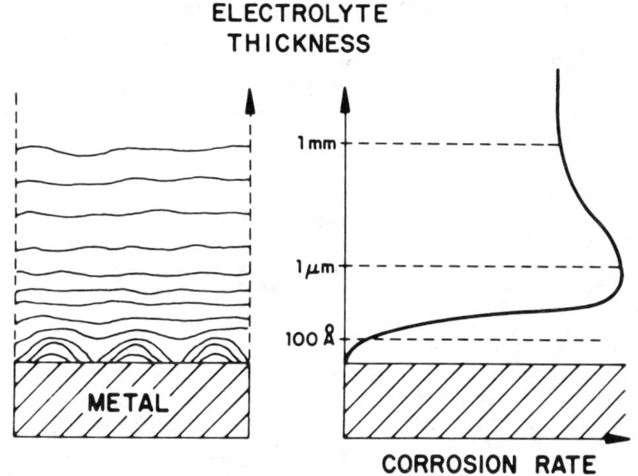

FIG. 7 - Humidity layer thickness as controlling parameter of corrosion rate of steel in the atmosphere.

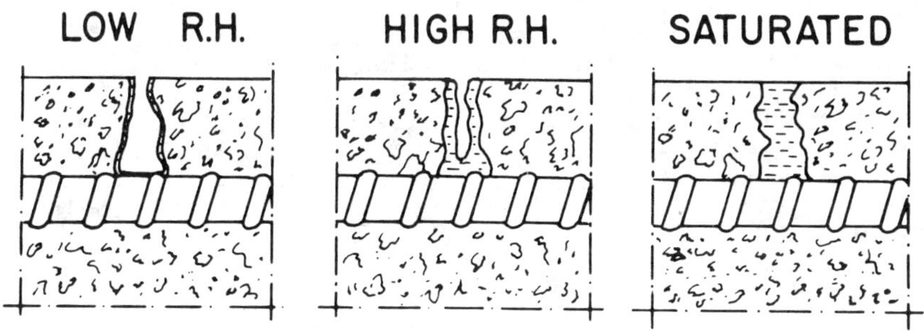

FIG. 8 - Schematic illustration of water layer on concrete pores in function of ambient humidity.

If the steel corrodes actively or is depassivated, the corrosion rate value established will depend on concrete humidity content, oxygen availability, temperature and presence of aggressive substances (chlorides, sulfates or others), as illustrated in Fig. 5. Nevertheless, if steel corrodes actively, the propagation period will be short in comparison with the service life of design, unless oxygen availability is controlled by water saturation, either of concrete or soil, where the LLW facilities are buried.

ALTERNATIVE SOLUTIONS TO MAINTAIN THE STEEL PASSIVE

A final comment will be emphasized for those cases in which the LLW facility is located in places where corrosion of steel cannot be avoided during the storage life, and alternative protective solutions are needed.

To face this situation is very difficult, because if the reinforced concrete record of experience is shorter than 100 years, the experience with methods like the use of inhibitors, epoxy coated, galvanized rebars or cathodic protection, is even shorter.

It should be repeated that it is best to try to select disposal locations, as non aggressive to reinforced concrete as possible, because an extrapolation of good performance for any of the aforementioned complementary protective measurements is merely intuitive, due to the few practical experiences and the short period of use recorded to date. With the present knowledge, cathodic protection looks to be the most theoretically efficient, although its permanent need of maintenance gives an additional difficulty. Solutions like galvanized rebars or inhibitors, may be underdesigned for future aggressivity or protection needs, and the durability of an organic material such as epoxy, is also at present, very controversial.

Therefore as a general rule, only well known and stable materials should be used in LLW disposal if longevities as long as 500 years are required. The use of new materials, either admixtures, aggregates, or fibers not well experimented and with no experience records longer that 50-100 years, should be avoided in this type of structures. In addition, places in principle non aggressive for concrete should be selected.

CONCLUSIONS

The present experience with modern reinforced concrete is much shorter that the expected lifes of LLW concrete disposals. This fact requires extrapolation of short term experience to 300-500 years life. However, non reinforced concrete has lasted 2000 years.

In the design phase, these concrete disposal facilities should be planned to avoid initiation of corrosion during the life time foreseen, because, when the corrosion initiates the propagation period will be in general very short (1-30 years for cover cracking) compared to the total life expected. Only in very particular cases not still well controlled, will the active steel corrode at very low corrosion rates. In permanently buried structures it seems too risky to base the design on the probability of occurrence of active (corrosion) periods.

If the steel remains passive, it is then possible to expect suitable long term behaviour of the structure. Corrosion rates of 0.01 $\mu A/cm^2$ (0.1 μm/year attack penetration) are expected and therefore negligible diameter losses are expected in 500 years.

Deductions of future corrosion rates based on the linear dependence of oxygen diffusion rate through cover thickness can lead to erroneous prediction of service life. Corrosion rate may be controlled by oxygen diffusion in water saturated concrete while, in not fully saturated concrete, oxygen arrives in enough quantity to support high rates of corrosion, making the concrete resistivity the controlling rate parameter.

Only very well experienced materials and methodologies should be used in LLW structures if long longevities are designed.

REFERENCES

[1] TUUTTI, K., "Corrosion of steel in Concrete" , Swedish Cement and Concrete Institute-Stockholm (1982).

[2] VESIKARI, E., "Service Life Design of Concrete Structures with regard to Corrosion of Reinforcements" Technical Research Centre of Finland, Espoo - Research Report nº 553 (1988).

[3] FRONSDORFF, G., MASTER, L.W., MARTIN, J.W., "An approach to improved durability test for building materials and components" - Nat.Bur.Stds. Technical Note 1120, Luly - Washington D.F.USA (1980)

[4] CLIFTON, J.R., "Predicting the Remaining Service Life of Concrete" - NISTIR 4712 - Washington D.F.USA - Nov.1991.

[5] WALTON, J.C., SEITZ, R.R., "Performance of Intact and partially degraded Concrete Barriers in Limiting Fluid Flow" - NUREG/CR-5614-E66-2614-RW, CC, CJ, CO, CY. Office of Nuclear Regulatory Research - Washington D.C. - NRC FIN A6858.

[6] CLIFTON, J.R., KNAB, L.I., "Service life of Concrete" - NUREG/CR-5466-NISTIR 89-4086-RW.

[7] WALTON, J.C., PLANSKY, L.E., SMITHR. W., "Models for estimation of Service Life of Concrete Barriers in low level Radioactive Waste Disposal" - NUREG/CR-5542 EGG-2597- Office of Nuclear Regulatory Research - Washington D.C.- NRC FIN A6858. - Sep.1990.

[8] SCHIESSL, P., "Relation between the crack width and the amount of corrosion at the reinforcement", Betonwerk + Fertigteil-Technik, vol.12 (1975) pg.594-598.

[9] PARROTT, L., "Design for avoiding damage due to carbonation-induced corrosion". 3rd. Canmet - ACI Int. Conf. on Durability of Concrete - Nice - France - May 1994.

[10] BROWNE, R.D., GEOGHEGAN, M.P., BAKER, A.F., "Analysis of structural condition from durability results" Corrosion of Reinforcement in concrete construction - Crane Ed. London (1983) 193. Ellis Horwood Ltd. Publishers - Society of Chemical Industry.

[11] ANDRADE, C., ALONSO, C., GONZALEZ, J.A., "An initial effort to use the corrosion rate measurements for estimating rebar durability" - ASTM STP 1065 (1990) p.25. Philadelphia - USA.

[12] ESCALANTE, E., WHITENTON, E., QUIU, F., "Measuring the rate of corrosion of Reinforcing steel in concrete". Final report NBSIR 86-3456, National Bureau of Standards, Oct.(1986). Washington D.C. - USA.

[13] HARDON, R.G., LAMBERT, P., PAGE, C.L., "Relationship between electrochemical noise and corrosion rate of steel in salt contaminated concrete" - Brit. Corrosion J. 22,4 (1988) 225.

[14] SAGUES, A., "Critical issues in electrochemical corrosion measurement techniques for steel in concrete" CORROSION 91 - NACE, National Ass. Corrosion Eng. Paper nº 141 - Cincinnati (Ohio) USA.

[15] GONZALEZ, J.A., ALGABA, S., ANDRADE, C., "Corrosion of reinforcing bars in carbonated concrete" - British Corrosion J. 3(1980) 135.

[16] ALONSO, C., ANDRADE, C., GONZALEZ, J.A., "Relation between concrete resistivity and corrosion rate of the reinforcements in carbonated mortar made with several cement types". Cement and Concrete Research 18 (1988) 687.

[17] STERN, M., GEARY, A.L., "A theoretical analysis of the shape of Polarization Curves" - J. Electrochemical Soc. Jan (1957) 56.

[18] FELIU, S., GONZALEZ, J.A., ANDRADE, C., FELIU, V., "On-site determination of the Polarization Resistance in a reinforced concrete beam" - Corrosion Engineering 44, 10 (1988) 761.

[19] ANDRADE, C., ALONSO, C., GARCIA, M.A., "Oxygen availability in the corrosion of reinforcements" - Advances in Cement Research 3, 1 (1990) 127.

[20] ANDRADE, C., ALONSO, C., GONZALEZ, J.A., FELIU, S., "Similarity between atmospheric/underground corrosion and reinforced concrete corrosion" - Corrosion of Reinforcement in concrete - C.L.Page, K.W.J. Treadaway, P.B. Bamforth Ed - SCI London (1990) 39.

Experimental Techniques

Yuan Xu[1] and Howard W. Pickering[2]

A NEW INDEX FOR THE CREVICE CORROSION RESISTANCE OF MATERIALS

REFERENCE: Xu, Y. and Pickering, H. W., "**A New Index for the Crevice Corrosion Resistance of Materials,**" Application of Accelerated Corrosion Tests to Service Life Prediction of Materials, ASTM STP 1194, Gustavo Cragnolino and Narasi Sridhar, Eds., American Society for Testing and Materials, Philadelphia, 1994.

ABSTRACT: Recent studies have revealed the crucial role played by the macro corrosion cell (potential coupling between the inside and outside of a cavity) in crevice and pitting corrosion. It was found that acidification and the existence of chloride ions in the local cell are not the sole and necessary conditions for localized corrosion to occur, and that their accelerating effects on crevice corrosion and pit growth can be explained within the framework of the macro cell (the IR voltage mechanism). Upon analysis of the results of the experiments, quantitative modeling, and the literature, a new characteristic parameter - the critical distance into the crevice, d_c - has been suggested for indexing the crevice corrosion resistance of a material under specified conditions. The advantages of using d_c as the index of the crevice corrosion resistance are: (1) it may be obtained through experiment and may also be estimated through computational approaches; (2) it has a distinct and straightforward physical meaning; (3) it may be employed in engineering design and (4) it is a single parameter which can reflect the integrated influence of several factors known to affect the crevice corrosion resistance of a material from past practical experience and research work. Preliminary work has shown good agreement between the measured and the computed values of d_c. The experimental technique and the principle of the mathematical approach to obtain d_c are described.

KEYWORDS: quantitative test, predictive method, accelerated test, corrosion testing, crevice corrosion susceptibility

---

[1] Singapore Institute of Standards and Industrial Research (SISIR), Metal and Advanced Materials Center, 1 Science Park Drive, Singapore 0511

[2] Department of Materials Science and Engineering The Pennsylvania State University, University Park, PA 16802, USA

Introduction

Numerous crevice corrosion testing methods have been proposed and used with varying success. American Society for Testing and Materials (ASTM). Specification G78 provides guidance in the conduct of crevice corrosion tests for stainless steels and related nickel-base alloys in sea-water and other chloride-containing environments [1], although it does not provide any particular test technique. The large amount of testing methods may be divided into two major groups. The first includes those methods which involve the use of artificial crevices. Samples with artificial crevices are immersed into the testing solution for a period of time. The crevice corrosion resistance is evaluated by the number of crevices which are found to have been corroded during the test. The spool specimen test racks [2], Ferric chloride tests [3], the Materials Technology Institute Tests (MTI-1 to MTI-5) [4] and Multiple-Crevice Assembly Testing [5] fall in this group. Among these, the Multiple-Crevice Assembly Testing [5] is the most often used. The other group employs electrochemical techniques. It includes two ASTM standard methods, ASTM G61 for iron-nickel- and cobalt-based alloys [6] and ASTM F746 for metallic surgical implant materials [7], as well as other methods, for example, potentiostatic test [8]; potentiodynamic test [9] and remote crevice assemblies test [10]. Although each of the techniques has its own merits and has been used with varying success, the currently used testing methods have several common problems. For instance, the data obtained through one of these tests can only serve the purposes of comparison and screening but can not be used directly for quantitative engineering design. Another problem is that data obtained by the different methods are not convertible to each other. Therefore, no method can be claimed to be the best. The reason for this may lie in the lack of a generally recognized theory on crevice corrosion so that it is difficult to link the method itself and the obtained data with the crevice corrosion mechanism.

In the present paper, a brief introduction on the progress of the mechanistic study of crevice corrosion is given first. Based on the new progress, a characteristic parameter, d_c - the critical distance into the crevice - is proposed to index the crevice corrosion resistance of a metal. The advantages for using d_c are discussed. The suggested experimental technique and the mathematical approach for obtaining d_c are described.

Progress on the Understanding of Crevice Corrosion

Two major mechanisms on crevice corrosion have been proposed. The one based on the solution composition change within crevices [for example 11] suggests that the hydrolysis of dissolved metal ions increases the acidity (lowers the pH) and the resulting autocatalytic effect increases appreciably the metal dissolution rate (causing crevice corrosion). However, how the so called autocatalytic effect can increase the corrosion rate within a crevice has never been explained. As a matter of fact, the actual metal dissolution rate measured in an acidified solution (equivalent to the hydrolysed crevice electrolyte solution) is far less than the observed crevice corrosion current density. Furthermore, crevice corrosion sometimes occurs in the absence of a pH decrease. Another mechanism, the IR voltage mechanism [12,13] focuses on the macro corrosion cell between the active crevice wall and the passivated sample's outer surface where the (cathodic) reduction of oxygen occurs. Or, in a potentiostatic test, the cathodic reduction occurs at the counter electrode. The crevice corrosion process may be explained by the potential distribution along the crevice wall (Fig. 1). Due to the electric shield effect, the electrode potential at some distance into a sufficiently deep crevice always remains at the mixed or equilibrium

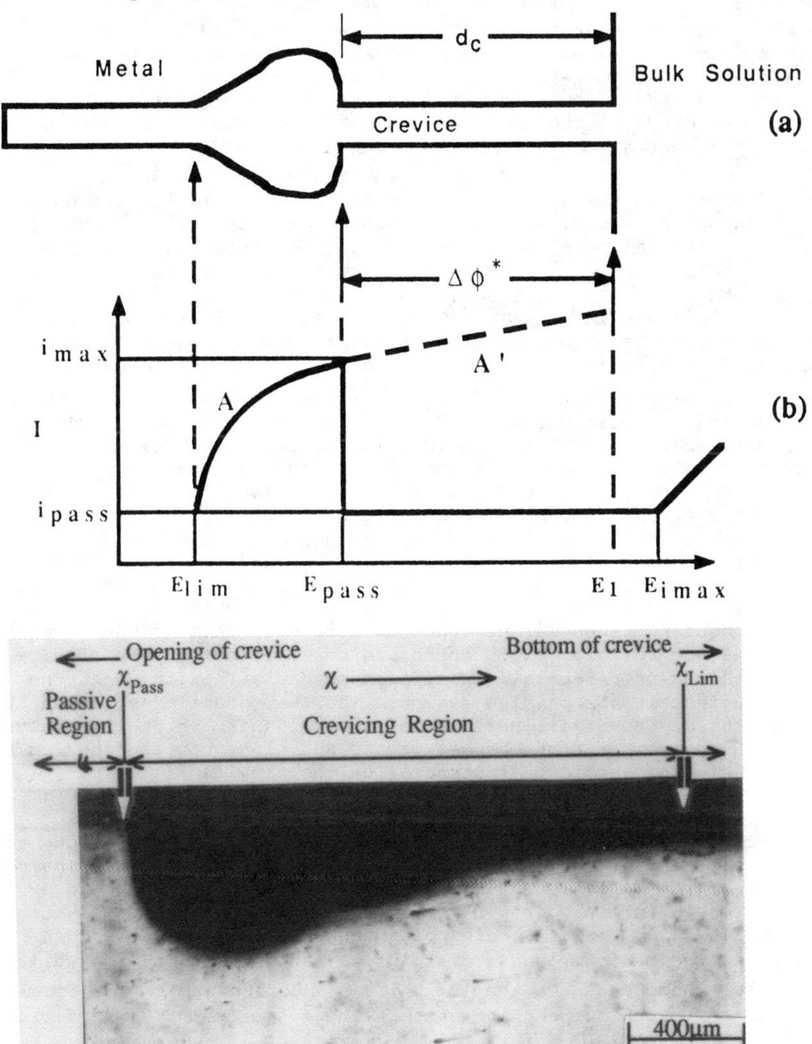

Fig. 1 (a) A crevice undergoing corrosion below d_c where $E < E_{pass}$ since $IR > \Delta\phi^*$. (b) anodic polarization curve of the sample in the crevice solution. A' is the extension of the active portion A. It is used in the mathematical modeling [15]. (c) a cross section micrograph of the crevice wall below d_c where crevice corrosion occurred [19].

potential in the crevice electrolyte, referred to as the limiting potential [12,13], x_{lim} in Fig. 1c, irrespective of the more oxidizing potential at the outer surface. The outer surface potential is established either by a potentiostat or by the reduction of dissolved oxygen in the bulk solution at the passivated outer surface. Thus, along the crevice wall, the local electrode potential decreases gradually from the value at the crevice opening (the more oxidizing potential) to a less oxidizing potential inside the crevice. At a certain distance, d_c, also referred to as x_{pass} (Fig. 1c), where the electrode potential equals the passivation potential of the anodic polarization curve of the crevice electrolyte, the crevice wall changes its state from passive to active. Thus, active dissolution occurs beyond d_c to the distance where the potential decreases to the limiting potential, E_{lim} (Fig. 1b). The crevice wall dissolution current is highest at distances slightly greater than d_c, corresponding to the peak current in the anodic polarization curve. This can be seen in cross sections of the crevice wall as shown in Fig. 1c where the penetration of the crevice corrosion is deepest just to the right of x_{pass}. Since this potential distribution is controlled by the IR voltage of the dissolution current flowing out of the crevice (that is, the IR voltage within the crevice must exceed the difference between the sample's outer surface potential and the passivation potential, E_{pass}, of the crevice electrolyte polarization curve, IR > $|\Delta\phi^*|$ (Fig. 1), in order that crevice corrosion occurs), this process is referred to as IR induced crevice corrosion. The IR voltage mechanism suggests: (1) Crevice corrosion is due to the macro corrosion cell involving well separated anodic and cathodic reactions occurring at very different electrode potentials. This is consistent with the generally accepted understanding of crevice corrosion. (2) The highest metal dissolution rate within a crevice corresponds to the peak current of the crevicing electrolyte polarization curve. That is, the crevice corrosion current is the anodic dissolution rate at high anodic overpotentials within the active region, which is hundreds or even a thousand times larger than the corrosion rate at the limiting potential due to local micro corrosion cells deeper in the crevice or at the corrosion potential (or applied potential) existing at the outer surface in the passive region. The IR mechanism overcomes the difficulty of the traditional hydrolysis mechanism which is solely based on the composition changes of the solution, not considering the distribution of anodic overpotential on the crevice wall.

The IR voltage mechanism can satisfactorily explain the accelerating effect on crevice corrosion of the pH decrease (due to hydrolysis) and of the chloride ion build up in the crevice. The pH decrease of the crevice electrolyte and the chloride ion build up may always result in a higher peak current and a more noble passivation potential, E_{pass}, in the anodic polarization curve, which in turn increases the maximum anodic dissolution rate on the crevice wall (IR increases) and decreases the $|\Delta\phi^*|$ value. Consequently, the total crevice corrosion current increases. The increase in IR and decrease in $|\Delta\phi^*|$ both lead to a decrease in the d_c value. In many situations these composition changes, by decreasing passive film stability in the low overpotential region, facilitate the IR > $|\Delta\phi^*|$ condition by producing an active region and, thus, an active/passive transition in the polarization curve where none existed in the corresponding curve of the bulk electrolyte.

Critical Distance Into the Crevice, d_c

The critical distance into the crevice, d_c, at which the crevice wall changes its state from passive into active, has a specific meaning for crevice corrosion. When the depth of a crevice is less than the maximum critical distance under the specified conditions, the crevice wall will be fully passivated and no crevice corrosion occurs. On the other hand, when the depth is larger than the maximum critical distance, the crevice will be active at distances into the crevice greater than the d_c distance. It has been found, in both experiments and computations [14-19], that the critical distance into the crevice, d_c, is affected by several parameters as shown in Figs. 2 and 3. The crevice gap dimension has a remarkable influence on d_c. The larger the crevice gap, the larger the d_c (Fig. 2a). On the other hand, the crevice depth, d_o, has a weak influence and d_c, with d_c increasing slightly as d_o decreases to the d_c value, for example, the maximum d_c occurs for $d_c = d_o$ [14]. The magnitude of the passive current density affects d_c to a less extent than the gap dimension, with a larger passive current giving rise to a smaller d_c (Fig. 2b). A high solution conductivity decreases the resistance and so increases d_c (Fig. 2c). Also the higher (more oxidizing) the electrode potential (produced by an oxidant or power supply) at the sample's outer surface, the larger the d_c. An almost linear relationship exists between d_c and the outer surface electrode potential (Fig. 3). In addition the anodic behaviour of the metal, especially the peak current in the anodic polarization curve, has been found to significently influence d_c. To have the same critical distance for a larger crevice gap requires a higher peak current density for otherwise identical conditions [18]. All these relations are fully compatible with the IR voltage mechanism.

The above factors affecting d_c are surprisingly consistent with those factors which are known, by experience and past research work, to influence the crevice corrosion resistance of materials. In the work by Fitzgerald and his predecessors [20-23], the crevice corrosion resistance has been found to increase with increasing crevice gap, passive potential range and solution conductivity but with decreasing peak current density of the anodic polarization curve and the active potential range. These are exactly the same factors affecting d_c described above. Therefore, it may be suggested to use the critical distance into the crevice, d_c, as the index of the crevice corrosion resistance of a material, for example, the larger the d_c in a specified situation, the more resistant the material is to crevice corrosion.

The advantages of using d_c are apparent. First, it has a distinct and straightforward meaning. That is, d_c is the distance into a crevice beyond which crevice corrosion occurs. Secondly, d_c may be used in corrosion prevention design. As stated before, when the depth of an existing crevice d_o, is less than d_c for the specified conditions, the crevice will be in the passive state and so it can be tolerated. But when d_o is larger than d_c, the crevice wall beyond d_c will be in the active state and crevice corrosion will occur down to the distance of the limiting potential, E_{lim}. A nomograph or table of the relationship between d_c and the crevice geometry in a specified environment/metal combination may be prepared by computations (as explained later) and experiments. By referring to the graph or table, one may find out whether a crevice is safe (no crevice corrosion) or not. Thirdly, d_c is a single parameter to represent the crevice corrosion resistance. It combines the comprehensive effects of the several factors affecting the crevice corrosion resistance (the peak current density, the solution conductivity, the passive potential, and so forth).

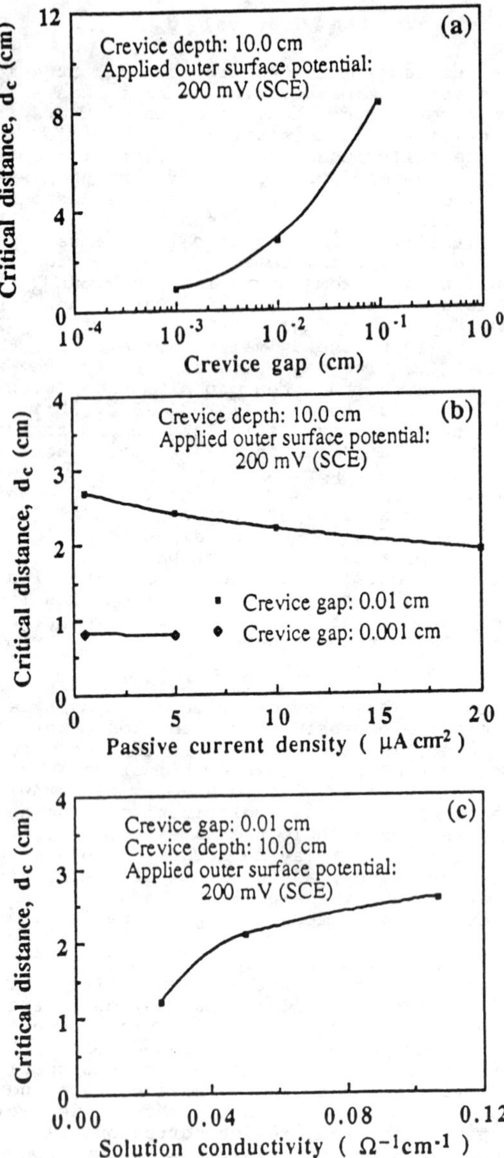

Fig. 2 The critical distance into the crevice, d_C, as a function of the (a) crevice gap dimension, (b) passive current density, and (c) solution conductivity.

Experiment for Measuring d_c

The experimental set up for measuring d_c is given in Fig. 4a. It consists of a three electrode corrosion cell, a potentiostat and a current recording device. The working electrode is made of the material to be tested. An artificial crevice is made on it. Two kinds

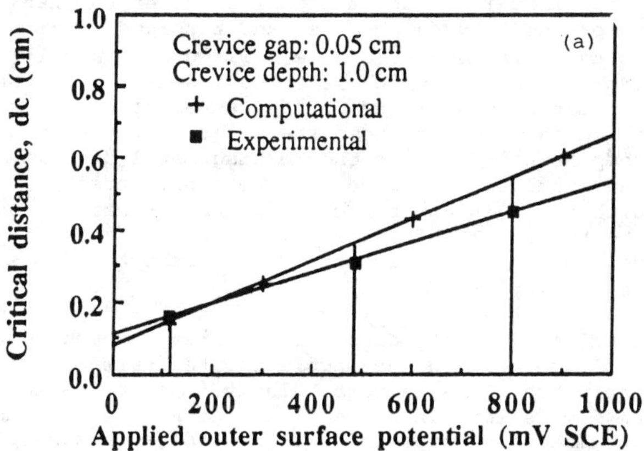

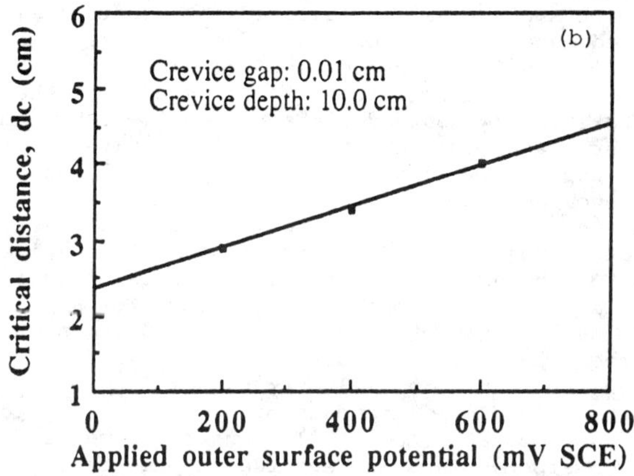

Fig. 3 Effect of the applied potential at the sample's outer surface on the critical distance into the crevice, d_c. (a) High purity iron in a 0.5 M acetic acid + 0.5 M sodium acetate solution. Computation results are in good agreement with the experiments. (b) High purity iron in 1 M ammonia hydroxide + 1 M ammonia nitrate.

of artificial crevice may be used: the straight crevice (Fig. 4b) and the cylindrical crevice (Fig. 4c). The former has been used for a long time in the Corrosion Laboratory of the Department of Materials Science and Engineering, The Pennsylvania State University. The straight crevice is made of a flat pyrex sheet on which a groove, 0.5 cm x 1.0 (or longer, say 2.0) cm x g, was cut, where g is the crevice gap dimension which can be made as small as 25 mm. The grooved face of the pyrex sheet is pressed against the flat surface of the sample. Thus, an artificial crevice of dimensions 0.5 cm x 1.0 cm (or longer) x g cm is formed. One advantage of using the straight crevice is that a camera may be installed in front of the pyrex sheet so that the crevice corrosion process may be observed and recorded in situ [17]. The cylindrical crevice is made of a hole drilled into a bulk component and a core component which is to be inserted into the hole. The core component may be positioned in the center of the hole by an orifice. The crevice is formed between the walls of the hole and the core. The bulk component is made of the testing metal or alloy and the core is made of an insulating high polymer, or the opposite. Alternatively, both the bulk and the core components may be made of the material to be tested. This device is similar to that used by France and Greene [23]. By adjusting the diameters of the hole and the core, the crevice gap may be adjusted.

The Luggin probe is placed close to the crevice opening to minimize the IR voltage between the sample's outer surface and the opening of the capillary. This is particularly important when the solution conductivity is low. The area of the counter electrode should be large enough to make the potential distribution at the sample's outer surface more uniform so that the electrochemical test may best simulate the actual crevice corrosion situation in the natural environment. The test solution can be the one in which the metal is immersed during its service or a specially prepared corrosive solution for accelerating the test. The applied potential at the sample's outer surface must be within the passive region of the anodic polarization curve of the tested material. For comparison of different materials, the applied potential and the crevice gap dimension should be the same. The total current of the circuit is monitored from the beginning of the test. It takes a few hours for the current (which is mainly the current flowing out of the crevice) to reach a relatively stable value. After a period of time (one or two days if the peak current is large, or one or two weeks if the peak current is low), the experiment is terminated and the crevice wall or cross section is checked in an optical microscope to measure the critical distance. Sometimes the current declines to a very low value after an initial high reading, indicating that the crevice has become passivated. For some material/environment combinations, for example, 304 stainless steel in sea water, the anodic polarization curve does not exhibit an active-passive transition and so the initial current reading is low. It takes longer time for the acidification and chloride ion buildup to occur (either of which can decrease the stability of the passive film especially at low overpotentials and enlarge or produce an active potential region, enabling $IR > |\Delta\phi^*|$), and then the crevice corrosion current increases with time.

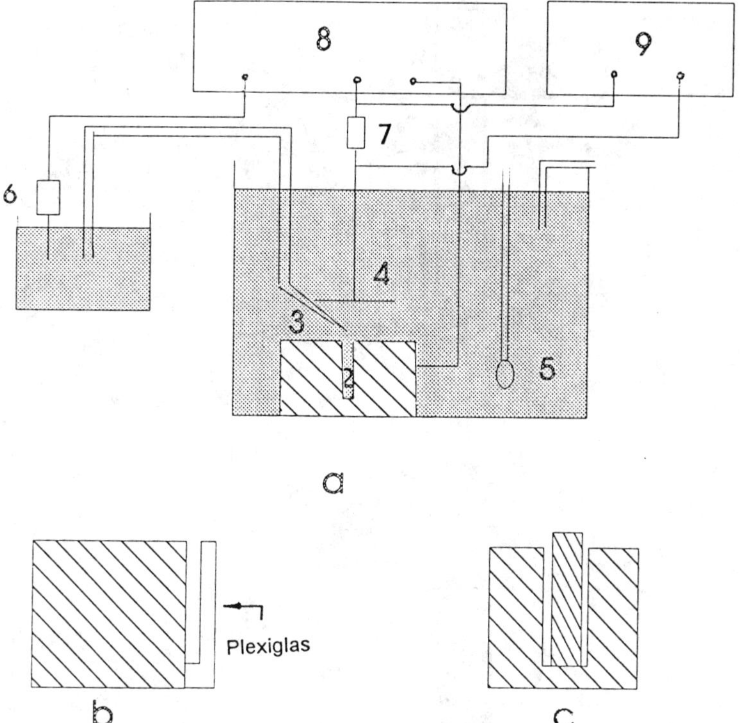

Fig. 4 (a) Experimental set up for measuring the critical distance into the crevice, d_c: 1.working electrode (the sample); 2. crevice; 3. Luggin capillary; 4. counter electrode; 5. gas purger; 6. reference electrode; 7. precision resistor for current measurement; 8. potentiostat; 9. pen recorder. (b) straight crevice with transparent (Plexiglas) wall. (c) cylindrical crevice.

Estimation of d_c by Computation

Another advantage of using d_c is that it can be estimated by computation. The coordinate system for the computation is shown in Fig.5. In the early stage of crevice corrosion, after the initial current transient due to double layer charging and passive film formation, the potential distribution in the system observes the Laplace equation. For a straight crevice, in Cartesion coordinates, the equation is written as:

$$\frac{\partial^2 \varphi}{\partial x^2} + \frac{\partial^2 \varphi}{\partial y^2} = 0$$

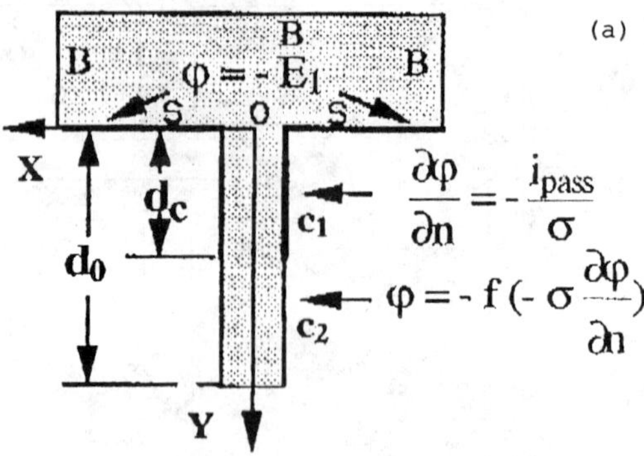

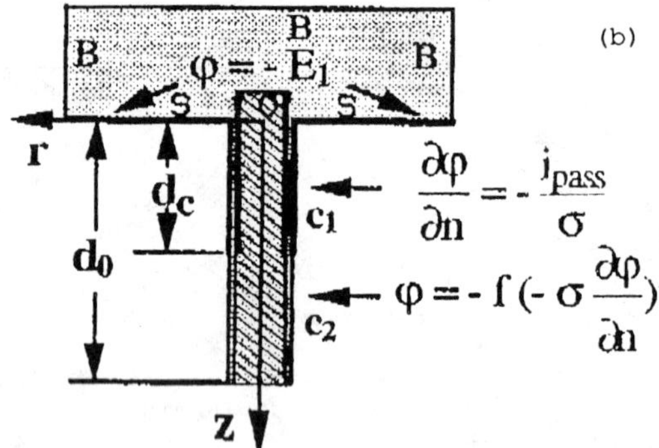

Fig. 5. (a) Boundary conditions for the computation of d_c (Cartesian coordinates, straight crevice). (b) Boundary conditions for the computation of d_c (cylindrical coordinates, cylindrical crevice).

For the cylindrical crevice, in cylindrical coordinates, the Laplace equation has the following form:

$$\frac{\partial^2\varphi}{\partial r^2}+\frac{1}{r}\frac{\partial\varphi}{\partial r}+\frac{\partial^2\varphi}{\partial z^2}=0$$

The boundary condition at the sample's outer surface (S in Fig. 5) is:

$$\varphi_s=-E_1$$

where E_1 is the applied potential at the sample's outer surface. The reference potential of φ is given in Appendix II of reference [15]. Along boundaries encircling the system (B in Fig. 5), the boundary condition is:

$$\left(\frac{\partial\varphi}{\partial n}\right)_B = 0$$

In the upper part of the crevice wall (passivated part, C_1 in Fig. 5), the boundary condition is:

$$\left(\frac{\partial\varphi}{\partial n}\right)_{c1} = -\frac{i_{pass}}{\sigma}$$

where i_{pass} is the passive current and σ is the solution conductivity. In the bottom part of the crevice (active dissolution part, C_2 in Fig. 5), the boundary condition becomes:

$$\varphi_{c2} = -E_{c2} = f(i_{c2}) = f\left(-\sigma\frac{\partial\varphi_{c2}}{\partial n}\right)$$

where $E_{c2} = -f(i_{c2})$ is the active loop portion of the anodic polarization curve. This boundary condition is non linear with respect to the potential and its gradient.

The above Laplace equation contains a non linear boundary condition and so can not be solved by conventional numerical methods. A boundary variation and trial and error technique may be used to solve the problem. The full details of the computation method have been given elsewhere [14-16]. The solution of the Laplace equation is the potential distribution in the system, including that on the crevice wall. Thus the critical distance into the crevice, d_c , where

TABLE 1. Comparison between the computational [15,16] and experimental data [16,18] of the critical distance into the crevice, d_c, and the total crevicing current, I, for pure iron in 0.5 M acetic acid + 0.5 M sodium acetate solution. Crevice gap: g = 0.05 cm; applied potential at the sample's outer surface: E_1 = 115 mV (SCE); crevice length: l = 0.5 cm; crevice depth: d_o = 1.0 cm.

Critical distance, d_c (cm)		Total crevicing current, I (mA)	
Computational	Experimental	Computational	Experimental
0.14	0.12-0.20	1.2	1.2-1.4

the potential is decreased to the passivation potential, E_{pass}, may be obtained. From the potential distribution and the polarization curve, the current distribution along the crevice may be determined. Then by integration, the total crevice corrosion current, I, may also be calculated. Preliminary computation has shown good agreement between the computed and the experimental values. Tables 1 and 2 and Fig. 2a are comparisons of the computation results with experiments for high purity iron in an acetic acid buffer solution [15,16].

It is noted here that the above computation can only give an estimation of the d_c because of three reasons: (1) the polarization curve of the crevice electrolyte should be used in the computation. If the polarization data for the bulk solution is used, the obtained d_c is accurate only at the beginning of the crevice corrosion process when the composition of the crevice solution has not changed appreciably. With the progress of crevice corrosion, the composition and pH of the crevice electrolyte change with time and so does d_c. (2) The shape of the crevice changes with time due to corrosion so that the boundary condition on the crevice wall should also be changed with time. Thus, using the original boundary condition results in additional errors in d_c. Therefore, although the computation can sometimes give a good estimation of d_c, it can not replace the experiment. (3) The model does not consider the mass transfer effect (the so called tertiary distribution). In the later stage of crevice corrosion, the mass transfer effect becomes important, and the model results thus contain larger errors.

TABLE 2. Comparison between the computational and experimental data of the critical distance into the crevice, d_c, for high purity iron on 0.5 M acetic acid + 0.5 M sodium acetate solution at different applied potentials, E_1, at the sample's outer surface. Crevice gap: $g = 0.05$ cm; crevice depth: $d_o = 1.0$ cm.

Applied potential, E_1 (mV SCE)	Critical distance, d_c (cm)	
	Computational	Experimental
115	0.14	0.12-0.20
485	0.37	0.29-0.31
801	0.55	0.41-0.45

Summary

A new index for the corrosion resistance of materials - the critical distance into the crevice, d_c - has been proposed. At d_c the local potential of the crevice wall is the passivation potential in the anodic polarization curve of the material in the crevice electrolyte. At greater distances, the crevice undergoes active anodic dissolution (crevice corrosion) down to the distance of the limiting potential. The larger the d_c, the better the crevice corrosion resistance of the material. Two artificial crevice designs have been introduced which can be used to measure the critical distance into the crevice. The d_c may also be estimated through computation. The advantages in using d_c as the index of crevice corrosion resistance have been discussed.

Acknowledgement

The authors are indebted to their colleagues and the research students in the Corrosion Laboratory, Department of Materials Science and Engineering, The Pennsylvania State University, for their collective work in the theoretical development and experimental justification of the IR induced crevice corrosion mechanism. The work is in part sponsored by the Office of Naval Research, Contract No. N00014 - 91 - J - 1189 (Dr. A. J. Sedriks).

References

[1] "Standard Guide for Crevice Corrosion Testing of Iron-Base and Nickel-Base Stainless Alloys in Seawater and other Cloride-Containing Aqueous Environments", American Society for Testing and Materials, Annual Book of ASTM Standards, **03.02**, G78

[2] A. H. Tuthill, "Resistance of Highly Alloyed Materials and Titanium to Localized Corrosion in Bleach Plant Environments", Mater. Perform., Vol. 24, October, 1985, pp. 43-49

[3] "Standard Test Methods for Pitting and Crevice Corrosion Resistance of Stainless Steels and Related Alloys by use of Ferric Chloride Solution", American Society for Testing and Materials, Annual Book of ASTM Standards, **03.02**, G48

[4] R. S. Treseder and E. A. Kachik, Laboratory Corrosion Tests and Standards, STP 866, American Society for Testing and Materials, 1985, 373

[5] D. B. Anderson, "Statistical Aspects of Crevice Corrosion in Seawater", Galvanic and Pitting Corrosion - Field and Laboratory Studies, STP 576, American Society for Testing and Materials, Philadelphia, Pa., 1976, 231-242.

[6] "Standard Test Method for Conducting Cyclic Potentiodynamic Polarization", American Society for Testing and Materials, Annual Book of ASTM Standards, **03.02**, G61

[7] "Standard Test Method for Pitting or Crevice Corrosion of Metallic Surgical Implant Materials", American Society for Testing and Materials, Annual Book of ASTM Standards, **03.02**, F746

[8] S. Bernhardson, Paper 85, presented at Corrosion/80, Houston TX, NACE, 1980

[9] J. W. Oldfield and W. H. Sutton, "Crevice Corrosion of Stainless Steel", Br. Corros. J., Vol. 13, 1978, pp. 104-111.

[10] T. S. Lee, Electrochemical Corrosion Testing, STP 727, American Society for Testing and Materials, 1981, 43

[11] M. G. Fontana, Corrosion Engineering, 3rd Ed., McGraw-Hill, New York, 1986.

[12] H. W. Pickering, "Significance of the Local Electrode Potential Within Pits, Crevices and Cracks", Corrosion Sci. Vol. 29, 1989, pp. 325-341.

[13] H. W. Pickering, "On the Roles of Corrosion Products in Local Cell Processes", Corrosion, Vol. 42, March, 1986, pp. 125-140.

[14] Yuan Xu, Minghua Wang and H. W. Pickering, "A Mechanism of Pitting Corrosion", Oxide Films on Metals and Alloys, Electrochemical Soc., Pennington, NJ, in press.

[15] Yuan Xu and H. W. Pickering, "The Initial Potential and Current Distribution in the Crevice Corrosion Process", J. Electrochem. Soc., Vol. 140, 1993, pp. 658-668.

[16] Yuan Xu and H. W. Pickering, "A Model of the Potential and Current Distributions Within Crevices and Its Application to the Iron-Ammoniacal System", Critical Factors in Localized Corrosion, G. S. Frankel and R. C. Newman, eds., The Electrochem. Soc., Princeton, NJ, 1991, pp. 389-406; K. Cho and H. W. Pickering, ibid., pp. 407-419.

[17] K. Cho and H. W. Pickering, "The Role of Chloride Ions in the $IR>IR^*$ Criterion for Crevice Corrosion, J. Electrochem. Soc., Vol. 138, 1991, pp. L56-L58.

[18] K. Cho, MS Thesis, The Pennsylvania State University, 1991.

[19] K. Cho, PhD Thesis, The Pennsylvania State University, 1992.

[20] B. J. Fitzgarald, Thesis, University of Connecticut, 1976

[21] C. Edeleanu and J. G. Gibson, Chem. & Ind., 1961, p. 301.

[22] M. N. Folkin and V. A. Timonin, Dokl. Akad. Nauk. SSSR, Vol. 164, 1965, pp. 150-153.

[23] W. D. France and N. D. Greene, "Passivation of Crevices During Anodic Protection", Corrosion, Vol. 24, 1968, pp. 247-251.

Masatsune Akashi[1]

ACCELERATION OF STRESS-CORROSION CRACKING TEST FOR HIGH-TEMPERATURE, HIGH-PURITY WATER ENVIRONMENTS BY MEANS OF ARTIFICIAL CREVICE APPLICATION

REFERENCE: Akashi, M., **"Acceleration of Stress-Corrosion Cracking Test for High-Temperature, High-Purity Water Environments by Means of Artificial Crevice Application,"** Application of Accelerated Corrosion Tests to Service Life Prediction of Materials, ASTM STP 1194, Gustavo Cragnolino and Narasi Sridhar, Eds., American Society for Testing and Materials, Philadelphia, 1994.

ABSTRACT: In order to study the effects of introducing an artificial crevice to the specimen on the stress-corrosion cracking (SCC) test for stainless alloys in the boiling water reactor (BWR) primary coolant environments, uniaxial constant load tests were performed in a high-temperature, high-purity water environment on Type 304 stainless steel and Alloy 600 type Ni-base alloys, both sensitized. On analyzing the lifetime distribution data by fitting to the exponential distribution model, it was found: (1) the lifetime to initiation of SCC can be accelerated by introducing an artificial crevice to the specimen; (2) the acceleration efficacy of graphite-fiber wool as a crevice-forming material is confirmed; (3) the acceleration effect of artificial crevice is stronger for Alloy 600 type Ni-base alloys than the Type 304 stainless steel by a factor of 5 to 6; and (4) by comparison of the performance data in actual environments and the results of analysis conducted on the laboratory acceleration test data, the use of artificially-creviced specimen for the SCC lifetime test has been justified because it effectively shortened the SCC initiation lifetime without losing the factual correspondence to the SCC that occurs in actual service environments.

KEYWORDS: stress-corrosion cracking, stainless steel, Ni-base alloy, BWR environment, artificial crevice, lifetime distribution

In the primary coolant environment of operating boiling water nuclear reactors (BWR), the weld-sensitized Type 304 stainless steel [*1*-*3*] and Alloy 600 type Ni-base alloys [*3*] have been reported to have failed by stress-corrosion cracking (SCC). In the meantime, the fact that initiation of SCC in these systems can be accelerated greatly in the laboratory test on introduction of artificial crevice to the specimen has become well established since Copson and Dean [*4*]. The

[1] Senior research engineer, Research Institute, Ishikawajima-Harima Heavy Industries Co., Ltd. (IHI), 3-1-15 Toyosu, Koto-ku, Tokyo 135, Japan

author and his associates have also been studying the material and other factors that exert influence on the SCC liabilities of various alloys in these environments, mostly with a piece of graphite-fiber wool (GFW) fabric as the artificial crevice former in the creviced bent-beam test method (the CBB test method) [*5,6*].

The purpose of this paper is the quantitative examination of the extent of SCC acceleration achievable on application of an artificial crevice to the specimen in the SCC susceptibility assessment test. The experiments were conducted in a high-temperature, high-purity water environment simulating the actual BWR primary coolant environment using many specimens of Type 304 stainless steel and two Ni-base alloys of Alloy 600 and Alloy 182, and in the uniaxial constant load mode. The SCC initiation lifetime data thus acquired were analyzed for their distribution characteristics by using the exponential distribution as a model.

THE EXPONENTIAL DISTRIBUTION MODEL

For SCC occurring in BWR primary coolant environments, it has been shown that the distribution of initiation lifetime can be described quite well, as for other systems, by the exponential distribution model [*7-10*]. In this model, $F(t)$, the cumulative probability of SCC occurring in a testing time t is expressed as:

$$F(t) = 1 - \exp[-(t - a)/\theta] \quad \text{for } t \geq a \text{, or} \\ = 0 \quad \text{for } t < a, \tag{1}$$

where a is the location parameter, which is the distribution lower limit, and θ is the scale parameter, which is the standard deviation of the distribution. Since the cumulative probability $F(t)$ and the cumulative hazard $H(t)$ are mutually related as:

$$F(t) = 1 - \exp[-H(t)], \tag{2}$$

$H(t)$ can thus be expressed as:

$$H(t) = (t - a)/\theta \quad \text{for } t \geq a \text{, or} \\ = 0 \quad \text{for } t < a. \tag{3}$$

This means that by plotting the data on exponential probability paper shown (Fig. **1**), -namely, taking $H(t)$ on one coordinate axis in an arithmetic scale, then taking the corresponding $F(t)$ on an axis drawn parallel to it, while taking t on the other coordinate axis in an arithmetic scale-, one should obtain a linear array of data points, from which one should be able to determine the location parameter a as the t-intercept of the linear line and the scale parameter θ as the inclination.

EXPERIMENTAL

The samples were two heats of Type 304 stainless steel and one

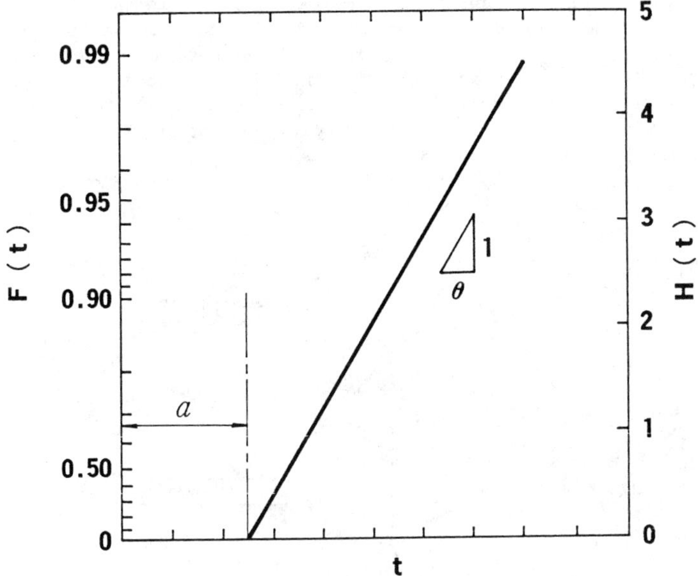

FIG. 1 -*Construction of the exponential probability paper.*

heat each of two Alloy 600 type Ni-base stainless alloys, Alloy 600 and Alloy 182. Their chemical compositions are given in Table I. Heat PA was a seamless pipe of 12 inch in the outer diameter; Heat PB, a welded joint (weld heat-affected zone) of the same type of pipe; Heat BA, a 60 mm thick plate, and Heat WM, a weld overlay metal. They were all sensitized prior to specimen fabrication as follows: for PA, 750 C/1 h + 500 C/24 h; for PB, 500 C/24 h; for BA, 1200 C/24 h + 615 C/10 h + 450 C/200 h; and for WM, 615 C/10 h + 450 C/200 h.

TABLE I -*Chemical compositions of alloys investigated (mass %).*

Heat No.	Alloy Type	C	Si	Mn	P	S	Ni	Cr	Fe	Ti	Nb
PA	304	0.048	0.47	1.48	0.024	0.004	9.25	18.61	Bal.	-----	-----
PB	304	0.062	0.59	1.64	0.020	0.005	9.20	18.27	Bal.	-----	-----
BA	600	0.054	0.12	0.33	0.004	0.004	72.6	16.29	8.12	0.38	0.18
WM	182	0.037	0.44	7.30	0.004	0.003	67.2	13.64	8.41	0.42	1.90

Figure 2 illustrates the design of specimen (a flat tensile bar of an 11 mm long parallel portion) and the testing configuration. The specimens were machined out of the respective samples, polished to #100 emery finish, thoroughly degreased, and cleansed in an ethanol and a distilled water. Some of them were artificially creviced with either GFW or PTFE filter paper as shown in Fig. **2**. The test environment was 250 C high-purity water containing 20 ppm dissolved oxygen. All the SCC lifetime tests were performed by applying periodically the trapezoidal stress waveform shown in Fig. **3**.

In analyzing the fracture lifetime data for the two parameters of the exponential distribution model, the linear unbiased estimator method [*11*] was employed.

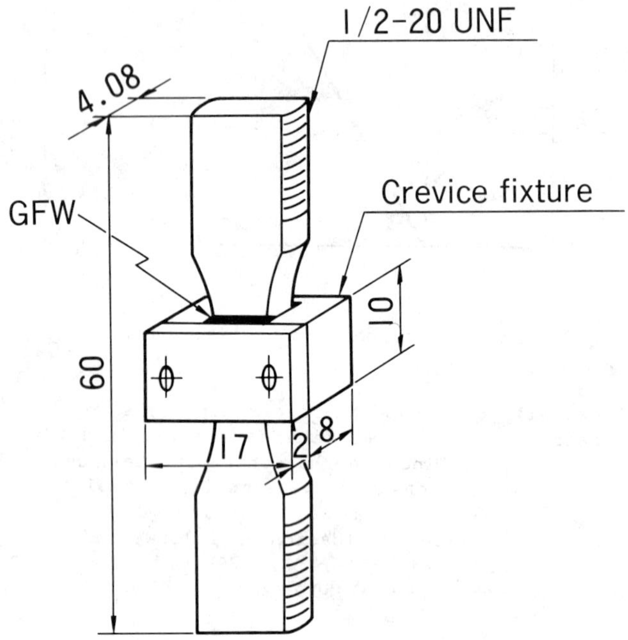

FIG. **2** -*Creviced specimen setup (in mm).*

RESULTS

The SCC lifetime distribution of sensitized 304 stainless steel (Heat PA) is presented in Fig. **4** as an exponential probability plot, comparing the crevice-free case to the GFW-creviced case. On introduction of artificial crevice with GFW, the lifetime to initiating SCC is remarkably shortened, and the two distribution parameters have both become noticeably smaller. A similar response is shown in Fig. **5**, which summarizes the results obtained for sensitized Alloy 600 (Heat BA), except that the acceleration effect of GFW-crevice is more apparent for Alloy 600 than for Type 304 stainless steel. Here, SCC was observed to have progressed in the intergranular mode, both in Type 304 stainless

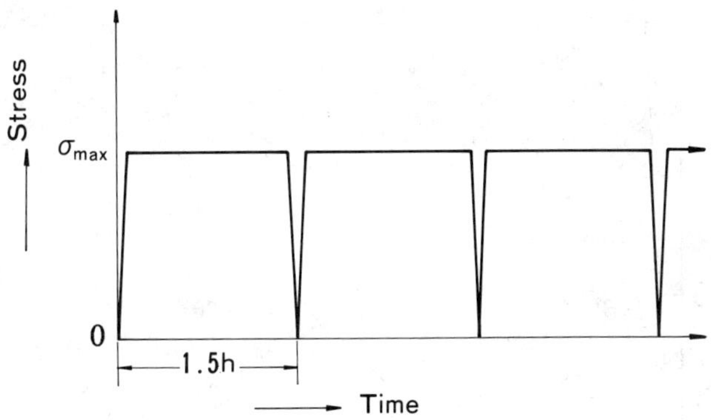

FIG. **3** -*Stress waveform employed during the test.*

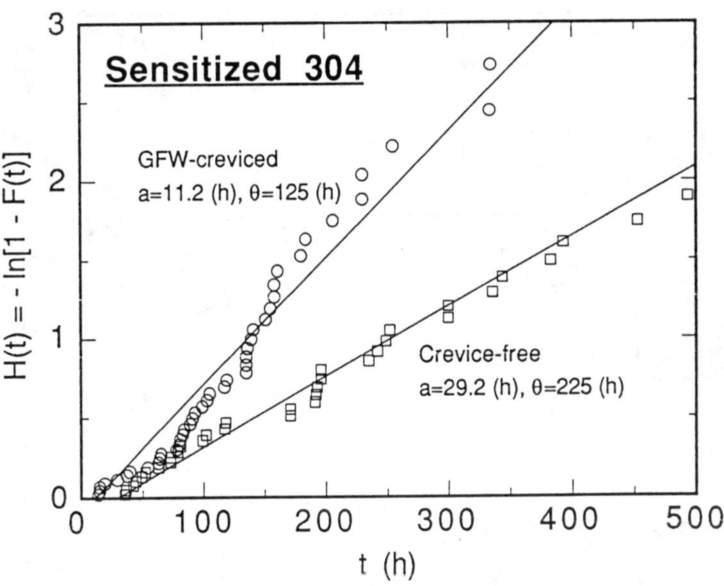

FIG. **4** -*Effect of GFW-crevice application on SCC lifetime distribution of sensitized Type 304 stainless steel ($\sigma_{max}=2S_y$).*

steel and Alloy 600.

Figure **6** compares the effect of GFW-crevice on SCC lifetime between weld-sensitized Type 304 stainless steel (Heat PB) and weld-sensitized Alloy 182 (Heat WM): no statistically significant difference is found. Here, the SCC of Alloy 182 was in the interdendritic mode.

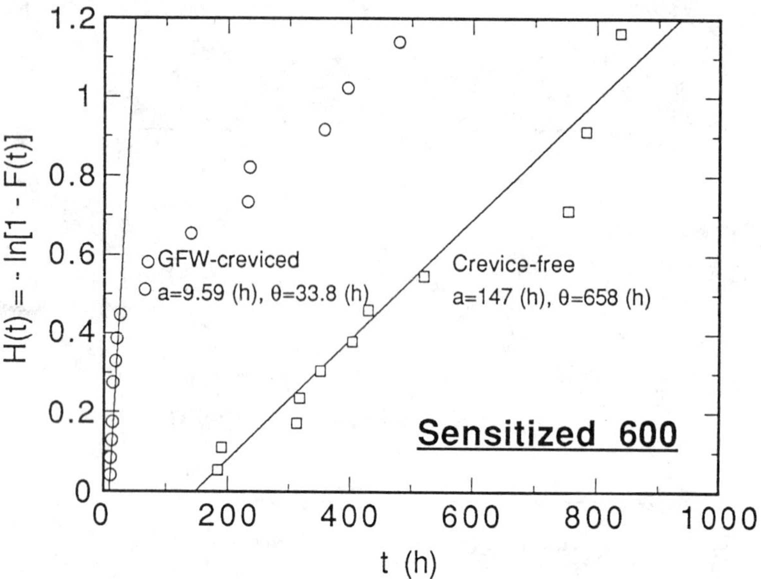

FIG. 5 -*Effect of GFW-crevice application on SCC lifetime distribution of sensitized Alloy 600* ($\sigma_{max}=2S_y$).

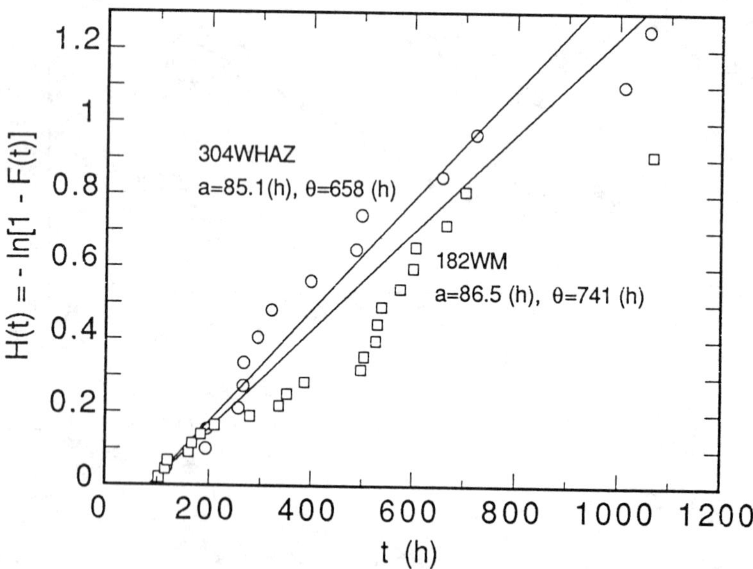

FIG. 6 -*Comparison of SCC lifetimes determined with GFW-creviced specimens between weld sensitized Type 304 stainless steel and weld sensitized Alloy 182* ($\sigma_{max}=2S_y$).

DISCUSSION

That GFW-crevice does accelerate SCC is thus clear. However, there is a possibility of the galvanic effect arising from the direct contact between GFW and metal contributing to the overall acceleration in addition to the crevice effect. To examine this, a filter paper made of PTFE was used as an insulative artificial crevice former. The effects of GFW-crevice and PTFE-crevice are compared in Fig. **7** for sensitized Type 304 stainless steel (Heat PA). The two SCC lifetime distributions are mutually overlapped so much that they are statistically indistinguish-able. This observation was further examined by the CBB test for Type 304 stainless steel sensitized for 24 h at 600 C in terms of the maximum crack depth distribution. The results are shown in Fig. **8** as fit to the Gumbel distribution model [*11,12*]. Here again, the two distributions are statistically indistinguishable. Finally, the free electrode potentials were determined for GFW and Alloy 182 in high-temperature, high-purity water environments. As seen in Fig. **9**, there is no significant difference between the two. These observations should be considered sufficient to conclude that any effect that may arise from the galvanic contact between GFW and specimen can safely be ignored in the present study.

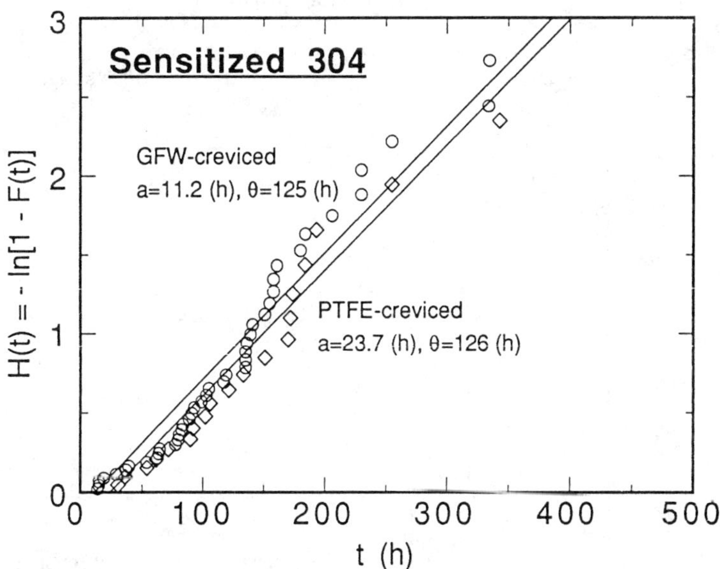

FIG. **7** -*Comparison of effects of GFW-crevice vs. PTFE-crevice for SCC lifetime distribution of sensitized Type 304 stainless steel* ($\sigma_{max}=2S_y$).

As for direct comparison with field experiences, there available are an analysis [*13*] conducted for the distribution of SCC initiation lifetimes of weld-sensitized Type 304 stainless steel in the piping systems of US BWR plants using the exponential distribution model and a

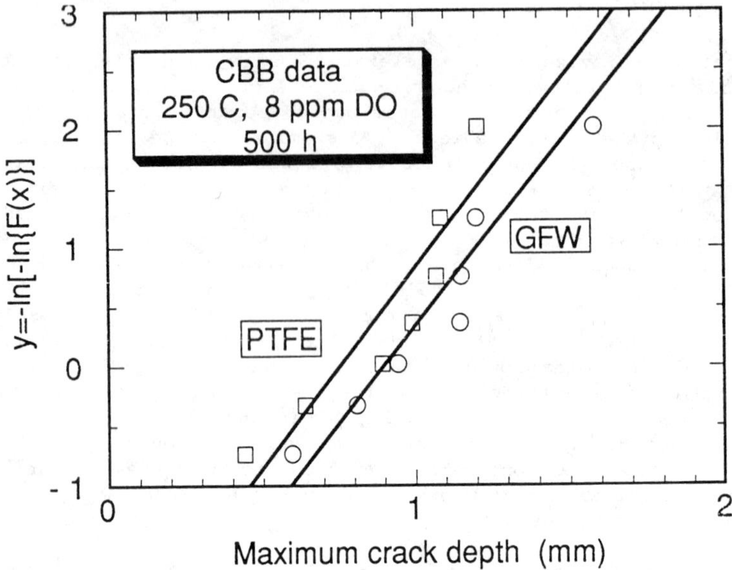

FIG. 8 -Comparison of effects of GFW-crevice vs. PTFE-crevice for maximum crack depth distribution of sensitized Type 304 stainless steel as determined by the CBB test.

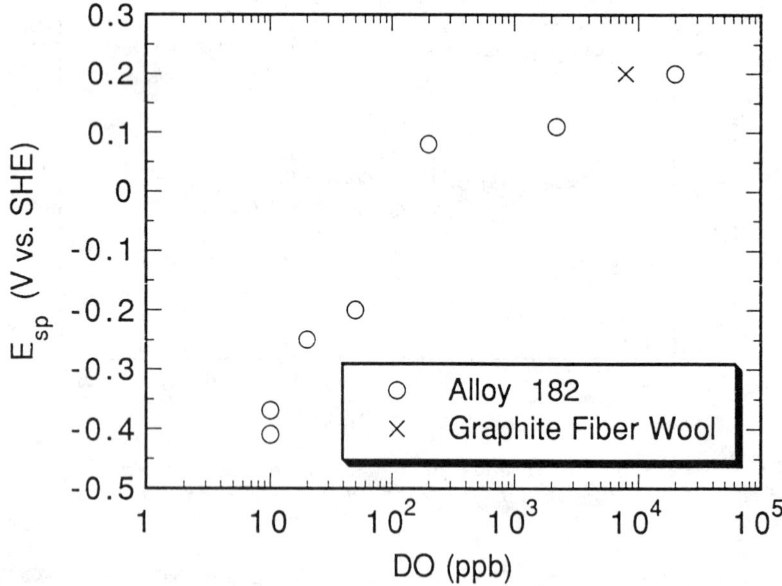

FIG. 9 -Free electrode potential of GFW and Alloy 182 specimens determined in a high-temperature, high-purity water environment.

paper [*3*] reporting an SCC failure found in a weld metal of Alloy 182 at the Duane Arnold-1 reactor. They are represented in Fig. **10**, which was plotted using the reported failure time (37,000 h) and the fact that there are 23 other plants that are operating without inducing similar SCC failure. Here, the straight line for Alloy 182 has been drawn by extending an empirical observation [*5,8-10*] that the ratio between the two distribution parameters a/θ remains approximately the same as long as they pertain to the SCC lifetime distributions, to an assumption that a/θ would remain the same between the Type 304 stainless steel and the Alloy 182.

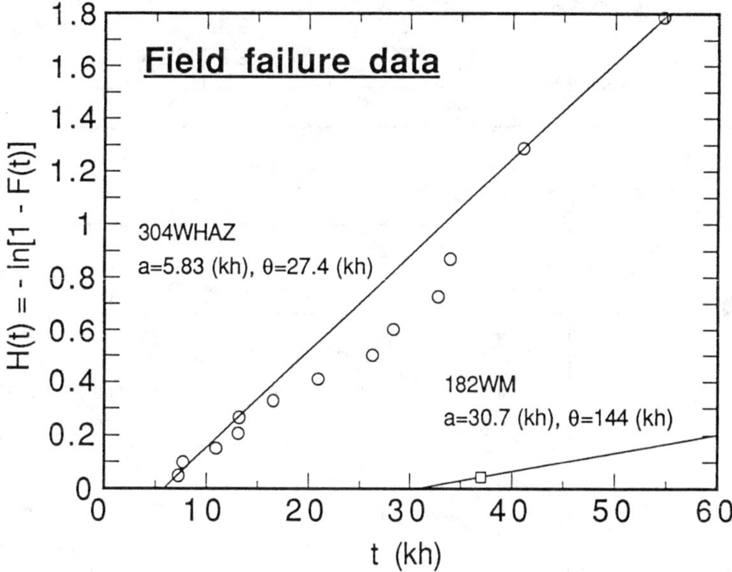

FIG. **10** -*Exponential probability plot of the distribution of SCC failure occurrence in US BWR plants.*

The analytical results on all the SCC lifetime distribution data presented and discussed heretofore are summarized in Table **II**. Of the two distribution parameters, the location parameter *a*, which stands for the distribution lower limit, is of a particular technological importance. Therefore, the effect of introducing an artificial crevice on the parameter *a* has been examined in Fig. **11** as follows.

The results for sensitized Type 304 stainless steel and sensitized Alloy 600 comprise both the crevice-free cases and GFW-creviced cases (Table **II**). Comparing these two, it can be seen that the acceleration of SCC due to an artificial crevice is more effective for Alloy 600 than Type 304 stainless steel by a factor *F* of 5.99. Next, in the case of weld-sensitization, the only experimental data available are the GFW-creviced ones for Type 304 stainless steel and for Alloy 182. Yet, the

TABLE II -*Summary of SCC lifetime distribution analyses.*

Specimen	Crevice	Applied stress	a (h)	θ (h)
Sensitized 304	Free	2.00 Sy	29.2	225
	PTFE	2.00 Sy	23.7	126
	GFW	2.00 Sy	11.2	125
304 Weld HAZ	GFW	2.00 Sy	85.1	658
	Field failure data		5830	27400
Sensitized 600	Free	2.00 Sy	147	658
	GFW	2.00 Sy	9.59	33.8
182 Weld Metal	GFW	2.00 Sy	86.5	741
	Field failure data		30700	144000

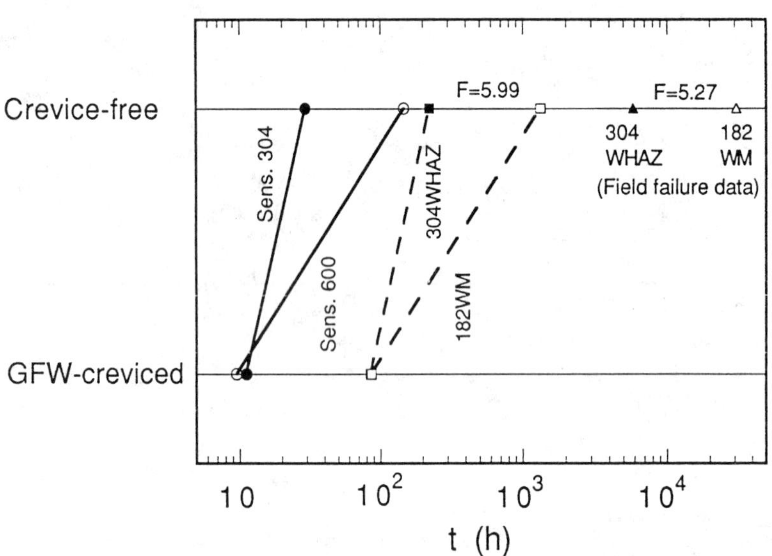

FIG. **11** -*Estimation of the lower limits of SCC lifetime distribution and evaluation of acceleration on introduction of artificial crevice.*

lower limit *a* of the SCC lifetime distribution in the crevice-free case can be estimated reasonably by assuming the parallel translation law to hold true between the furnace sensitized Type 304 stainless steel and weld-sensitized Type 304 stainless steel and between furnace sensitized Alloy 600 and weld-sensitized Alloy 182. This is shown in Fig. **11** with the pair of broken lines: since the *a*'s determined in the laboratory acceleration tests on the use of GFW-crevice for Type 304 stainless steel and for Alloy 182, both as weld-sensitized, agree quite well, the a's estimated as described above for the cases of crevice-free SCC should differ by a factor F of 5.99. Against this, the difference found for the parameter a on the field failure data between type 304 stainless steel and Alloy 182, both as weld-sensitized, is a factor F of 5.27. The agreement between these two values should be taken as supporting the use of an artificially creviced specimen in the SCC lifetime assessment test because it not only effectively accelerates the SCC lifetime but reflects the conditions encountered in the actual service environments.

This observation demonstrates that the use of a laboratory acceleration test performed with artificially creviced specimens effectively shortened the SCC initiation lifetime without losing the factual correspondence to the SCC that occurs in actual service environments. Also, the SCC initiation lifetime of Alloy 600 type Ni-base alloys is more prone to be accelerated by artificial crevice application than the Type 304 stainless steel by a factor of 5 to 6.

SUMMARY AND CONCLUSIONS

In order to study the effects of introducing an artificial crevice on the stress-corrosion cracking (SCC) test for stainless alloys in boiling water reactor (BWR) primary coolant environments, uniaxial constant load tests were conducted for sensitized Type 304 stainless steel and two sensitized Alloy 600 type Ni-base stainless alloys, Alloy 600 and Alloy 182. On analyzing the lifetime distribution data by fitting to the exponential distribution model, the following conclusions are drawn.

1. The SCC initiation lifetime can be accelerated by introducing an artificial crevice to the specimen, wherein the efficacy of graphite fiber wool as a crevice forming material has been well confirmed.
2. The Alloy 600 type Ni-base alloys are more prone to acceleration by artificial crevice application than the Type 304 stainless steel by a factor of 5 to 6.
3. On comparison of the performance data in actual service environments and the results of analysis conducted on the laboratory test data, the use of artificially creviced specimen for the SCC lifetime test has been justified as effectively shortening the SCC initiation lifetime without losing the factual correspondence to the SCC that occurs in actual service environments.

Acknowledgment

The author gratefully acknowledges the aid extended for him in performing the stress-corrosion cracking tests by Messrs. T. Kenjyo, T. Sudo, S. Matsukura, and K. Miyokawa.

REFERENCES

[1] Bush, S.H., and Dillon, R.L., "Stress Corrosion in nuclear Systems," USAEC Report, Washinton, D.C., 1973.

[2] Klepfer, H.H., *et al.*, "Investigation of Cause of Cracking in Austenitic Stainless Steel Piping," General Electric Company Report, No. NEDO 21000-1, 1975.

[3] U.S. Nuclear Regulatory Commission Pipe Crack Study Group, "Investigation and Evaluation of Stress-Corrosion Cracking in Piping of Light Water Reactor Plants," U.S. Nuclear Regulatory Commission, Washington, D.C., No. NUREG-0531, 1979.

[4] Copson, H.R., and Dean, S.N., *Corrosion*, Vol. 21, No. 1, 1965, pp. 1-8.

[5] Akashi, M., "CBB Test Method for Assessing the Stress Corrosion Cracking Susceptibility of Stainless Steels in High-Temperature, High-Purity Water Environments," *Localized Corrosion -Current Japanese Materials Research 4*, F. Hine, K. Komai, and K. Yamakawa, Eds., Soc. Mat. Sci. Jap., Elsevier Applied Science, London and New York, 1988, pp. 175-196.

[6] Akashi, M., "Alloy Chemistry Effects on the Stress-Corrosion Cracking Susceptibility of Nickel-Base Weld Metals in High-Temperature, High-Purity Water environments," presented at *International Conference on Advances in Corrosion and Protection*, G.C. Wood, Ed., Manchester, UK, 1992.

[7] Akashi, M., Kenjyo, T., and Kawamoto, T., *Corros. Eng. (Jpn.)*, Vol. 33, No. 11, 1984, pp. 628-634.

[8] Akashi, M., and Ohtomo, A., *J. Soc. Mater. Sci., Jpn.*, Vol. 36,No. 400, 1987, pp. 59-64.

[9] Nakayama, G., Ohtomo, A., and Akashi, M., *ISIJ International*, Vol. 31, No. 2, 1991, pp. 223-228.

[10] Akashi, M., "An Exponential Distribution Model for Assessing the Stress Corrosion Cracking Lifetime of BWR Component Materials," presented at *3rd NACE International Relations Committee Symposium on Life Prediction of Corrodible Structures*, R.N. Parkins, T. Shibata, and R.W. Staehle, Eds., Cambridge, UK, 1991.

[11] Kowaka, M., Tsuge, H., Akashi, M., Masamura, K., and Ishimoto, H., *Introduction to Life Prediction of Plant Materials -Application of Extreme-Value Statistics to Corrosion*, Japan Soc. Corros. Eng., Ed., Maruzen, Tokyo, 1984.

[12] Tsuge, H., Akashi, M., Ikematsu, K., Nakajima, H.,Saitoh, N., and Miyake, Y., *Proc. 73rd JSCE Symp.*, Japan Soc. Corros. Eng., Tokyo, 1988, pp. 93-125; Micro-Computer Code "EVAN," Japan Soc. Corros. Eng., Ed., Maruzen, Tokyo, 1989.

[13] Okamoto, A., Akashi, M., and Kitagawa, M., "Probabilistic Approach for the Defect Identification in BWR Pipings," *The Pressure Vessel and Piping Conf.*, ASME, 1987, Paper No. 87-PVP-30.

Pertti A. Aaltonen,[1] Pekka K. Pohjanne,[1] Seppo J. Tähtinen,[1] Hannu E. Hänninen[2]

APPLICATION OF A MULTIPOTENTIAL TEST METHOD FOR RAPID SCREENING OF AUSTENITIC STAINLESS STEELS IN PROCESS ENVIRONMENTS

REFERENCE: Aaltonen, P. A., Pohjanne, P. K., Tahtinen, S. J., and Hanninen, H. E., **"Application of a Multipotential Test Method for Rapid Screening of Austenitic Stainless Steels in Process Environments,"** Application of Accelerated Corrosion Tests to Service Life Prediction of Materials, ASTM STP 1194, Gustavo Cragnolino and Narasi Sridhar, Eds., American Society for Testing and Materials, Philadelphia, 1994.

ABSTRACT: The advantage of using MultiPotential Test, MPT, method together with, for example, double U-bend specimens is that in a single exposure a wide variety of corrosion phenomena can be screened in any environment. The MPT method has been applied in several tests both in simulated environments and in real process conditions. The measurement of threshold potential for stress-corrosion cracking of cold deformed and sensitized stainless steels in simulated boiling water reactor environment (BWR) is an example of successful applications of the MPT method. The MPT method applies a current source combined with potentiostat and resistors to polarize different specimens to desired potentials. The electrochemical potential range, which can be obtained, depends on the electrical instrumentation and also on the conductivity of the electrolyte.

KEYWORDS: Corrosion, Stainless Steels, Environment sensitive cracking, Potential Control

[1]Senior Research Scientist, Research Scientist, Senior Research Scientist, respectively, Technical Research Centre of Finland (VTT), Metals Laboratory, P.O. Box 26, SF-02151 Espoo, Finland.

[2]Professor, Helsinki University of Technology, Laboratory of Engineering Materials, Puumiehenkuja 3, SF-02150 Espoo, Finland.

INTRODUCTION

Material selection for process industries widely applies immersion tests in real environments. Immersion tests are best for eliminating from further testing materials that obviously can not be used. However, the corrosion resistance of passivating materials such as stainless steels is additionally influenced by the presence or absence of oxidizing species, that is, by corrosion potential. In this kind of situation, local corrosion reactions, for example, stress-corrosion cracking, pitting and/or crevice corrosion as well as hydrogen embrittlement may affect materials compatibility.

Conducting traditional immersion tests with coupons, U-bend or double U-bend specimens make it possible to evaluate materials sensitivity to corrosion. To evaluate the risk of localized corrosion of materials in a specific environment, separate immersion tests should be carried out in several modified and carefully controlled environments. However, double U-bend specimens together with applied electrochemical polarization to various potentials using the MultiPotential Test (MPT) method make it possible to examine all corrosion processes during a single exposure.

The MPT technique is based on the fact that localized corrosion phenomena take place at different electrochemical potentials. By using polarization with the MPT technique, different alloys in various metallurgical conditions can be tested for general corrosion, crevice and pitting corrosion as well as stress-corrosion cracking and hydrogen embrittlement, Fig. 1.

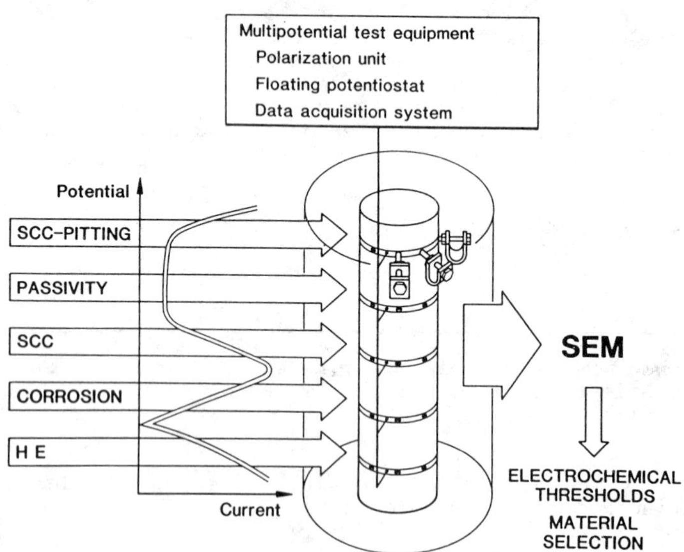

FIG. 1--The principle of MPT method to evaluate corrosion processes taking place at different potentials.

The MPT method applies a DC source (~10 mA) combined with a potentiostat and resistors (~30 Ω) to polarize different specimens to desired potentials. The potential range which can be covered depends on the electrical instrumentation and on the conductivity of the electrolyte. After exposure, the specimens are studied to evaluate the degree of different forms of corrosion.

The purpose of this paper is to present the MPT technique for corrosion testing which can be used both in the laboratory and in the field.

EXPERIMENTAL PROCEDURE

Principle of the MPT method

MultiPotential Test employs several test specimens exposed at the same time in the same test device, but polarized to different potentials. Therefore, corrosion test variables which are temperature, solution, or test history dependent in any way, are common to all the specimens, and the only significant variable is the applied potential of the test specimens. Fig. 2 shows the electrical configuration used for MPT method. A potentiostat is used to control the applied potential of one set of test specimens. The remaining sets of specimens are biased to other potential levels, relative to the potentiostatically controlled set of specimens by adjusting a combination of current level and resistor settings to obtain the desired potential levels. The adjused potential levels should be checked by using separate reference electrode to verify, that the target potentials have been achieved.

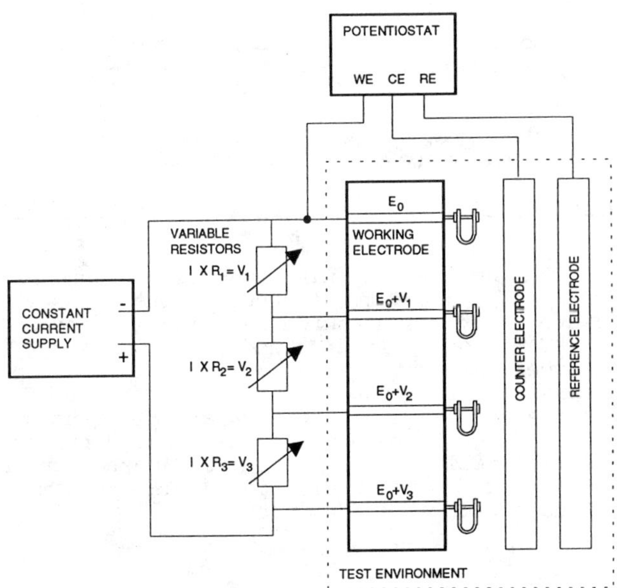

FIG. 2--Schematic of the MPT set-up to obtain 4 different potential levels.

MultiPotential Tests can be performed at ambient temperatures, but they can also be carried out in high temperature, high pressure autoclaves. By adding one set of samples, which are not polarized, the open circuit corrosion potential of the test materials during the test period can also be measured. Based on the measured open circuit corrosion potentials, and observations of periodically conducted specimen inspections, information concerning the kinetics of the reactions can even be obtained. Usually, the first screening tests cover a wide potential range, which can then be refined in the following tests.

The MPT experiment should be carried out in the real process streams to ensure that all chemical species are included in the environment. In simulated environments special attention should be paid to careful control of the environment. To conduct electrochemical tests in process equipment or autoclaves it is necessary to use floating potentiostats and proper insulation of specimens from the test chambers.

RESULTS

MPT method has been successfully applied at ambient as well as at elevated temperatures in material studies for nuclear and process industries to find, for example:

- threshold corrosion potentials for intergranular stress-corrosion cracking (IGSCC) in sensitized and cold deformed stainless steels in pure BWR water,

- SCC susceptibility of compound tubes in black liquor recovery boiler environments,

- hydrogen embrittlement susceptibility of ferritic heat exchanger tube materials in sea water under applied cathodic potentials, and

- critical potentials for various forms of corrosion of austenitic stainless steels in boiling NaCl solutions.

Critical potential for IGSCC of austenitic stainless steel in BWR-environments

The critical stress-corrosion cracking potential, below which IGSCC does not occur, for austenitic stainless steel in high temperature water was measured by using MPT and SSRT methods [1,2,3]. The MPT method applied U-bend specimens for IGSCC testing in high temperature water, Fig. 3, taken from a pipe, which had been in service for 45,000 h at 230 °C in Ringhals I BWR plant. The degree of sensitization and carbide structure in the heat affected zone (HAZ) of the weld was studied by the electrochemical potentiokinetic reactivation (EPR) method and by transmission electron microscopy (TEM), Fig. 3. The test material was welded stainless steel, AISI 304, pipe.

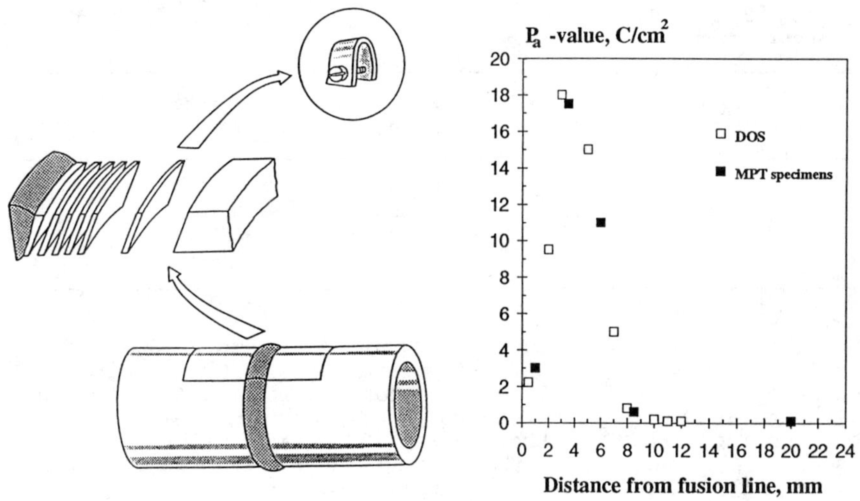

FIG. 3--Manufacturing of specimens for MPT experiment (a). The degree of sensitization (DOS) as a function of distance from the weld fusion line and the location of cut U-bend specimens (b).

The MPT exposure in the low conductivity pure BWR water (no water clean-up during the test) at 288 °C in a static (low O_2 content) autoclave was conducted first for 50 days at the following potentials: -300, -275, -250, -225 and -200 mV_{SHE}. When the optical microscopy inspection after this test period revealed no cracking, the test potentials were changed as follows: -250, -200, -150, -100 and 0 mV_{SHE}. The second exposure time was 60 days. The observed stress-corrosion cracking is summarized in Table 1.

TABLE 1--MultiPotential Test results for U-bend stainless steel specimens exposed to pure water at 288 °C for 60 days. (N= no cracking, P=pitting, and C=cracking).

Test potential mV_{SHE}	Distance from the fusion line (mm)				
	1.0	3.5	6.0	8.5	20.0
0	N+P	N+P	C	N	N
-100	N+P	C	C	N	N
-150	N	N	N	N	N
-200	N	N	N	N	N
-250	N	N	N	N	N

The SSRT testing at controlled electrode potentials was performed in high purity water at 250 °C. The inlet conductivity of the water was less than 0.08 µS/cm. The tests were performed in a 0.8 l autoclave in air-saturated deionized water with a flow rate of 5 l/h. The results of the SSRT tests are summarized in Table 2.

TABLE 2--SSRT test results of austenitic stainless steel weld specimens in BWR water at 250 °C.

Specimen No.	Electrode potential mV_{SHE}	Elongation rate mm/min	Elongation to fracture %	Fracture surface
NKA 2	-250	5.8×10^{-4}	50	Dimple
NKA 3	-250	5.8×10^{-4}	60	Dimple
NKA 4	Air-sat	7.5×10^{-5}	4	IGSCC
NKA 5	Air-sat	7.5×10^{-5}	12	IGSCC

SSRT tests of specimens from a pipe weld removed from Ringhals 1 BWR-plant in high purity water at 250 °C at controlled potentials of -250 and -200 mV_{SHE} did not result in IGSCC. Testing in air-saturated water gave entirely intergranular fracture surfaces. Testing with MPT method showed IGSCC at 288 °C at potentials of 0 mV_{SHE} and -100 mV_{SHE}. Cracking was observed in specimens taken from the distance of 3.5 to 6 mm from the fusion line.

Both tests revealed IGSCC at the potentials corresponding high oxygen content. No cracking was observed at -200 mV_{SHE} independent of test method. MPT exposure showed cracking at the location of the highest DOS near the original inner surface of the pipe, which was not included in SSRT specimens, since round specimens were used.

Stress-corrosion cracking of cold worked austenitic stainless steel in BWR environment

Surface cold work due to grinding, machining, scratching or cold bending operations has been shown to cause stress-corrosion cracking failures of austenitic stainless steel tubes in BWR water environment [4,5]. Cracks initiate typically transgranularly and propagate intergranularly. Cold work results in major changes in the mechanical properties due to work hardening and in the microstructure due to strain-induced martensite structures. Cold work is also known to enhance sensitization.

The test material had been 11 years in service in Ringhals 1 BWR-plant at the operating temperature of 240 °C. The material was 88.9 x 8.8 mm, solution annealed, seamless AISI 304 steel tube, which had been cold bend to angles up to 90 degrees over a mandrel without any subsequent heat treatment. The microstructure and stress-corrosion properties were determined as a function of the tube wall thickness emphasizing the effect of heavily deformed surface layers on the crack initiation. Hydrogen changing was used to simulate the effect of the mechanically cold worked surface layers on the crack initiation.

The MultiPotential Test results after 93 days exposure in a BWR environment at four different potential levels -300, -100, 0 and +150 mV_{SHE} and SSRT results are summarized in Tables 3 and 4. Specimens from the unscratched, smooth inner surface of the bend tube showed ductile behavior in slow strain rate tests and no cracks were detected in MPT experiment. Specimens from the middle of the tube wall showed TGSCC in slow strain rate tests with lowest strain rate used, 2 x 10^{-7} s^{-1}, but no cracks were found in the MPT experiment.

IGSCC was found in MPT exposure only in those inner surface specimens, which had either the original mechanically scratched surface by the mandrel contact or hydrogen charged surfaces. Shallow transgranular surface cracking was present in all specimens with pretreated surfaces already after bending of the U-bend specimens.

During the MPT exposure the environment was monitored using an additional sample for open circuit corrosion potential measurement. The variation of the measured potential during 93 days is shown in Fig. 4.

The microstructure of the bend tube wall varied through the tube wall thickness as can be seen in Fig. 5. The main difference in deformation structures between the inner and outer surfaces of the bend tube was the amount of **α**'-martensite. The amount of **α**'-martensite due to bending was high near the inner surface and gradually decreased to the outer surface of the bend tube. Intergranular corrosion tests, oxalic acid and EPR-tests, showed grain boundary attack near the outer surface of the bend tube wall indicating that the material was slightly sensitized. This was also confirmed by transmission electron microscopy observations of chromium rich precipitates on some of the austenite grain boundaries.

These results indicated that stress-corrosion crack initiation is difficult and material in the middle of the tube wall is somewhat more susceptible to TGSCC as compared to the unscratched, smooth inner surface material. The MPT and SSRT results showed that surface pretreatment, for example, mechanical scratching by mandrel contact or hydrogen charging enhanced crack initiation. It can be concluded, that the heavily deformed martensitic surface layer is needed for crack initiation. Once the cracks have initiated they can grow intergranularly even in the very slightly sensitized material in BWR-water.

TABLE 3--Summary of the MPT results of cold bent AISI 304 steel tube in BWR water at 288 °C with U-bend specimens. (C = IGSCC, SC = surface cracking, N = no cracking).

Potential mV_{SHE}	Surface, scratch	Surface, no scratch	Middle	Surface hydrogen charged	Middle hydrogen charged
+150	SC+C	N	N	SC	SC
0	N	N	N	SC	SC+C
-100	SC	N	N	SC	SC
-300	SC	N	N	SC	SC

TABLE 4--Summary of the SSRT results of cold bent AISI 304 steel tube in BWR water at 288 °C.

Specimen	Strain rate 10^{-6} s^{-1}	Failure time h	Total elong. %	Maximum stress MPa	Applied potential mV_{SHE}	TG fracture %
smooth surface	1.0	78.8	27.8	575	0	0
smooth surface	0.2	412.9	26.6	580	-191	0
smooth surface	1.0	75.8	26.6	575	-214	15
middle	1.0	71.2	25.0	570	-151	13
middle	0.2	225.3	16.0	550	-142	54
middle	1.0	74.7	26.4	560	-140	18
smooth surface hydrog.	1.0	58.6	20.3	570	-25	47

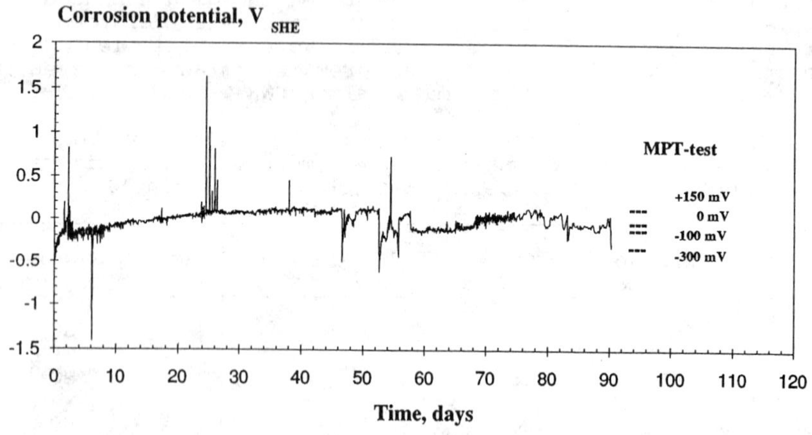

FIG. 4--Measured corrosion potential of cold worked austenitic stainless steel during 93 days MPT test in pure water with 200 ppb O_2 at 288 °C. Note also the applied MPT potentials.

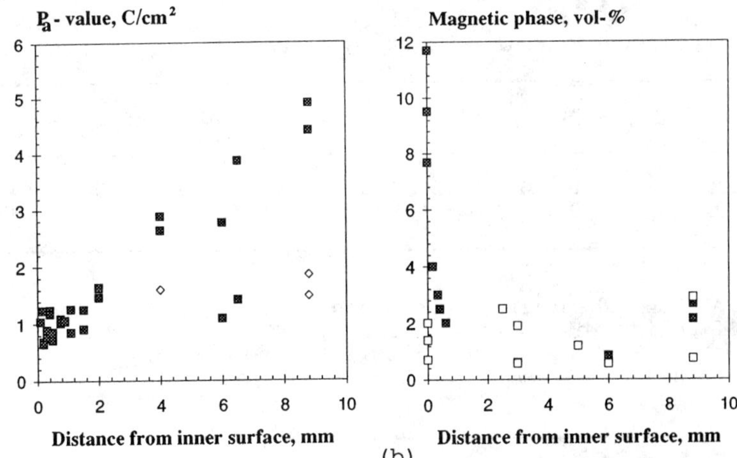

FIG. 5--Results of the EPR measurements through the bend (closed symbols) and straight (open symbols) tube wall (a). The ferrite detector measurements through the bend tube wall on the scratched (closed symbols) and smooth (open symbols) surfaces (b).

Cracking of compound tubes in black liquor recovery boilers

Compound tubes are employed in recovery boilers for protecting the tubes against fireside corrosion in the lower furnace. Compound tubes manufactured by hot extrusion consist typically of 1.5 mm thick AISI 304 or 304L outer layer on 3.5 mm to 5 mm thick carbon steel (ASTM A210 Gr. A1) base material. Recently cracks have widely been found in the austenitic cladding of the compound floor tubes of the recovery boilers. The susceptibility of compound tubes to stress-corrosion cracking was studied in the laboratory by immersion tests in simulated boiler washing water conditions [6]. Test material was a commercially produced compound tube (Table 5). The test solution was made by dissolving 193 g solidified smelt in 1 liter of distilled water. The solidified smelt was received from two Finnish recovery boilers. The chemical compositions of test solutions are presented in Table 6. Tests were performed with two pH-values: pH = 13 and pH = 10. Hydrochloric acid (HCl) was used to adjust the pH-value to 10.

Stress-corrosion tests were conducted using C-ring specimens, cut from the compound tube. The side surfaces of the specimens were ground with 600-grit emery paper and the outer surface was left in as-received condition. The side surfaces and the inner surface of the specimens were coated with a silicon coating in order to prevent corrosion of carbon steel. Specimens were stressed to a desired stress level (300 MPa) according to the ASTM standard G 38-73. Stress level was also checked with X-ray diffraction measurements on some samples. A MPT method allowing testing of several samples in different potentials at the same time was employed. The test duration was 168 hours and the temperature was 80 °C.

TABLE 5--Chemical composition of the compound tube used in the stress-corrosion cracking tests (wt.-%).

Material	%C	%Cr	%Ni	%Mn	%Si	%S	%P
Stainless steel	0.042	18.5	10.0	1.24	0.51	0.004	0.014
Low alloy steel	0.21	-	-	0.70	0.21	0.015	0.013

TABLE 6--Chemical compositions of the test solutions, 193 g of solidified smelt was dissolved in 1 liter of distilled water. Test solution A = solidified smelt from recovery boiler A and test solution B = solidified smelt from recovery boiler B.

Element	Solution A	Solution B
Cl^- (g/l)	0.66	1.65
SO_4^{2-} (g/l)	15	24
S^{2-} (g/l)	8.4	7.1
CO_3^{2-} (g/l)	35	57
Na^+ (g/l)	69	64
K^+ (g/l)	5.7	4.1
pH (25 °C)	12.5	13
Conductivity (mS/m, 25 °C)	11400	12800

Results from the stress corrosion tests performed in simulated water washing solutions are presented in Table 7. Stress-corrosion cracking was observed only in washing solutions where the pH-value was adjusted to 10 with an HCl addition. In solution "A", stress-corrosion cracking and pitting corrosion were observed at potentials -20 mV_{SHE} and 180 mV_{SHE}. In other specimens, potentials -220 mV_{SHE} and -420 mV_{SHE}, no corrosion or cracking was observed. Tests performed in solution "B" at pH 10 showed indications of stress-corrosion cracking and pitting at much lower potentials than in solution "A". In these tests cracks were observed only at potential -260 mV_{SHE} and pitting corrosion was observed at potentials above -160 mV_{SHE}. All observed large cracks were branched, transgranular cracks and their fracture surfaces were similar to those of stress-corrosion cracks found in austenitic stainless steels in the floor tubes of recovery boilers.

In tests without lowering the pH-value, solution "B" at pH 13, neither stress-corrosion cracking nor pitting corrosion was observed at potentials -360 to 140 mV_{SHE}. All samples were almost uncorroded and only slight general corrosion was observed.

TABLE 7--Stress-corrosion tests performed with compound tube in simulated water washing conditions: C-ring specimens, T = 80 °C, solidified smelt 193 g/l. (P = pitting, C = stress-corrosion cracking, N = no corrosion, A = solidified smelt from boiler A, B = solidified smelt from boiler B).

Test	Sample no.	pH	Potential mV_{SHE}	Duration h	Type of corrosion
1^A	1	10	180	145	P+C
	2		-20	145	P+C
	3		-220	145	N
	4		-420	145	N
2^B	19	10	140	168	P
	20		40	168	P
	21		-60	168	P
	22		-160	168	P
	23		-260	168	C
	24		-360	168	N
3^B	36	13	140	168	N
	35		40	168	N
	34		-60	168	N
	33		-160	168	N
	32		-260	168	N
	31		-360	168	N

<u>Threshold potentials for corrosion of stainless steels in saturated NaCl solution at 100 °C</u>

The effect of potential on corrosion behavior of austenitic and duplex stainless steels was studied in hot chloride solutions [7]. Test materials were two austenitic AISI 304 and 316 type steels and one cast duplex stainless steel (22% Cr, 5.5% Ni, 3% Mo) in mill annealed condition. Tests were performed in saturated NaCl solution with two pH-values, pH = 3 and pH = 6, at temperature of 100 °C using U-bend and double U-bend specimens. Test duration in all tests was 168 hours.

The summary of tests performed with these different stainless steels are presented in Tables 8, 9 and 10. The AISI 304 type stainless steel was susceptible to stress-corrosion cracking in saturated NaCl solution at pH = 6, when the potential was between -150 and -50 mV_{SHE}. No stress-corrosion cracking was observed at potential -200 mV_{SHE}. At pH-value 3 cracking was also observed at potentials from -150 to -50 mV_{SHE}. No stress-corrosion cracking was observed at potential -200 mV_{SHE}. In both solutions cracks were observed both in the inner specimens and in the outer specimens of the double U-bend specimens. In addition to cracks also pitting and crevice corrosion were observed.

The AISI 316 type stainless steel was not susceptible to stress-corrosion cracking at potentials between -200 and -50 mV_{SHE} in a saturated NaCl-solution at pH 6. In the solution with lowered pH-value, pH = 3, cracking was observed at potentials from -100 to -20 mV_{SHE}. In all cracked specimens, cracks were observed only in the inner specimens of the double U-bend specimens, and in addition to cracks, crevice corrosion was observed.

Also the cast duplex stainless steel was immune to stress corrosion cracking at potentials between -200 and -50 mV_{SHE} in saturated NaCl-solution at pH = 6. In the solution with lowered pH-value, pH = 3, cracking was observed at potentials from -30 to 50 mV_{SHE} and intensive pitting corrosion was present at potential 100 mV_{SHE}. In addition to cracks, pitting and crevice corrosion were observed.

TABLE 8--Effect of potential on corrosion behavior of AISI 304 type stainless steel in saturated NaCl solution at temperature of 100 °C. (Double U-bend specimens, N = no corrosion, P = pitting, C = crevice corrosion, G = general corrosion, SCC = stress-corrosion cracking).

Material	pH	Potential mV_{SHE}	Type of corrosion
AISI 304	6	-200	N
	6	-150	P+C+SCC
	6	-100	P+C+SCC
	6	-50	P+C+SCC
	3	-200	N (+C)
	3	-150	P+C+SCC
	3	-100	P+C+SCC
	3	-50	P+C+SCC
	3	0	P+C+SCC

TABLE 9--Effect of potential on corrosion behavior of AISI 316 type stainless steel in saturated NaCl solution at temperature of 100 °C. (Double U-bend specimens, N = no corrosion, P = pitting, C = crevice corrosion, G = general corrosion, SCC = stress-corrosion cracking).

Material	pH	Potential mV_{SHE}	Type of corrosion
AISI 316	6	-200	N
	6	-150	N
	6	-100	N
	6	-50	N
	3	-150	N
	3	-100	C+SCC
	3	-50	C+SCC
	3	-20	C+SCC

TABLE 10--Effect of potential on corrosion behavior of cast duplex stainless steel in saturated NaCl solution at temperature of 100 °C. (U-bend specimens, N = no corrosion, P = pitting, C = crevice corrosion, G = general corrosion, SCC = stress-corrosion cracking).

Material	pH	Potential mV_{SHE}	Type of corrosion
Cast duplex stainless steel	3	-100	N
	3	-50	SCC
	3	-30	P+SCC
	3	50	P+SCC
	3	100	P+G

DISCUSSION

The MPT technique applied with constant displacement specimens, for example, U-bend, C-ring, bent beam specimens gives information on potential dependence of corrosion phenomena and crack initiation [8]. An additional advantage of the use of constant displacement specimens is the possibility to define the effect of surface condition, for example, as-received, machined, ground or scratched surfaces on crack initiation. The SSRT as such is one of the most severe test method for stress-corrosion cracking and it requires numerous time consuming experiments giving only limited information, for example, on initiation of environmental assisted cracking.

The most reliable MPT results are obtained by conducting immersion tests in real process streams. In the laboratory tests, it is essential to control the chemical species, pH and conductivity in addition to temperature carefully. However, it has been shown that it is not the presence of oxygen, per se, but corrosion potential which controls the initiation and propagation of SCC in passivating materials [9]. This makes it possible to conduct MultiPotential Tests in simulated environments where oxygen is consumed during testing.

The MultiPotential Test system has been shown to give reliable and reproducible threshold potential values for pitting and stress-corrosion cracking for stainless steels in various environments. Until now the MPT technique is most often used with U-bend or double-U-bend specimens, but also other commonly used constant displacement specimens, for example, C-ring, bent beam specimens or precracked fracture mechanical specimens, such as wedge-opening-loaded (WOL) specimens can be used. The full applicability of the MPT technique can be realized by utilizing both constant displacement and precracked fracture mechanical specimens to obtain detailed design criteria and life predictions in well defined stress and environment conditions.

MPT technique has also been successfully applied in studies of ferritic stainless steels, titanium and high strength steels in sea water under applied cathodic potentials [10]. Specimens with thickness of 25 mm are polarized in present hydrogen embrittlement tests.

ACKNOWLEDGEMENT

The authors acknowledge following organizations and companies for support of this work. The work has partly been financed by the Technical Research Centre of Finland (VTT) , Swedish Power Inspectorate (SKI) and Ministry of Trade and Industry in Finland (KTM). Also the funding of The Finnish Black Liquor Recovery Boiler Committee, the Nordic Liaison Committee for Atomic Energy (NKA) and Outokumpu Co. is gratefully acknowledged.

REFERENCES

[1] Aaltonen, P., Molander, A., "Critical Potential for Intergranular Stress-Corrosion Cracking of Austenitic Stainless Steel", 11th Scandinavian Corrosion Congress, Stavanger, June 19-21, 1989, paper F5.

[2] Jargelius, R., Hertzman, S., Symniotis, E., Hänninen, H., Aaltonen, P. "Evaluation of the EPR Technique for Measuring Sensitization in Type 304 Stainless Steel." Corrosion, Vol. 47, No. 6, 1991, pp 429-435.

[3] Hänninen, H., Aaltonen, P.,Nenonen, P., Jargelius, R., Lehtinen, B.,"Characterization of Austenitic Stainless Steel Pipe Welds After Prolonged Power Plant Exposure." Corrosion, Vol. 48, No. 2, 1992, pp 114-123.

[4] Danko, J., Smith, R., Grandy, D., "Effect of Surface Preparation on Crack Initiation in Welded Stainless Steel Piping". Proceedings of the 5th Int. Symp. Environmental Degradation of Materials in Nuclear Power Systems- Water Reactors, August 25-29, 1991. Monterey, CA, American Nuclear Society Inc., pp 372-377.

[5] Hänninen, H., Aho-Mantila, I., "Effects of Sensitization and Cold Work on Stress Corrosion Susceptibility of Austenitic Stainless Steels in BWR and PWR Conditions", Report 88, Technical Research Centre of Finland, Espoo, May, 1981.

[6] Pohjanne, P., Hänninen, H., Mäkipää, M., Ehrnsten, U., "Cracking of Conpound Tubes in Black Liquor Recovery Boilers", To be presented in the 7th Int. Symp. on Corrosion in the Pulp and Paper Industry, 1992.

[7] Pohjanne, P., Hänninen, H., "Stress-Corrosion Cracking and Corrosion Fatigue of Austenitic and Duplex Stainless Steels in Hot Chloride Solution", VTT-MET B-208, Technical Research Centre of Finland, Espoo, June, 1992. 37p. (in Finnish)

[8] Cullen, H.,W., Partridge, M.,J., Gorman, J.,A. & Paine, J. P. N., "IGA/IGSCC of Alloy 600 in Sulfate and Chloride Solutions", Proceedings of the Int. Symp. on Environmental Degradation of Materials in Nuclear Power Systems-Water Reactors, August 25-29, 1991. Monterey, CA, American Nuclear Society Inc., pp.780-788.

[9] Park, H., Cragnolino, G., Macdonald, D., " Stress-Corrosion Cracking of Sensitized Type 304 Stainless Steel in Borate Solutions at Elevated Temperatures", Proceedings of the Int. Symp. on Environmental Degradation of Materials in Nuclear Power Systems-Water Reactors, August 22-25, 1983. Myrtle Beach, South Carolina, American Nuclear Society Inc., pp 604-622.

[10] Henrikson, S., "Corrosion in Sea Water Systems", Final report of the NKA Project MAT510. Nordic Liaison Committee for Atomic Energy. Stockholm, Sweden, Sept. 1988. 21 p.

S. Tsujikawa [1], T. Shinohara [1], and Wen Lichang [2]

SPOT-WELDED SPECIMEN MAINTAINED ABOVE THE CREVICE-REPASSIVATION POTENTIAL TO EVALUATE STRESS CORROSION CRACKING SUSCEPTIBILITY OF STAINLESS STEELS IN NaCl SOLUTIONS

REFERENCE: Tsujikawa, S., Shinohara, T., and Lichang, W., **"Spot-Welded Specimen Maintained Above the Crevice-Repassivation Potential to Evaluate Stress Corrosion Cracking Susceptibility of Stainless Steels in NaCl Solutions,"** Application of Accelerated Corrosion Tests to Service Life Prediction of Materials, ASTM STP 1194, Gustavo Cragnolino and Narasi Sridhar, Eds., American Society for Testing and Materials, Philadelphia, 1994.

ABSTRACT: A test is proposed to evaluate the susceptibility of austenitic stainless steels to SCC, particularly in low-concentration chloride solutions. A spot-welded specimen with both a crevice and residual stresses was immersed in the NaCl solutions, and kept potentiostatically for 72 hours at a potential slightly above the crevice repassivation potential with no additional applied stress. This method was also adopted in the development of an improved SCC resistant type 304 steel. The SCC critical temperature was measured for a group of 52 highly purified steels in which the P, Cu, Mo and Al contents were systematically varied. A 14Cr-16Ni-0.013P-2Cu-1Al-(0.3~1) Mo steel was found to have a high critical temperature of 110C, which is markedly higher than that of the conventional type 304 steel (50C) and therefore, it is a good candidate for industrial applications.

KEYWORDS: stress corrosion cracking, spot-welded specimen, NaCl solution, type 304, repassivation potential, crevice corrosion, alloying element.

---

[1]Professor and associate professor, respectively, Department of Metallurgy, Faculty of Engineering, University of Tokyo, 7-3-1 Hongo, Bunkyo-ku, Tokyo, 113 Japan.

[2]Professor, Department of Metallic Corrosion and Protection, Shanghai Institute of Metallurgy, Chinese Academy of Sciences, 865 Changning Road, Shanghai, 200050 China.

INTRODUCTION

The service life of materials, when limited by stress corrosion cracking (SCC), can be divided into two periods: the time for initiation and the time for propagation of cracks. It is considered that cracks will not grow when the SCC growth rate is lower than a threshold value of 10^{-10} m/s or 3 mm/yr.[1]. As a consequence, the time for propagation is then short compared to the service life of thousands of years demanded for the package materials used in the geological disposal of high-level nuclear wastes. Since a longer service life implies a longer time to initiation, in order to analyze the possible susceptibility of a material to SCC, any proposed test must accelerate the crack initiation time.

SCC in aqueous solutions containing chloride (Cl^- SCC) is frequently encountered in austenitic Fe-Cr-Ni stainless steels. Two necessary conditions for Cl^- SCC initiation were determined in previous studies [2,3,4]:

1. Cracks occur at a dissolving surface,
2. A microcrack can grow to a macrocrack (growing crack) only when the crack growth rate, **C'**, is faster than the dissolution rate, V, at the surface.

Therefore, the $V < C'$ condition is a prerequisite for the initiation of Cl^- SCC.

In neutral solutions, the dissolving surface specified in the first condition consists of large pits (B in Table 1) or a corroding crevice (C in Table 1), depending on the Cl^- concentration. By diluting the bulk Cl^- concentration, it becomes difficult for cracks to start at pits; and, as the $V < C'$ condition is readily satisfied with a lower

TABLE 1 -- Initiation site for SCC cracks in chloride solutions

	solution		laboratory test solution	initiation site
	pH	Cl^- concentration		
A	acidic		boiling 42, 45%$MgCl_2$ [5] 25C, $H_2SO_4+Cl^-$ [4] [6] 25C, HCl [7]	uniformly corroding plain surface
B	neutral	high	80C, 25~30% $MgCl_2$ [2]	large pits
C	neutral	low	80C, 0.3~3% NaCl [3] 90C, 100ppm Cl^- [8]	corroding crevice

dissolution rate, a corroding crevice becomes a good initiation site in low Cl^- solutions [2]. A circumferential notch, introduced into a rounded type 310S steel bar as a corrosion crevice and also as a stress intensifier, extended the SCC range to both the less noble potential region and the lower concentrated range of $MgCl_2$ solutions at 80C when compared to that corresponding to SCC initiated from pits on a smooth specimen [2].

In NaCl solutions with a concentration of 3, 0.3, 0.03 wt.% at 80C, an artificial slit crevice, introduced into a tapered double cantilever beam specimen of type 316 steel, induce cracks when the electrode potential is more noble than the repassivation potential for the crevice, E_R, and the stress intensity is above K_{ISCC}, 4.0 ~ 4.3 MPa-$m^{1/2}$ [3].

In this paper, SCC tests were performed with a spot-welded specimen at various temperatures [8]. The specimen had both a crevice condition and residual stresses, thus, no additional external loading is needed. By conducting the test at several temperatures, the SCC susceptibility of a steel can then be evaluated in terms of the critical temperature (CT). The steel will not exhibit SCC when the temperature is equal to or less than the critical temperature.

The CT is defined as the temperature at which C' and V are equal. The C' value is mostly influenced by temperature, and far less by chemical parameters, such as electrode potential and bulk Cl^- concentration [3]. In contrast, the dissolution rate, V_{II}, depends strongly on the electrode potential, and only slightly on temperature [2,3,4]. The dissolution rate, V_{II}, is the steady-state rate in the growth stage of crevice corrosion, stage II; and, it can be determined via the Moire technique for all locations of the specimen under the glass. Thus, it is chosen as the dissolution rate, V, at the initiation condition [9,10,11]. Hence, C' can be enhanced by increasing the temperature, and the V_{II} < C' condition is then easily satisfied.

The conditions for Cl^- SCC in terms of temperature and the localized corrosion (crevice corrosion) are summarized in Fig.1. In the potential range equal to and below E_R, the steel is immunized for both localized corrosion and SCC at any temperature [4]. When a localized corrosion is introduced at the potential E_R^+, the criteria is readily satisfied by the minimum dissolution rate, V^*_{II}; thus, SCC can be initiated whenever the critical temperature is exceeded.

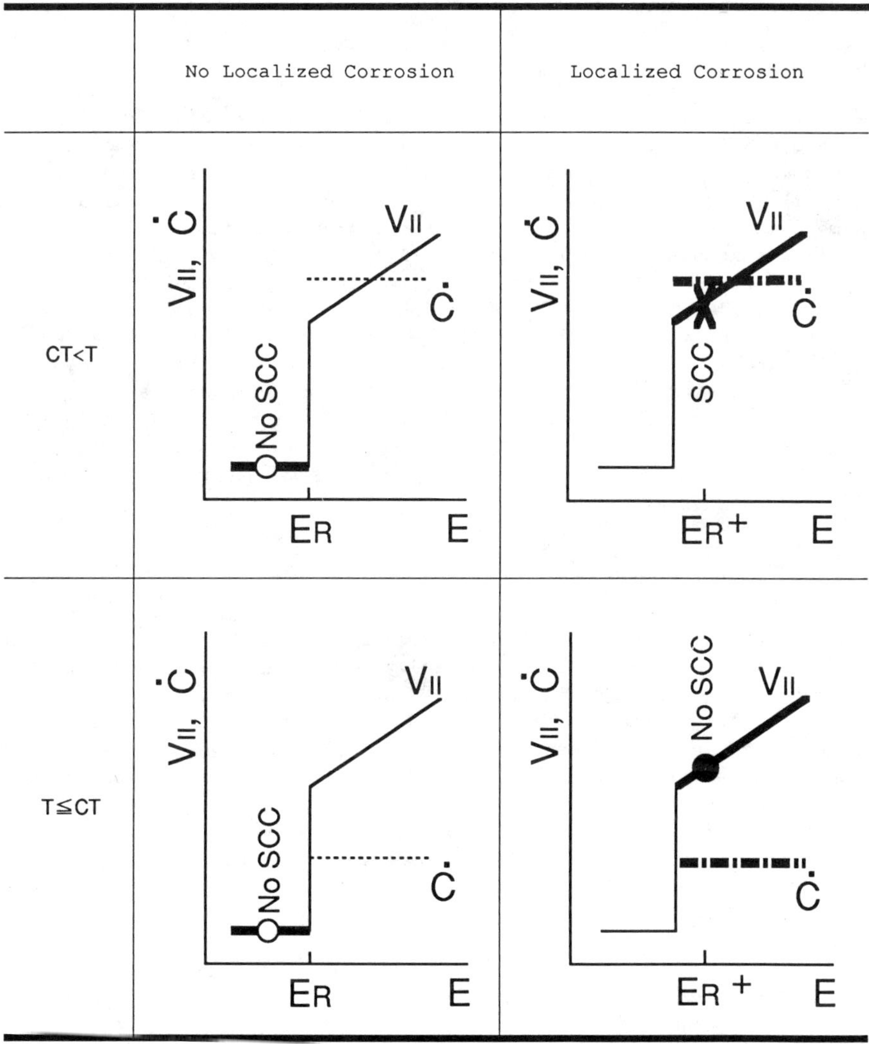

CT : critical temperature for SCC

E : electrode potential

VII : dissolution rate at local anode

ER : repassivation potential for localized corrosion

Ċ : crack growth rate

ER+: electrode potential slightly above ER

FIG. 1 -- Chloride SCC conditions in terms of localized corrosion and temperature

EXPERIMENTAL METHOD

Specimens

Fifty-two kinds of steels with different chemical compositions, as shown in Tables 2 through 4, were investigated. With few exceptions, the composition of the base steels (4M-1 and 4M-9) were Si (<0.1%), Mn (<0.1%), P (20ppm), Mo (<0.01%) and N (<0.006%). With approximately the same composition, type 304 steel has a greater resistance to SCC in boiling concentrated $MgCl_2$ solution [5]. A 14% Ni concentration is required for the 18Cr-14Ni steels to remain fully austenitic at 3% Si. Since the 14Cr-16Ni steels, containing 14% Cr and 16% Ni, can remain fully austenitic up to 3% Al, they were used in studies of Al effects [12]. The 18Cr-14Ni steels were heat-treated for 15 minutes at 1050C and the 14Cr-16Ni steels were heat-treated for 30 minutes at 1200C before water quenching. After cutting a 1mm thick steel sheet in two pieces of 40mmx20mm and 20mmx10mm, these pieces were ground flat to a 600 grit finish before being spot-welded to each other, as shown in Fig.2.

SCC test

The spot-welded specimen was maintained at a potential above the repassivation potential for crevice corrosion, E_R, to induce localized corrosion in NaCl solution at several temperatures [13], and specimen current was recorded during the test. After a test duration of 72 h., the specimen was removed from the test cell and a hole was drilled through each of the welded portion (nugget); then, the two sheets were separated manually and the inner surfaces were examined for cracking using a scanning electron microscope. At temperatures above 100C, the SCC tests were carried out in an autoclave that was equipped with a Ag/AgCl/0.1M KCl external pressure-balanced reference electrode. The electrode potentials are given with respect to the saturated calomel electrode (SCE) at room temperature [14].

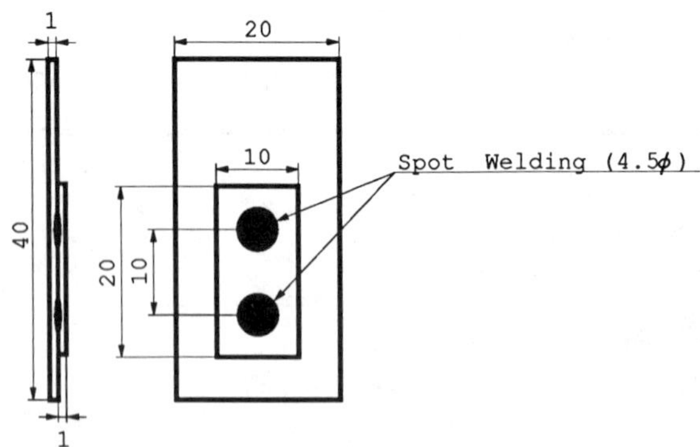

FIG. 2 -- Spot-welded specimen for SCC test (mm).

TABLE 2 -- Chemical compositions in wt% of the steels 4M-1 through 4M-14.

	C	Si	Mn	P	S	Cr	Ni	Cu	Mo	Al	N
4M-1	0.019	<0.10	<0.10	0.002	0.0018	17.05	13.77	<0.05	<0.01	0.003	0.0015
4M-2	0.020	<0.10	<0.10	0.006	0.0019	17.32	13.87	<0.05	<0.01	0.005	0.0018
4M-3	0.018	<0.10	<0.10	0.030	0.0097	18.00	14.29	<0.05	<0.01	0.003	0.0050
4M-4	0.020	<0.10	<0.10	0.108	0.0096	17.81	14.26	<0.05	<0.01	0.002	0.0064
4M-5	0.022	0.52	<0.10	<0.002	0.0093	17.90	14.14	<0.05	<0.01	0.002	0.0052
4M-6	0.023	1.01	<0.10	<0.002	0.0093	17.54	14.02	<0.05	<0.01	0.003	0.0049
4M-7	0.017	2.03	<0.10	<0.002	0.0069	17.85	14.15	<0.05	<0.01	0.006	0.0029
4M-8	0.019	3.05	<0.10	<0.002	0.0070	18.01	14.20	<0.05	<0.01	0.006	0.0038
4M-9	0.019	<0.10	<0.10	<0.002	0.0095	17.72	14.10	<0.05	<0.01	0.006	0.0051
4M-10	0.022	<0.10	1.07	<0.002	0.0099	17.69	14.13	<0.05	<0.01	0.005	0.0046
4M-11	0.019	<0.10	4.93	<0.002	0.0102	17.68	14.10	<0.05	<0.01	0.003	0.0050
4M-12	0.019	<0.10	9.57	<0.002	0.0111	17.50	14.06	<0.05	<0.01	0.003	0.0058
4M-13	0.019	<0.10	<0.10	<0.002	0.0091	18.14	14.15	0.98	<0.01	0.004	0.0050
4M-14	0.020	<0.10	<0.10	<0.002	0.0089	17.97	14.06	2.03	<0.01	0.005	0.0046

TABLE 3 -- Chemical compositions in wt% of the steels TS-1 through TS-9.

	C	Si	Mn	P	S	Cr	Ni	Cu	Mo	Al	N
TS-1	0.020	<0.10	<0.100	0.005	0.0053	17.72	13.89	0.99	<0.010	0.031	0.0063
TS-2	0.022	<0.10	<0.100	0.013	0.0051	18.00	14.05	0.99	<0.010	0.031	0.0047
TS-3	0.017	<0.10	<0.100	0.028	0.0052	17.65	14.06	1.00	<0.010	0.037	0.0046
TS-4	0.020	<0.10	<0.100	0.030	0.0049	17.89	14.13	2.00	<0.010	0.033	0.0055
TS-5	0.021	0.51	<0.100	0.030	0.0047	17.99	14.36	1.02	<0.010	0.033	0.0040
TS-6	0.022	1.00	<0.100	0.028	0.0049	17.84	14.08	0.99	<0.010	0.033	0.0062
TS-7	0.021	1.99	<0.100	0.029	0.0045	17.99	14.18	1.00	<0.010	0.042	0.0055
TS-8	0.020	0.50	0.740	0.006	0.0048	17.66	14.02	1.98	<0.010	0.032	0.0053
TS-9	0.020	0.58	0.760	0.030	0.0128	17.87	13.74	<0.05	<0.010	0.002	0.0022

TABLE 4 -- Chemical compositions in wt% of the steels TSA-1 through TSA-29.

	C	Si	Mn	P	S	Cr	Ni	Cu	Mo	Al	N
TSA-1	0.019	<0.10	<0.10	<0.002	0.0028	14.12	16.30	<0.005	<0.005	0.005	-
TSA-2	0.020	<0.10	<0.10	<0.002	0.0016	14.14	16.33	<0.005	<0.005	0.564	-
TSA-3	0.019	<0.10	<0.10	<0.002	0.0043	14.13	16.18	<0.005	<0.005	0.964	-
TSA-4	0.022	<0.10	<0.10	<0.002	0.0024	14.32	16.40	<0.005	<0.005	1.955	-
TSA-5	0.023	<0.10	<0.10	0.002	0.0020	14.42	16.50	<0.005	<0.005	2.981	-
TSA-6	0.020	<0.10	<0.10	<0.002	0.0036	14.33	16.52	2.056	<0.005	3.376	-
TSA-7	0.021	<0.10	<0.10	0.005	0.0026	14.31	16.48	2.053	<0.05	3.369	-
TSA-8	0.018	<0.10	<0.10	0.013	0.0035	14.06	16.49	2.093	<0.005	3.440	-
TSA-9	0.020	<0.10	<0.10	0.031	0.0035	14.11	16.54	2.089	<0.005	3.433	-
TSA-10	0.019	<0.10	<0.10	<0.002	0.0045	14.35	16.43	1.996	0.029	2.964	-
TSA-11	0.020	<0.10	<0.10	<0.002	0.0042	14.38	16.49	1.995	0.058	2.972	-
TSA-12	0.022	<0.10	<0.10	<0.002	0.0037	14.33	16.41	1.980	0.307	2.951	-
TSA-13	0.021	0.41	1.010	0.002	0.002	14.12	16.18	1.97	1.04	2.920	0.0268
TSA-14	0.021	0.38	1.000	0.004	0.002	14.26	16.32	1.96	2.04	2.940	0.0272
TSA-15	0.021	0.36	0.973	0.004	<0.001	14.51	15.85	1.94	3.04	2.840	0.0253
TSA-16	0.018	0.41	0.986	0.008	0.002	14.07	15.92	1.94	1.00	0.006	0.0288
TSA-17	0.020	0.37	0.980	0.010	<0.001	14.60	15.90	1.92	0.99	0.463	0.0266
TSA-18	0.020	0.42	1.010	0.012	<0.001	14.24	16.30	1.98	1.02	0.982	0.0277
TSA-19	0.020	0.39	0.990	0.011	<0.001	14.14	16.18	1.93	0.99	1.930	0.0235
TSA-20	0.021	0.39	1.000	0.012	<0.001	14.19	16.29	1.98	1.02	2.910	0.0177
TSA-21	0.021	0.40	0.90	0.011	0.002	14.27	16.02	1.93	0.27	2.99	0.0025
TSA-22	0.018	0.39	0.90	0.028	0.002	14.10	15.91	1.96	0.30	2.48	0.0010
TSA-23	0.024	0.38	0.91	0.011	0.002	14.01	16.05	2.02	2.05	2.32	0.0012
TSA-24	0.021	0.40	0.95	0.011	0.002	14.27	16.16	2.00	0.28	1.01	0.0016
TSA-25	0.018	0.40	0.94	0.029	0.002	14.32	16.06	2.01	0.30	0.82	0.0012
TSA-26	0.024	0.39	0.95	0.011	0.002	14.14	16.11	2.02	2.05	0.89	0.0013
TSA-27	0.023	0.40	0.95	0.011	0.002	14.14	16.11	2.02	0.03	<0.01	0.0014
TSA-28	0.021	0.41	0.97	0.012	0.002	14.19	16.21	2.03	0.29	0.01	0.0015
TSA-29	0.017	0.40	0.96	0.005	0.002	14.27	16.08	2.00	0.32	<0.01	0.0013

RESULTS

Potential range for SCC

The potential range for SCC was determined for numerous highly purified 18Cr-14Ni steels (4M-1 through 4M-14), each with a different percentage of P in 0.03%, 0.3% and 3% NaCl solutions at 80 C (Fig. 3).

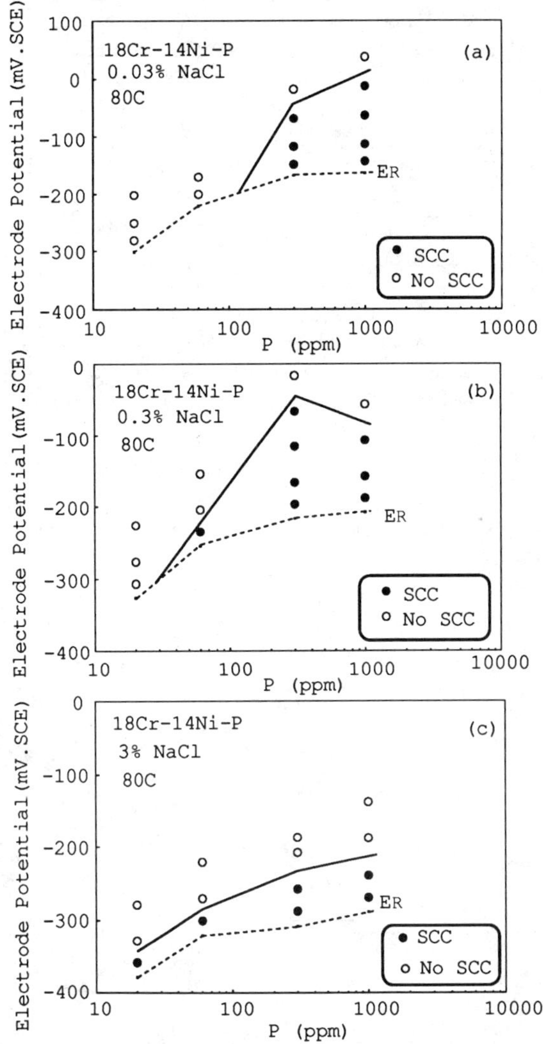

FIG. 3 -- Potential range for SCC versus P content in (a) 0.03%, (b) 0.3% and (c) 3% NaCl solutions.

It was demonstrated that all steels with SCC susceptibility in chloride solutions will have SCC cracks at a potential which is 20mV above E_R. If a steel does not crack at this potential, no SCC will be initiated at any other potentials. SCC can be induced at a less noble potential when V_{II} is slower than C', which is estimated as shown in Fig. 4 as the value corresponding to that of V_{II} at the boundary potential between SCC and no SCC ranges [9]; but, when V_{II} is faster than C', initiation becomes impossible even at an increased potential. Fig.4a shows the dependence of V_{II} on potential for the 20ppm P steel in 3% NaCl and Fig.4b those for the 300ppm P steel in 0.3% and 3% NaCl where the rate V_{II} at E_R, $V_{II}*$, is slower than C'. However, as shown in Fig.4a for the 20ppm P steel in 0.3% NaCl $V_{II}*$ can be faster than C' and the SCC initiation condition, $V_{II} < C'$, is never satisfied for potentials above E_R. A potentiostatic test, based upon this unique property associated with E_R^+, was adopted in the evaluation of the susceptibility of austenitic stainless steels to Cl^- SCC.

SCC test at the E_R^+ potential

The results of SCC tests for 4M-1 through 4M-14 steels at the E_R^+ potential are plotted in Fig.5. The effects of P, Si, Mn and Cu when added to the base steel are shown in Fig.5a, Fig.5b, Fig.5c and Fig.5d respectively. While increasing contents of P, Si or Mn do not suppress the tendency to cracking, the addition of 1~2 % Cu was found to improve the SCC resistance[14].

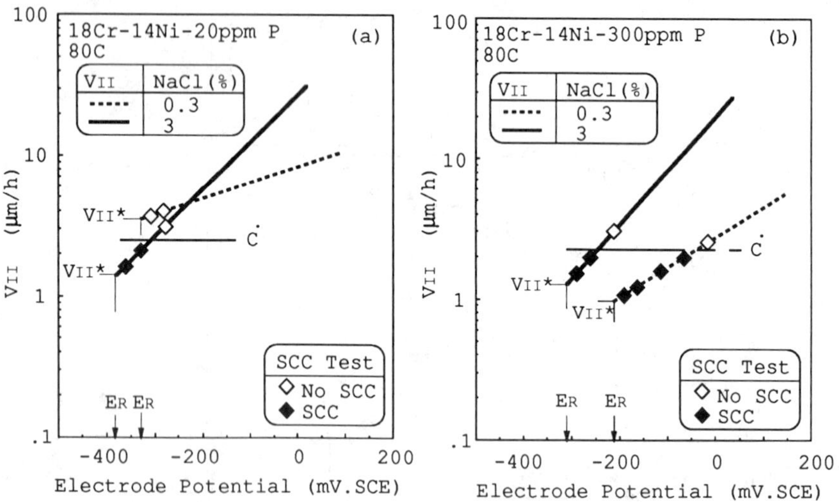

FIG. 4 -- Effect of electrode potential on dissolution rate of V_{II} for the steels with 20ppm P (4M-1) (a) and 300ppm P (4M-3) (b) in 0.3%(·····) and 3%(——) NaCl solutions at 80C with occurrence of SCC(◆) or no SCC(◇).

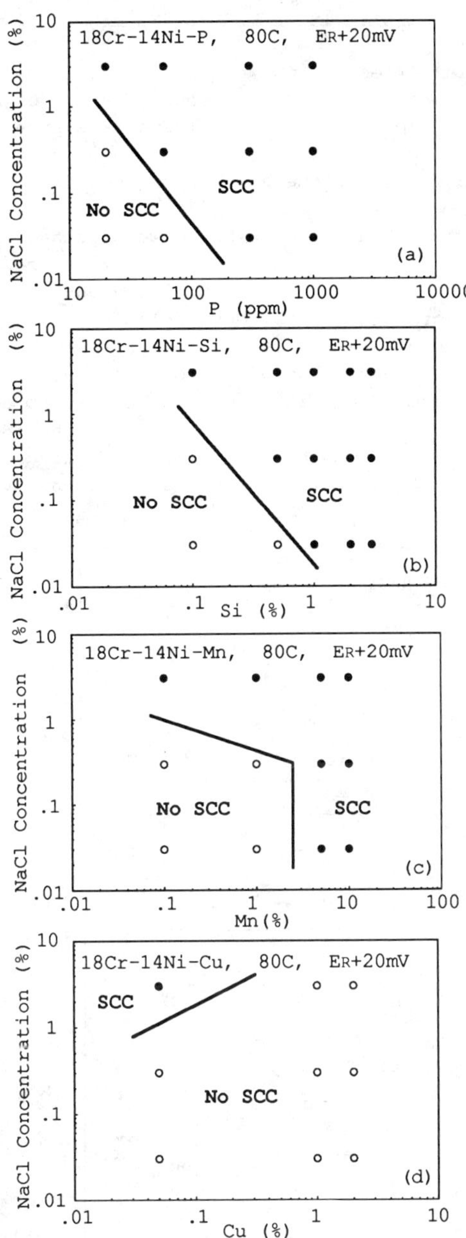

FIG. 5 -- SCC and no SCC ranges determined with spot-welded specimens at a potential of ER+20mV in terms of NaCl concentration and alloying content of (a) P, (b) Si, (c) Mn or (d) Cu.

Susceptibility to SCC in terms of temperature

The beneficial effect of Cu was observed for the highly purified steels with 20ppm P and <0.01% Mo. A concentration of P less than 60~80 ppm is currently unpractical in the industrial steel making, and concentration of Mo up to 0.3%, as present in conventional type 304 steel, is accepted to reduce the cost in the melting process by using return steels.

To better examine the trade-offs between Cu and P, the SCC tests were performed for steels with assorted Cu and P concentrations. The results of the highly purified 18Cr-14Ni-2Cu-0.006P steel tested at various electrode potentials in 3% NaCl solution at 130C and 150C are plotted in Fig.6; the E_R value is -340mV.SCE at both 130C and 150C which is very close to the value of -345mV measured at 80C. SCC was observed at -300mV and -320mV at 150C, but it was not detected at the critical temperature of 130C. The currents for the TS-8 steel specimen are plotted in Fig.7 as a function of time for various electrode potentials. The differences for the currents at -300mV and -320mV are

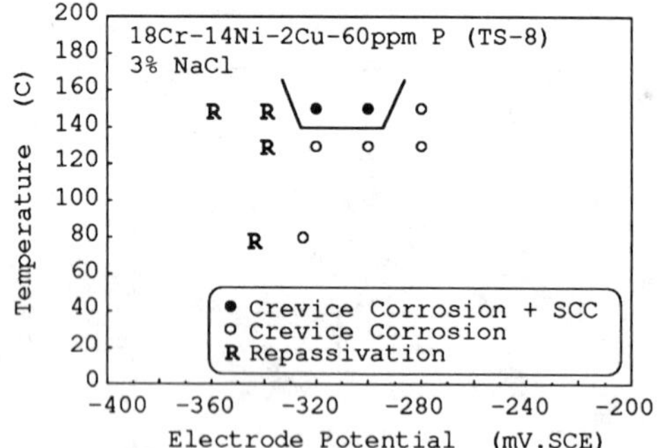

FIG. 6 -- SCC test results for TS-8 (2Cu-60ppm P) steel in 3% NaCl solution in terms of temperature and electrode potential.

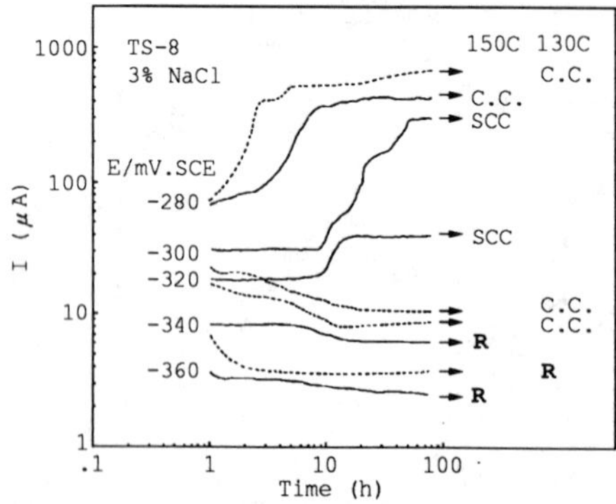

FIG. 7 -- Variation of current with time during SCC tests for TS-8 steel specimens shown in FIG. 6. Full line: 150C; dotted line: 130C. "C.C." denotes crevice corrosion with no SCC and "**R**" repassivation.

significant between 150C and 130C, which are shown as full lines and dotted lines, respectively. At 150C, the currents increased rapidly after 10h leading to SCC, but at 130C, the current decreased gradually leading to crevice corrosion without SCC.

The results of SCC tests conducted with TS-1 through TS-9 steels at E_R^+ are shown versus P and Cu contents in Fig.8a and Fig.8b. A maximum CT of 140C was attained for the 2% Cu and 20ppm P steel; but, as the concentrations of P raises to 300ppm P, CT was reduced to 100C.

The CT of the 2% Si, 1% Cu and 300ppm P steel was found to be 100C [15]. The CT of a similar steel with 3% Si, 2% Cu and 200~300ppm P was 120C; but, the CT did not decrease when the content of Mo was increased to 0.8% [14]. The ability of Si to compensate the deleterious effect of increasing P content may be utilized in strengthening the SCC resistance. However, increasing the Si concentration also brings new problems. For example, intergranular corrosion may be enhanced because Si increases the activity of carbon which favors the precipitation of carbides inducing sensitization. Therefore, it is necessary to explore the beneficial effect of other alloying elements.

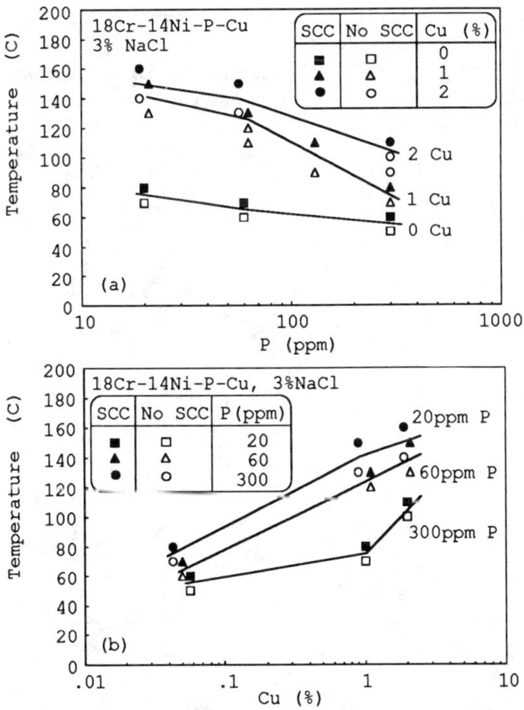

FIG. 8 -- Occurrence (●,▲,■) or no occurrence (○,△,□) of SCC in terms of temperature, P and Cu contents for 18Cr-14Ni-P-Cu steels.

Effect of Al combined with 2% Cu on the critical temperature for SCC

According to the condition V_{II} < C', those elements that accelerate the dissolution rate, V_{II}, are effective in improving the SCC resistance. Among them, Al is selected as an alloying element because of its capacity to enhance the dissolution rate [17,18] and lengthen the time-to-failure [19]. However, when Al is added to a highly purified 14Cr-16Ni steels without Cu, the CT is unaffected and remained at 60C up to 3% of Al additions. The CT of many 14Cr-16Ni-2Cu-P-Mo-(0~3)Al steels (TS-1 through TS-29), kept at the E_R^+ potential in 3% NaCl solution, are summarized in terms of their P and Mo contents in Fig.9.

A 2% Cu addition improved the SCC resistance of 14Cr-16Ni steels containing 50ppm P and <0.01% Mo, elevating the critical temperature to 140C. However, the high CT obtained by Cu addition diminished to 100C with 300ppm P and to 60C with 0.3% Mo. Additions of 1 or 3% Al, combined with 2% Cu, were found to attenuate the deleterious effect of P and Mo. A CT of 110C was attained for steels containing 0.3~1% Mo and a slightly reduced P content of 130 ppm. With 3% Al addition, a 14% Cr and 16% Ni composition is required for the treated steels to remain fully austenitic. Actually only 1% Al addition is needed and hence, the Cr and Ni contents can be adjusted to those of type 304 steel.

The would-be-developed type 304 stainless steel, modified with 2% Cu ,1% Al, 130ppm P and 0.3% Mo, will have a CT of 110C, which is higher than that of 50C for the conventional type 304 steel [9,15] and that of 70C [16] or 100C for the type 316 steel.

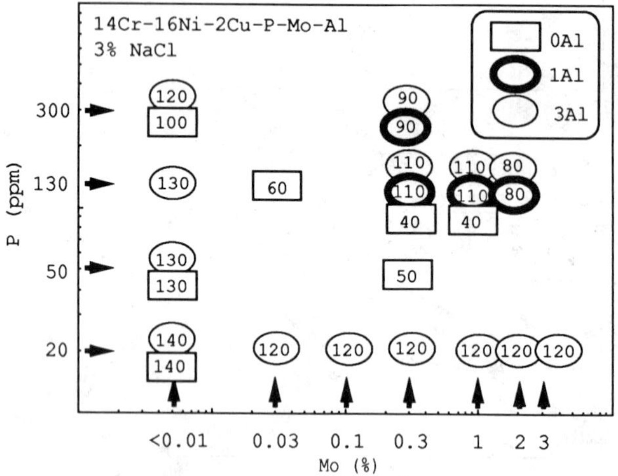

FIG. 9 -- Summary of critical temperature for SCC (in C) in terms of P content and Mo content for 14Cr-16Ni-2Cu-P-Mo steels with 0, 1 and 3% Al.

CONCLUSION

When the electrode potential of an austenitic stainless steel is maintained at a potential lower than the repassivation potential for localized corrosion, the steel will not undergo either localized corrosion or SCC in neutral chloride solution. However, stainless steels often experienced localized corrosion prior to SCC.

The initiation of SCC cracks from a local anode can be simply reproduced in an accelerated test using a spot-welded specimen that is held potentiostatically at a potential slightly above the repassivation potential for crevice corrosion. If the test is performed at various temperatures, the susceptibility to SCC can be analyzed in terms of the critical temperature for SCC. This accelerated test has been further adopted in the development of an improved SCC resistant type 304 steel.

ACKNOWLEDGEMENTS

This research for an improved chloride SCC resistant type 304 steel was a three years effort with support from NiDI (Nickel Development Institute, Toronto, Canada; Japan Liaison Office at Tokyo). We are grateful to Stainless/Titanium Research Laboratory of Nippon Steel Co., for the preparation of steel specimens.

REFERENCES

[1] Speidel,M.O. and Hyatt,M.V., Advances in Corrosion Science and Technology , Vol.2, Plenum Press, 1972, pp 115-337.

[2] Tsujikawa,S., Shinohara,T., and Hisamatsu,Y., Corrosion Cracking, American Society for Metals, Metals Park, OH, 1985, pp 35-42.

[3] Tamaki,K., Tsujikawa,S., and Hisamatsu,Y., Proceedings of the Second International Conference on Localized Corrosion, National Association of Corrosion Engineers, Houston, 1987, pp 207-214.

[4] Tsujikawa,S., Proceedings of International Conference on Stainless Steels, Iron and Steel Institute of Japan, Tokyo, 1991, pp 48-55.

[5] Kowaka,M., and Fujikawa,H., Journal of Institute of Metals of Japan, Vol.34, No.10, 1970, pp 1047-1054, 1054-1062; Localized Corrosion (NACE-3), National Association of Corrosion Engineers, Houston, 1974, pp 437-446.

[6] Acello,S.J. , and Greene,N.D. , Corrosion, Vol 18, No.8, August 1962, pp 286-290.

[7] Bianchi,G. , Mazza,F. , and Torchio,S. , Corrosion Science, Vol 13, No.3, March 1973, pp 165-173

[8] Masuo,M., Ono,H., and Ohashi,N., Boshoku Gijutsu, Vol.26, No.10, October 1977, pp 573-581.

[9] Tsujikawa,S., Shinohara,T., and Liang,C., Proceedings of International Conference on Stainless Steels, Iron and Steel Institute of Japan, Tokyo, 1991, pp 196-203.

[10] Shinohara,T., Tsujikawa,S., and Masuko,N., Corrosion Engineering,

Vol.39, No.5, May 1990, pp 255-269.
[11] Tsujikawa,S., Critical Factors in Localized Corrosion, Electrochemical Society, Pennington, NJ, 1992, pp378-388.
[12] Zaizen,T., Yamazaki,T., Otoguro,Y., Ogawa,T., Itoh,H., and Yamanaka,M., Materials for Energy Systems, Vol.2, September 1980, pp 21-29.
[13] Liang,C., Shinohara,T., and Tsujikawa,S., Boshoku Gijutsu, Vol.37, No.11, November 1988, pp 679-684; Corrosion Engineering, Vol.37, No.11, 1988, pp 613-617.
[14] Liang,C., Shinohara,T., and Tsujikawa,S., Boshoku Gijutsu, Vol.38, No.12, December 1989, pp 650-656; Corrosion Engineering, Vol.38, No.12, 1989, pp 709-720.
[15] Liang,C., Shinohara,T., and Tsujikawa,S., Boshoku Gijutsu, Vol.39, No.6, June 1990, pp 309-314; Corrosion Engineering, Vol.39, No.6, 1990, pp 343-352.
[16] Adachi, T., Proceedings of International Congress on Stainless Steels, Iron and Steel Institute of Japan, Tokyo, 1991, pp 189-195.
[17] Chen,W.Y.C., and Stephens,J.R., Corrosion, Vol. 35, No.10, October 1979, pp 443-451.
[18] Osozawa,K., Okato,N., Fukase,Y., and Yokota,K., Boshoku Gijutsu, Vol. 24, No.1, January 1975, pp 1-7.
[19] Staehle,R.W., Royuela,J.J., Raredon,T.L., Sarrate,E., Morin,C.R., and Farrar,R.V., Corrosion, Vol.26, No.11, November 1970, pp.451-486.

M. R. Louthan, Jr.[1] and W. C. Porr, Jr.[2]

CONSTANT EXTENSION RATE TESTING AND PREDICTIONS OF IN-SERVICE BEHAVIOR: THE EFFECT OF SPECIMEN DIMENSIONS

REFERENCE: Louthan, M. R., Jr. and Porr, W. C., Jr., **"Constant Extension Rate Testing and Predictions on In-Service Behavior: The Effect of Specimen Dimensions,"** Application of Accelerated Corrosion Tests to Service Life Prediction of Materials, ASTM STP 1194, Gustavo Cragnolino and Narasi Sridhar, Eds., American Society for Testing and Materials, Philadelphia, 1994.

ABSTRACT: Constant extension rate test measurements of a material's susceptibility to environmental degradation may depend on the dimensions of the test specimen. Long, slender samples will appear to be more susceptible to degradation than will short, thick samples of the same material. This sample size effect is due to the importance of the tearing modulus on the onset of unstable fracture. Therefore, if constant extension rate tests are used to qualify materials for in-service applications, the effect of specimen size on test results should be evaluated.

KEYWORDS: environmental degradation, unstable fracture, tearing modulus, free machining brass, mercury embrittlement

---

Slow strain rate testing (SSRT), also termed constant extension rate testing (CERT), is used to assess the susceptibility of materials to stress corrosion, hydrogen embrittlement, and other forms of environmentally induced cracking. These test techniques were refined in the mid 1960's and publications outlining the history of SSRT began

---

[1]Senior Advisory Engineer, Savannah River Technology Center, Westinghouse Savannah River Company, Aiken, SC 29808

[2]Graduate Assistant, School of Engineering and Applied Science, University of Virginia, Charlottesville, VA 22903

to appear in the 1970's [1-6]. The ASTM Standard G-49 (Preparation and Use of Direct Tension Stress Corrosion Test Specimens) is now available as a guide to applying SSRT/CERT techniques in investigating the susceptibility of metals and alloys to stress corrosion cracking [7]. This standard states that "axially loaded tension specimens provide one of the most versatile methods of performing a stress corrosion test because of the flexibility permitted in the choice of the type and size of test specimens" [7].

A "typical" test specimen is a smooth, round bar machined to a geometry similar to that shown in Figure 1.

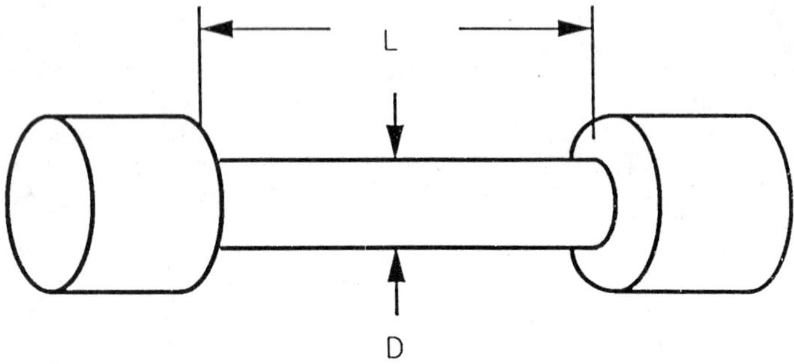

FIG. 1-- Schematic of Typical Tensile Specimen for SSRT.

Selection of the gage length, L, and the diameter, D, is controlled by ASTM Standard E8 which establishes sample size requirements through terms such as "whenever possible" and "product dimensions determine test specimen sizes" [8]. Although the smooth test specimens are typical, specimens that have been used for slow strain rate testing include: single edge notch [9], notched round [5], cracked plates [10], and smooth bars [11-13]. Rectangular bars [11], wires [12], and round bars [13] are examples of the type of smooth bar specimens that have been tested. After the specimen configuration and size have been selected and the tests completed, evaluation of the test data may be another multiple choice exercise.

Parameters that have been used to evaluate the stress corrosion susceptibility of metals and alloys in slow strain rate studies include: crack velocity [9,10], ultimate strength [7,9,10,12], time to failure [9], elongation at failure [9-14], crack length [10], reduction in area [7,10,14,15], hardness through the cross section [12,15], fraction of the fracture area that has a topography indicative of a stress corrosion fracture mode [11,12] and work or energy required to fracture the test specimen

[7,12,15]. The three most common SSRT/CERT measures of stress corrosion susceptibility are ultimate strength, elongation to failure and evaluation of fracture mode changes.

Material-environment-test parameter combinations that produce a susceptibility to stress corrosion generally cause the specimen to develop surface cracks during the testing process. Under this condition, even an initially smooth test specimen becomes a cracked sample. The test is continued until the cracked sample can no longer support the applied load. If the final fracture is by ductile processes, tearing instability is involved in the fracture. An analysis of the elastic-plastic fracture behavior of cracked tensile samples has shown that onset of tearing instability is very dependent on specimen dimensions [16]. This paper demonstrates that the dependence of tearing instability on specimen dimension has a significant influence on the measure of the stress corrosion susceptibility that is established by slow strain tensile test procedures.

BACKGROUND

A CERT study to quantify the susceptibility of metals and alloys to environmental degradation typically involves a consistent chronology of events. A test sample is strained in the test environment. A stress corrosion crack, or some other form of environmentally induced cracking, will initiate as the specimen is strained. The crack(s) will propagate by environmentally influenced processes until the specimen is completely fractured or until the conditions for mechanical instability are achieved. The onset of plastic tearing instability is controlled by material properties that can be characterized through the standard tensile test and the J-integral R-curve and by the geometry and dimensions of the test sample [16].

Tearing instability in a cracked round bar having a length, L, a diameter, D, and a uniform circumferential crack to a depth, a, will result when [16],

$$(dJ/da)(E/\sigma_0^2) = [(4/\beta)(\sigma_f/\sigma_o)][(L)(D-2a)/(D^2)], \qquad (1)$$

where the terms $(dJ/da)(E/\sigma_0^2)$ describe the tearing modulus, T, of the material. The tearing modulus can be determined from the J-integral R-curve as shown in Figure 2.

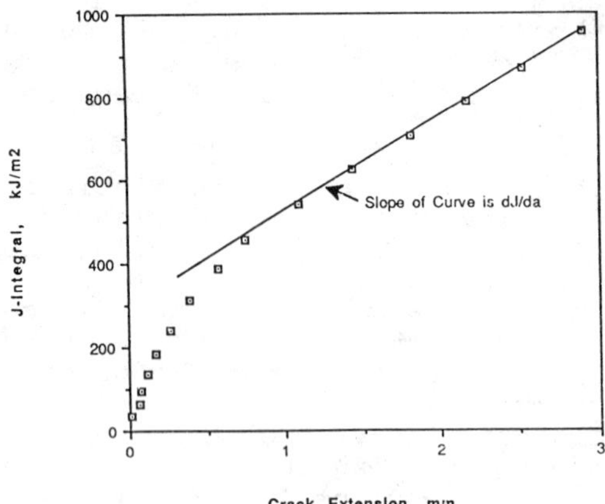

FIG. 2--A J-integral R-curve for a Tough, Ductile Metal

However, the characterization of the post blunting portion of the R-curve as linear, dJ/da = constant, is an over simplification because of the curvature apparent in many experimental R-curves [17].

The micromechanisms of ductile fracture, includes the involvement of dislocations, inclusions and other defects, thus making fracture a heterogeneous processes. This heterogeneity and the large scale plastic strains that frequently develop during the test, cause major difficulties in modeling experimental J-integral R-curves. However, even though Equation 1 is not strickly valid, the relationships expressed demonstrate that the conditions for instability during a tensile test of a notched or cracked bar are dependent on both the length and the diameter of the tensile specimen. Therefore, the tearing instability analysis for notched bars may be applicable to the fracture of CERT test specimens.

The instability analysis predicts that long thin test samples (samples which have large values of L/D) will become mechanically unstable under conditions that would not produce instability in short thick samples. The variation in the ease of mechanical instability with L/D ratio provides the basis for a hypothesis that the results of CERT studies will depend on sample dimensions. An experimental program to test this hypothesis was therefore developed.

EXPERIMENTAL

The importance of sample dimensions in the determination of the apparent susceptibility of a metal to environmental degradation was established by tensile studies of smooth cylindrical samples of free machining brass (65.1% Cu, 35.5% Zn, 3% Pb). The tensile tests were conducted in dilute (0.05m) aqueous solutions of hydrochloric acid which were saturated with $HgCl_2$ (some $HgCl_2$ remained undissolved in the test solution). Mercury was plated on the brass surfaces during the test and the combination of stress and exposure to the liquid mercury caused liquid metal embrittlement, (LME), of the brass test samples. All tests were made with samples machined from the same bar stock and were conducted at a constant crosshead displacement rate of 2.54 $\times10^{-3}$ cm/sec. Testing was conducted at room temperature or approximately 20°C.

Test conditions were selected to assure that material and test variables such as grain size, alloy content, temperature and metallurgical condition were constant. These variables are known to influence the susceptibility to mercury embrittlement [18] and were controlled to minimize that influence.

The microstructure of the test material is shown in Figure 3 and a fractograph illustrating the LME induced crack around the outer diameter of a test sample are shown in Figure 4.

FIG. 3--Microstructure of Test Material

FIG. 4--Fractrograph of SSRT Specimen

The metallographic observations demonstrated that the environmentally induced cracking caused a sharp notch to develop in the test specimen. The depth of the notch, a, is represented by the depth of the intergranular fracture in Figure 4. Under the conditions of these tests, the parameters given on the right hand side of Equation 1 can be directly determined from pre- and post test measurements.

Samples tested to establish a measure of the susceptibility of the free machining brass to mercury cracking included: specimen having pretest diameters which varied from 0.38 cm. to 0.76 cm. and pretest gage lengths which varied from 1.9 cm. to 13 cm. Crack or notch depths were determined on the samples after testing and varied from 1.5 mm to 4.6 mm. These specimen parameters provided test samples having values for the right hand side of Equation 1 {(L)(D-2a)/(D^2)} which varied from 2.7 to 18.4. Specimen parameters, including elongation-to-fracture in the tensile test, are summarized in Table 1. Control tests for each specimen size were conducted in air.

Table 1--Specimen Parameters and Test Results

Sample	Diameter, cm.	Length, cm.	Crack Depth, cm.	Elong.
12	0.38	1.90	0.030	0.0287
13	0.38	2.29	0.025	0.0280
14	0.38	2.67	0.018	0.0435
15	0.38	3.81	0.020	0.0250
16	0.38	4.57	0.025	0.0181
18	0.38	6.48	0.023	0.0180
19	0.38	7.62	0.023	0.0160
21	0.51	1.52	0.025	0.0578
22	0.51	2.54	0.046	0.0408
23	0.51	3.05	0.020	0.0458
24	0.51	3.56	0.030	0.0289
31	0.64	1.90	0.025	0.0580
32	0.64	3.18	0.025	0.0449
33	0.64	3.81	0.015	0.0400
34	0.64	4.45	0.025	0.0246
35	0.64	6.35	0.018	0.0234
36	0.64	7.62	0.020	0.0185
37	0.64	9.52	0.028	0.0182
39	0.64	12.70	0.025	0.0156
42	0.76	3.81	0.028	0.0395
43	0.76	4.57	0.020	0.0408
44	0.76	5.33	0.015	0.0441
45	0.76	7.62	0.025	0.0219
48	0.76	12.95	0.046	0.0192

Samples tested in air underwent approximately 6.5% elongation-to-fracture regardless of the length, diameter, or length-to-diameter ratio. Additionally, as expected, no surface cracks developed in any of the air tested samples. The effect of increasing {(L)(D-2a)/(D^2)} on the elongation-to-fracture of the samples tested in the mercury environment

(Figure 5) demonstrates that sample dimensions played an important role in determining the susceptibility of the free machining brass to LME. This observation is in general agreement with the tearing instability analysis summarized by Equation 1.

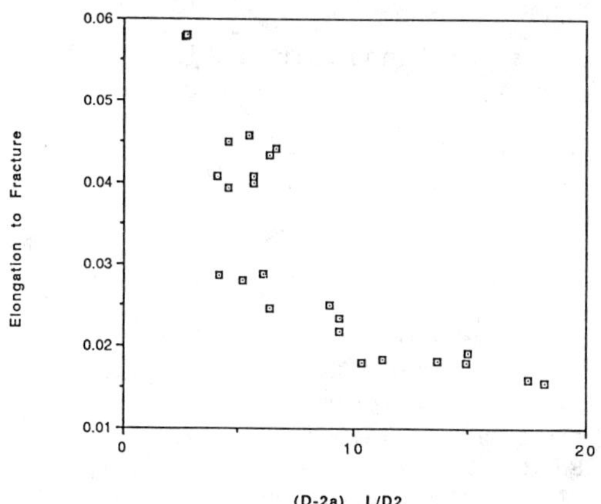

FIG. 5--Dependency of Elongation to Fracture on Specimen and Crack Dimensions

DISCUSSION

Inspection of Equation 1 reveals that value of left hand side of the equation is dependent on the properties of the material being tested while the right hand side is dependent on the term $[(4/\beta)(\sigma_f/\sigma_o)]$ and the dimensions of the test specimen. It has been shown that the term $[(4/\beta)(\sigma_f/\sigma_o)]$ approaches 6 for round tensile samples having external notches (16). Furthermore, as indicated previously, the term $\{(dJ/da)(E/\sigma_o^2)\}$ is the tearing modulus and may be approximated as a material constant, T. Therefore, Equation 1 may be rewritten as:

$$T = 6\ \{(L)(D-2a)/(D^2)\}. \qquad (2)$$

The length of a tensile specimen at instability, L_i, is the sum of the original specimen length, L_o, and the change in length, ΔL, that resulted from the testing process. Thus,

$$L_i = L_o + \Delta L. \qquad (3)$$

Equation 2 can be rewritten by replacing L with the actual specimen length at instability and multiplying by $(L_o)/(L_o)$ to give,

$$T = 6 \{(L_i)/(L_o)\}\{(L_o)(D-2a)/(D^2)\}. \quad (4)$$

Combination of Equations 3 and 4 gives

$$T = 6 (1 + \Delta L/L_o) [(L_o)(D-2a)/(D^2)]. \quad (5a)$$

However, $\Delta L/L_o$ is the fractional elongation (elong.) in a tensile test, thus Equation 5a can be written as

$$T = 6 (1 + elong.) [(L_o)(D-2a)/(D^2)]. \quad (5b)$$

or

$$elong = [T/6][(D^2)/\{(L_o)(D-2a)\}] -1. \quad (6)$$

Equation 6 is of the form *y = ax + b* where y is the elongation-to-fracture in the SSRT or CERT test and x is $(D^2)/[(L_o)(D-2a)]$. The magnitude of $(D^2)/[(L_o)(D-2a)]$ is determined by the specimen dimensions and the depth of cracking in the CERT test. Figure 6 is a replot of the test data summarized in Figure 5 but in the form predicted by Equation 6. The equation,

$$elong. = 0.135[(D^2)/\{(L_o)(D-2a)\}] + 0.01 \quad (7)$$

is a least squares fit to the experimental data. The empirical constants, however, were not those predicted through Equation 6.

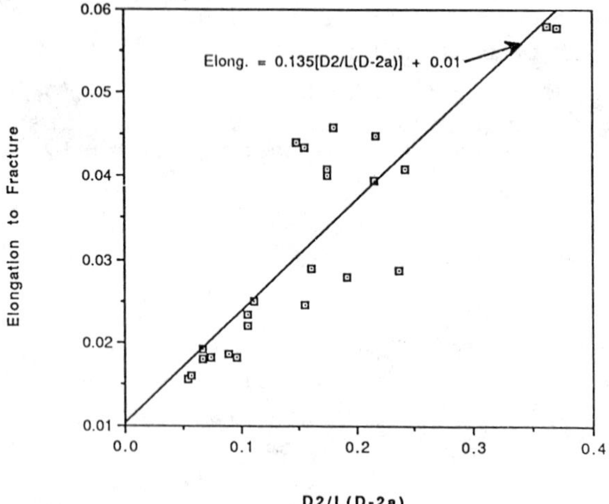

FIG. 6--Least Squares fit of Test Data to Equation of the Form Predicted by Equation 6.

A statistical analysis of the test data using Anderson-Bell AB stat™ computer software, an interactive statistical package, shows that the data fit an equation of the form $y = ax + b$ with b = 0. This analysis showed that within a 95% confidence (based on a normal distribution) the data for the 24 specimens were valid and described by a $y = ax + b$ equation to a probability exceeding 0.9999. This statistical analysis thus supports the observation that the degree of embrittlement measured in a CERT type experiment is significantly influenced by the choice of specimen dimensions.

The importance of specimen dimensions to a CERT evaluation of the susceptibility of a material to environmental effects results from the role of tearing instability in determining the elongation-to-fracture. The onset of tearing instability is also dependent on the tearing modulus. Experimental determinations of the tearing modulus have shown that T_{matl} is not a constant [19]. Curvature in the J-Resistance curve [20], specimen size effects [21], residual stresses [22] and material parameters such as strength, grain size and metallurgical structure influence T_{matl}. Because of such influences, the specific relationships developed in Equations 5 and 6 are only approximations of specimen size effects. However, the data presented demonstrate that the results of a CERT type study may be dependent on specimen size.

The size dependence should be most apparent when the combination of material, environment, specimen and test conditions produce surface cracks which propagate so that the final fracture develops before the environmental degradation process has propagated across the entire specimen diameter. In this regard, it should be noted that the observed specimen size effects were for cracks growing by liquid metal embrittlement processes and the crack velocities were quite high. For slowly growing cracks, such as cracks resulting from aqueous stress corrosion cracking, the results may be quantitatively different. Furthermore, the CERT test is a rising load test and the extent of crack opening displacement and the onset of unstable fracture may differ significantly from that observed in service produced cracking or in laboratory induced cracking under constant crack opening displacement conditions. This potenrial difference in crack opening displacement must also be carefully evaluated whenever CERT data are used to predict in-service performance.

Material to material differences in resistance to unstable tearing may cause an apparent difference in relative susceptibility of two different materials to environmental degradation. This apparent difference may reflect differences in T_{matl} rather than absolute differences in susceptibility to environmentally induced degradation. The material having the lower tearing modulus would reach instability at lower values of T_{app}, if specimen size and other considerations were identical. The earlier onset of unstable fracture would result in a reduced elongation-to-fracture and therefore in shorter times-to-fracture. It is also interesting to postulate that, under these conditions, the material showing the longest times-to-fracture and thus the least relative susceptibility to degradation, may also show fracture mode changes over the largest fraction of the fracture surface. The data and analysis developed during this study demonstrate that constant extension rate testing procedures to select and qualify materials for engineering service should include considerations of the role of specimen dimensions on the test results.

CONCLUSIONS

The results of this experimental study demonstrate that SSRT/CERT type measurements of a material's susceptibility to environmental degradation may depend on the dimensions of the test specimen. Long, thin samples will appear to be more susceptible to degradation than will short, thick samples of the same material. This sample size effect is due to the importance of the tearing modulus on the onset of unstable fracture.

REFERENCES

[1] Parkins, R. N., Stress Corrosion Research, Sijhoff and Noordoff International Publishers, The Netherlands, 1979, pp. 1-28

[2] Parkins, R. N., Stress Corrosion Cracking-The Slow Strain-Rate Technique, ASTM STP-665, American Society for Testing and Materials, 1979, pp. 5-25

[3] Payer, J. H., W. E. Berry and W. K. Boyd, Stress Corrosion Cracking-The Slow Strain-Rate Technique, ASTM STP-665, American Society for Testing and Materials, 1979, pp. 61-77

[4] Parkins, R. N., F. Mazza, J. J. Royqela and J. C. Scully, British Corrosion Journal, Vol. 7, 1972, pp. 154-167

[5] Parkins, R. N., F. Mazza, J. J. Royqela and J. C. Scully, British Corrosion Journal, Vol. 21, 1981, pp. 459-471

[6] Payer, J. H., W. E. Berry and W. K. Boyd, Stress Corrosion - New Approaches, ASTM STP-610, American Society for Testing and Materials, 1976, pp. 82-93

[7] American Society of Testing and Materials, Annual Book of ASTM Standards, Vol. 301, Designation: G49, pp. 284-291

[8] American Society of Testing and Materials, Annual Book of ASTM Standards, Vol. 201, Designation: E8, pp. 1047-1067

[9] Vgiansky, G. M. and C. E. Johnson, Stress Corrosion Cracking-The Slow Strain-Rate Technique, ASTM STP-665, American Society for Testing and Materials, 1979, pp. 113-131

[10] Kermani, M. and J. C. Scully, Corrosion Science, Vol. 19, 1979, pp. 89-110

[11] Holroyd, N. J. H. and G. M. Scammans, Environment-Sensitive Fracture: Evaluation and Comparison of Test Methods, ASTM STP-821, American Society for Testing and Materials, 1984, pp. 202-241

[12] Frignani, A. G., Trabenelli and F. Zucchi, Corrosion Science, Vol. 24, 1984, pp. 917-927

[13] Stoltz, R. E., Metallurgical Transactions A, Vol. 12A, 1981, pp. 543-545

[14] Abe, S., M. Kojima and H. Yuzo, Stress Corrosion Cracking-The Slow Strain-Rate Technique, ASTM STP-665, American Society for Testing and Materials, 1979, pp. 294-304

[15] Wei, R. P. and S. R. Novak, Environment-Sensitive Fracture: Evaluation and Comparison of Test Methods, ASTM STP-821, American Society for Testing and Materials, 1984, pp. 75-79

[16] Paris, P. C., H. Tada, A. Zahoor and H. Ernst, Elastic-Plastic Fracture, ASTM STP-668, American Society for Testing and Materials, 1979, pp. 5-36

[17] Howard, I. C. and A. A. Willoughby, Developments in Fracture Mechanics-2, Applied Science Publishers LTD, Essex, England, 1981, pp. 39-99

[18] Kamdar, M. H., Treatise on Materials Science and Technology, Embrittlement of Engineering Alloys, Vol. 25, Academic Press, 1983, pp. 362-465

[19] Loss, F. J., B. H. Menke, A. L. Hiser and H. E. Watson, Elastic-Plastic Fracture, Volume II, Fracture

Resistance Curves and Engineering Applications, ASTM STP-803, American Society for Testing and Materials, 1983, pp. 777-795

[20] Kobayashi, K., H. Nakamura and H. Nakazawa, Elastic-Plastic Fracture, Volume II, Fracture Resistance Curves and Engineering Applications, ASTM STP-803, American Society for Testing and Materials, 1983, pp. 420-438

[21] Chell, G. G., and I. Milne, Elastic-Plastic Fracture, Volume II, Fracture Resistance Curves and Engineering Applications, ASTM STP-803, American Society of Testing and Materials, 1983, pp. 179-205

[22] Davis, D. A., M. G. Vassilaros and J. P. Gudas, Elastic-Plastic Fracture, Volume II, Fracture Resistance Curves and Engineering Applications, ASTM STP-803, American Society for Testing and Materials, 1983, pp. 582-610

Michel Verneau,[1] Jacques Charles,[2] and François Dupoiron[3]

APPLICATIONS OF ELECTROCHEMICAL POTENTIOKINETIC REACTIVATION TEST TO ON-SITE MEASUREMENTS ON STAINLESS STEELS

REFERENCE: Verneau, M., Charles, J., Dupoiron, F., **"Applications of Electrochemical Potentiokinetic Reactivation Test to On-Site Measurements on Stainless Steels,"** Application of Accelerated Corrosion Tests to Service Life Prediction of Materials, ASTM STP 1194, Gustavo Cragnolino and Narasi Sridhar, Eds., American Society for Testing and Materials, Philadelphia, 1994.

ABSTRACT: The aim of the paper is to present some applications of the Electrochemical Potentiokinetic Reactivation (EPR) technique to the characterization of the corrosion resistance of stainless steels apparatus.
In particular, the use of the technique to the detection of chromium depletions on insufficiently pickled austenitic steels is explained. The application of the method to determine the degree of sensitization of duplex stainless steels by intermetallic phase precipitations is also discussed.
Then, the paper presents the possibility of use of the EPR test for on-site measurements, with a portable equipment.

KEYWORDS: intergranular corrosion, stainless steels, sensitization, on-site technique, electrochemical test.

INTRODUCTION

There are many specific applications which need a particular attention to the real surface state of stainless steels. This is the case when problems of contamination of the surface of the steels can occur. The presence of impurities, segregations or phases in the materials can favour selective attack, since the surface and the passive film will be modified. This may occur during all the process of fabrication of the products, or while assembling apparatus or equipment:

Sensitization of the materials often occurs when heat treatments

[1]Corrosion engineer, Materials Research Center, Creusot-Loire Industrie, 56 rue Clémenceau BP 56, 71202 Le Creusot Cedex, France.

[2]Stainless steels product manager, Creusot-Loire Industrie. Le Creusot

[3]Head stainless steels and corrosion department, Materials Research Center, Creusot-Loire Industrie, Le Creusot.

are badly chosen. Stainless steels need to be solution annealed and water quenched in order to put in solution all the different phases that they can contain and to homogenize their structure. In some cases, they need to be stress relieved after quenching in order to obtain the required mechanical properties. The latter treatment can favour precipitations in the material, if the temperature range is in the precipitation zone.

Welding operations may result in a thermal effect near the weld metal. The heat affected zone sometimes presents a lower corrosion resistance, due to carbide or phase precipitations. Then, whenever a post-welding heat treatment is not possible on the piece of metal or on the apparatus, a high corrosion resistance in the whole material must be maintained. Today, no efficient mean of control exists, and most of the equipments are built according to the "know-how" of the constructor. A quantitative method of testing would be appreciated.

The operation of pickling is necessary after heat treatments or after fabrication on austenitic and duplex (austenitic-ferritic) stainless steels, in order to give them a clean surface state. Indeed, oxides that have been formed on the surface of the steel during heat treatment need to be removed, in order to avoid the risks of corrosion that they could generate. The chromium depleted areas, linked with these superficial oxides may be at the origin of localized or intergranular corrosion, since the chromium level in these areas may be under the minimum level for passivation.

CLASSICAL INTERGRANULAR CORROSION TESTS

The best known intergranular corrosion tests are the ASTM Standard Practices for Detecting Susceptibility to Intergranular Attack in Austenitic Stainless Steels (A262, practices A to F). These techniques cover a wide range of different sources of susceptibility to intergranular corrosion : Precipitation of chromium carbides or intermetallic phases, even if they are not visible.

However, many deficiencies can be observed with these methods :

- All of them are destructive methods : they require to cut a sample off the material, before carrying the test out. This factor is detrimental for many applications, particularly when one wants to check the structure of the steel directly on assembled equipments.
- Except practice A, the minimum duration of the tests is 4 hours (for practice D), but the other practices require from 24 to 240 hours of immersion. Moreover, the time of preparation of the samples (i.e. machining and polishing) must be added to this time of immersion.
- Lastly, all these usual methods do not precisely quantify the degree of sensitization of the steels. Some of them are purely qualitative, such as methods A or E. Moreover, the result includes the whole surface of the sample that was immersed, and not only the surface that will be in contact with the environment. A more discerning method would often be desirable.

These different limitations explain why another method has been required to qualify the susceptibility of stainless steels to intergranular corrosion and the surface state of these materials.

Manufacturers or steel producers more and more need to reduce the delays of production including the operations of control. The reduction of the duration of the controls would be a satisfactory way to reach this objective.

When assembled apparatus are concerned, users cannot easily measure their real state, today, and the assembling procedures are usually based on a know-how that cannot be checked after the operations. A non-destructive method would be appreciated, to be able to verify the result of the work, and systematic controls after assembling would probably avoid many corrosion problems in service.

In conclusion, the main qualities that would be required for a better method of testing the susceptibility of stainless steels to intergranular corrosion would be :

1. rapidity
2. reproducibility
3. quantitative measurement
4. accuracy
5. non-destructive test.

AN ELECTROCHEMICAL TEST

Except ASTM A262 practice A, the methods previously described all consist of immersion tests. For all of them, cutting a sample is necessary.

Electrochemical methods have known a wide expansion during the recent past, and some techniques have been promoted to study the surface state of stainless steels. This is particularly the case of the Electrochemical Potentiokinetic Reactivation (EPR) technique [1][2].

The EPR test

Originally, the EPR method has been developed for the quantification of the susceptibility to intergranular corrosion of common austenitic stainless steels (AISI 304 or 304 L grades) [3][4]. The principle of the test is based on the drawing of the reactivation curve of the material in an appropriate chemical solution. The laboratory equipment that is used for this test consists in an electrochemical cell, which supports a working electrode (metal to be tested) a counter electrode (Platinum) and a reference electrode (saturated calomel electrode : SCE). The acquisition is realized with a potentiostat. A computer coupled with the potentiostat records the data, treats them and draws the polarization curves (Fig. 1).

For AISI 304 or 304L grades, the chemical solution that is generally used [2] is a mixture of sulfuric acid (H_2SO_4) and potassium thiocyanate (KSCN) : 0,5 M H_2SO_4 + 0,01 M KSCN. KSCN is a depassivating chemical species which enhances the reactivation of the material. The temperature of the solution is maintained at 30° C by a circulation of heated oil. At the beginning of the test, the samples are maintained

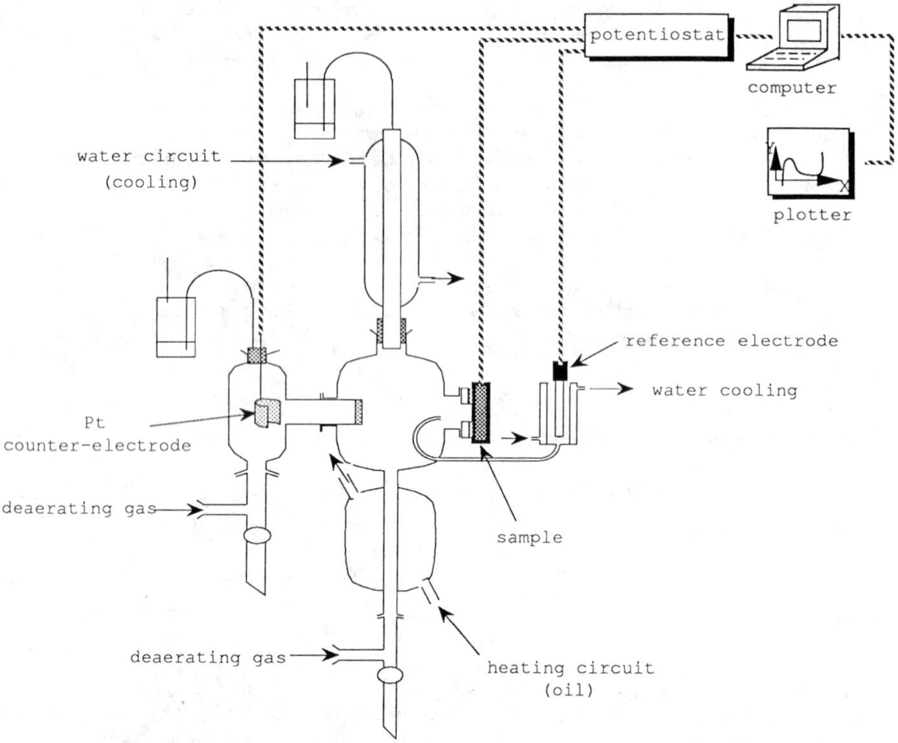

potentiostat
computer
plotter
water circuit
(cooling)
reference electrode
water cooling
Pt
counter-electrode
deaerating gas
sample
deaerating gas
heating circuit
(oil)

FIG. 1--Electrochemical set-up for EPR tests

Potential (mV/SCE)
Passive
potential
Passive area
Reactivation
peak
Active
area
Corrosion
potential
I (µA/cm 2)

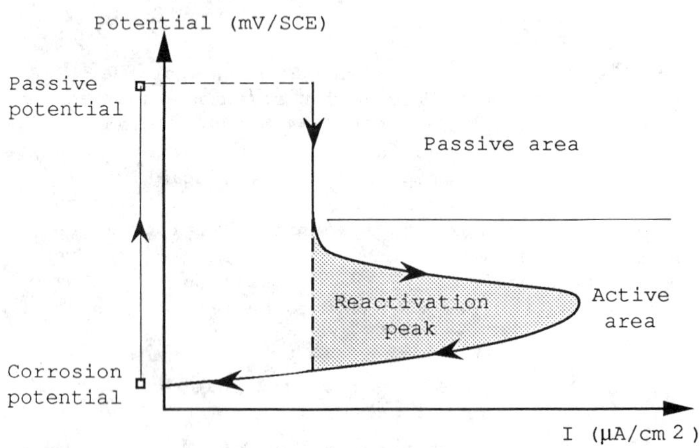

FIG. 2--Reactivation curve on stainless steels for the single-loop test

about 2 min at the free-corroding potential. Then, the potential is set in the passive area (+ 200 mV/SCE) for 2 min. This operation leads to a complete passivation of the surface of the steel. The time of passivation must be controlled, in order that the method keeps reproducible. After this, the potential back scan begins, using a sweep rate of 6V/h. In the single-loop test, the potentiodynamic reactivation curve is drawn, from the passive potential, towards the rest potential of the steel (Fig.2).

According to the degree of sensitization of the steel, the curve will or will not present a reactivation peak. Indeed, precipitations in the material or localized chromium depletions generally induce a weakening of the passive film. Conversely, materials which are not sensitized will not show any reactivation peak, since the passive film is not attacked.

The total weight of the species that have been dissolved from the material is directly proportional to the intensity of the dissolution reaction, itself proportional to the quantity of Coulombs exchanged during the reaction (Faraday's law):

$$m/M = Q/nF \tag{1}$$

where

m = weight loss (g)
M = atomic weight (g)f
Q = reactivation charge (C)
n = valency of the metal
F = Faraday's number (C)

This weight loss also directly depends on the total surface of the chromium depleted areas. Thus, the area under the reactivation peak, which is proportional to the reactivation charge is also proportional to the chromium depleted area. This means that the measurement of the total surface under the reactivation peak can be a quantification of the sensitization of the stainless steel.

Softwares have been developed to calculate the integral of the peak. The result, which takes into account the surface of the material in contact with the test solution, is expressed in mC/cm^2. Low integrals mean that the steel has not been sensitized. This is the case of water-quenched stainless steels, which do not have any phase or carbide precipitation. These have a "step structure" in the Oxalic Acid Etch Test, Practice A in ASTM A262. They usually present a reactivation curve without any peak or with only a small loop. Conversely, sensitized materials will produce high and wide reactivation peaks, that will depend on the degree of sensitization.

Discussion on the EPR method

The EPR method can be applied to a wide range of stainless steels or alloys [5][6]. However, the initial parameters of the test, that have been determined for AISI 304 and 304L stainless steels, have to be modified for each material. Indeed, they directly depend on the real characteristics of the materials, and their own corrosion resistance. This induces that most of the parameters of the test cannot be standardized, and that specific conditions must be found for each grade of stainless steel.

The main parameters which have to be calibrated for the test are the following :

1. The composition of the solution in H_2SO_4 and KSCN : the more H_2SO_4 concentration is high, the more the test is aggressive and the reactivation peak is high. The sensitivity of the test increases with H_2SO_4 concentration. KSCN has basically a similar effect. Nevertheless, pits may be initiated on the surface of the steel when a too high KSCN concentration is used. Therefore, the KSCN concentration will be limited to low values. Fig.3a presents the influence of H_2SO_4 concentration on the reactivation peak of a 23% Cr - 4% Ni - 0.1% N sensitized duplex stainless steel, after 30 minutes at 700 °C.

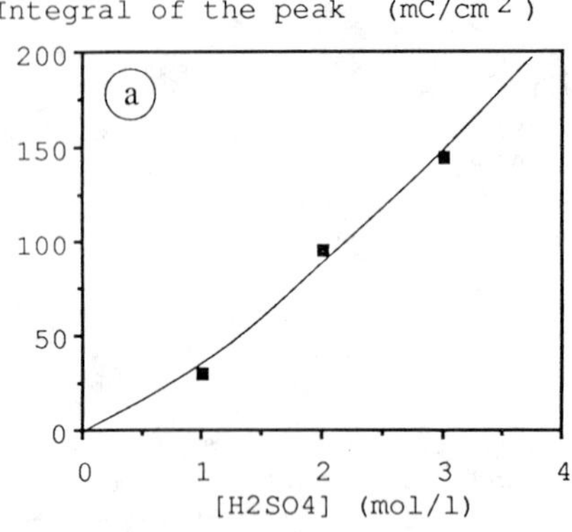

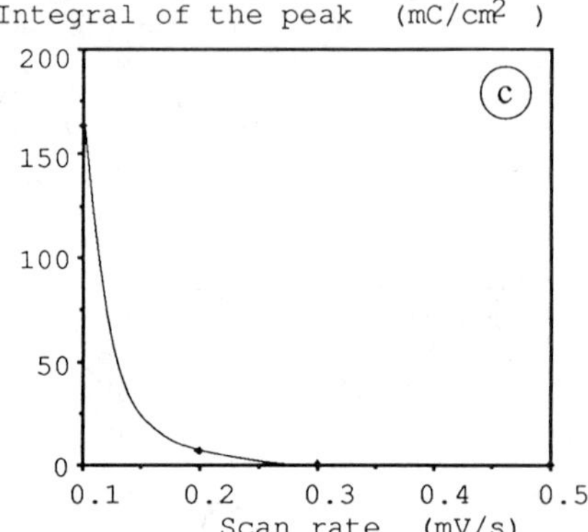

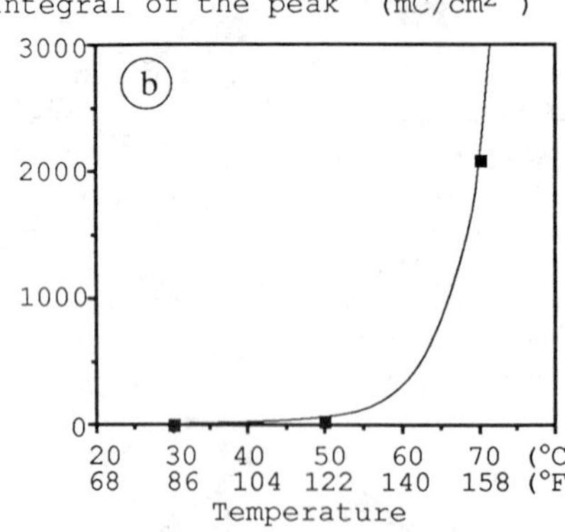

FIG. 3--Influence of
(a) the sulfuric acid concentration
(b) the temperature
(c) the scan rate
on the integral of the reactivation peak of a 23 %Cr-4 %Ni sensitized stainless steel

2. The temperature of the solution activates the electrochemical reactions and the corrosion resistance decreases when the temperature increases (Fig. 3b). However, the use of high temperatures may induce pits on the materials, though the sensivity would be better. So, it is

often necessary to limit the temperature of the test. The same result had been obtained by [4] on 304 stainless steels.

3. The surface state of the steel before the test is an important parameter. Indeed, the reproductibility of the chemical and electrochemical reactions requires to know perfectly the conditions of surface. Grinding samples to a 1200 - grit finish reduces the risks of preferential sites of attack. Moreover, it is always better to use freshly polished specimens. The effect of surface finish has been clearly observed by [4] on 304 stainless steels.

4. The choice of the passivation potential and the duration of the passivation is important for the success of the test. An optimum value of this potential is near the transpassivity. In these conditions, the sensitivity of the test is almost the best. However, it is sometimes difficult to avoid pits formation. Usually, the passive potential is maintained about 10 to 15 minutes. This time is sufficient for the construction of the passive film.

5. The sensitivity of the test decreases when the potential scan rate increases (Fig.3c). This explains why low scan rates are often used for the realization of the EPR test [4].

As can be seen, many parameters influence the EPR method, and no real standardization is possible. Each material requires a special calibration of the above parameters before applying the test. This calibration is realized step by step, parameter after parameter. Therefore, two specimens must be used :

The first one, usually waterquenched, free of any precipitations, will not reactivate during the test. One will measure no peak or only a small peak on the reactivation curve.

The second one is sensitized and the parameters of the test are determined in order that this specimen presents the highest peak, while no peak is observed on the other one. Then, each lightly or highly sensitized specimen will present a reactivation peak. The successive measurements of the integrals of these peaks allow to compare the degree of sensitization of each material.

Finally, as soon as the test has been calibrated for a special steel, its use becomes very simple and presents many advantages :

1. Rapidity : in most cases, the duration of the test does not exceed two hours. For some grades, it can be very rapid, less than 20 minutes. This is a real advantage in comparison with the usual chemical tests which last from 4 to 240 hours.

2. Sensitivity : the sensitivity of the EPR test is very good, and the beginning of sensitization can be measured on most of the steels. This only requires a fine calibration of the test. A quantitative measurement of the degree of sensitization of the materials can be obtained, with a very good precision. This is an important improvement with regard to some tests of the ASTM A262 series. However, in our tests, the maximum level of integrals of the reactivation peaks was under 5000 mC/cm^2. Some authors have observed a saturation of the EPR test, that becomes insensitive when the measurements are greater than 10 C/cm^2 [4].

3. Reproductibility : the EPR test shows a very good reproductibility on stainless steels, provided that their structure is homogeneous. This property has been verified on different steels, with different conditions of testing. As mentioned by [4], this good reproducibility can only be obtained if all the parameters of the test are carefully controlled.

4. Non destructive test : one of the major advantages of electrochemical tests is that they do not require to cut a sample off the material : Neither weight loss measurements, nor bending test are done. All the reactions are only surperficial. Thus, it is only necessary to fix a specific cell directly on the surface of the metal, in order to carry the test out. This can be done either in laboratories, or directly on-site, with the use of a specific equipment. Such an equipment will be presented at the end of this paper.

APPLICATION OF THE EPR METHOD TO THE CHARACTERIZATION OF PICKLING OPERATIONS

The EPR method has been applied to 317 LN stainless steel plates, to determine the influence of the duration of pickling on their corrosion behaviour. The chemical analysis of such plates is given in Table 1.

TABLE 1--Chemical composition of 317 LN (w/o)

C	S	P	Si	Mn	Ni	Cr	Mo	N
0.016	0.001	0.024	0.362	1.756	13.75	18.24	3.41	0.13

The test has been performed on hot rolled plates after a solution annealing heat treatment. Table 2 presents the results obtained for different conditions (time and temperature) of pickling in a solution containing hydrofluoric acid (HF) and nitric acid (HNO_3). The different parameters of the EPR test have been calibrated at the following values:

H_2SO_4 concentration : 0.5 M
KSCN concentration : 0.01 M
Temperature : 30 °C
Scan rate : 6 V/h
Passivation potential : + 200 mV/SCE
Passivation time : 2 min.

For each temperature, the results indicate that the reactivation charge decreases when the duration of pickling is increased. An increase of the temperature of the pickling solution also results in decreasing the reactivation peak obtained with the EPR test (Fig.4). These results are in total agreement with micrographic examinations. The surface

TABLE 2--Influence of the conditions of pickling on the reactivation peak

Duration of pickling, min	Reactivation charge, mC/cm^2		
	Pickling temperature		
	20°C (68°F)	40°C (104°F)	60°C (140°F)
0	1840	...	...
5	1763	800	26
10	...	52	0
15	575	10	0
20	93	3	0
25	...	0	0

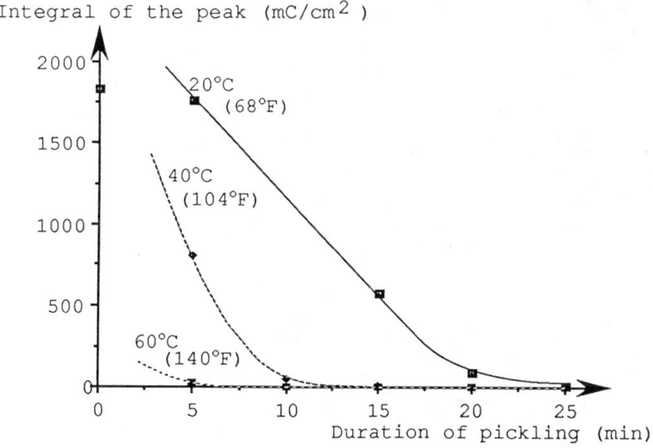

FIG. 4--Influence of the conditions of pickling on the integral of the reactivation peak for a 317 LN stainless steel

condition of the specimens appears to be directly linked with the result of the EPR test. High reactivation peaks indicate that pickling of the samples is not sufficient. Surface encrusted oxides, which have not been removed, create local preferential sites of attack and the reactivation is very high at these sites. Chromium depleted zones, around these oxides are also attacked during the test. In contrast, well pickled surfaces do not reactivate during the EPR test.

The EPR test appears to be an easy and reliable method to characterize the quality of pickling operations.

APPLICATION OF THE EPR TEST TO DETERMINE THE SENSITIZATION STATE OF DUPLEX STAINLESS STEELS

The chemical analysis of Type UNS 31803 duplex stainless steel is given in Table 3. The different parameters of the EPR test have been determined as following :

Polishing : 1200-grit paper
Solution : H_2SO_4 : 1.5 M and KSCN : 0.05 M
Passivation potential : +200 mV/SCE for 120 s
Temperature : 30 °C
Scan rate : 0.5 mV/s

TABLE 3--Chemical composition of UNS 31803 (w/o)

C	S	P	Si	Mn	Ni	Cr	Mo	N
0.023	0.001	0.019	0.419	1.720	5.74	22.20	2.76	0.13

In order to simulate different thermal sensitization treatments (intermetallic phase or carbide precipitations), several heat treatments were used :

- Water quenched
- Water quenched + 800°C/10 min - air
- Water quenched + 800°C/30 min - air
- Water quenched + 800°C/1h-air
- Water quenched + 800°C/5h-air
- Water quenched + 600°C/1h-air

For each treatment, the reactivation curve was measured (Fig. 5), and the integral of the peak has been calculated.

The water quenched UNS 31803 presents no reactivation peak. This indicates that the structure is completely free of sensitization. Micrographic examinations confirm this result (Fig. 6a).

After different treatments at 800°C, one can see an obvious progression in the amplitude of the reactivation peaks, that clearly shows the increase of the precipitation in the material with increasing time at temperature. Micrographic examinations are in complete agreement with this observation (Fig. 6b).

The five heat treatments that were tested on UNS 31803 were located on the phase diagram of the same grade (Fig. 7). On this diagram, one can notice different areas corresponding with some precipitations (σ or χ phase, carbides, etc).

The results of the EPR test and the phase diagram are in good accordance, since one can notice that the four heat treatments at 800°C that presented a reactivation peak agree with the probability of precipitations presented in the diagram. Conversely, the fifth treatment

(600°C/1h) does not give rise to precipitations, and no peak was detected after the EPR test. Thus, there is an obvious correlation between the EPR test and the microstructural investigations.

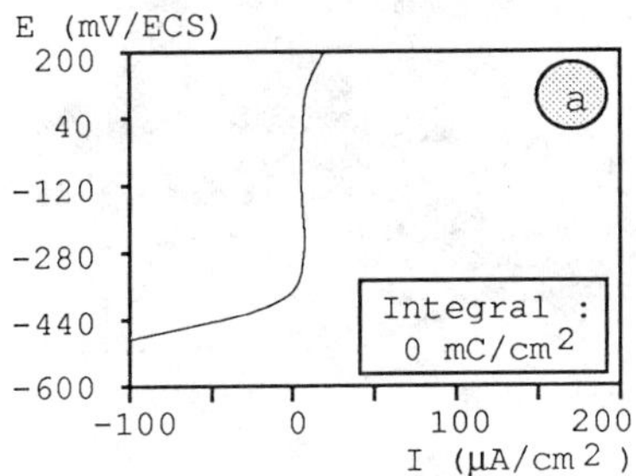

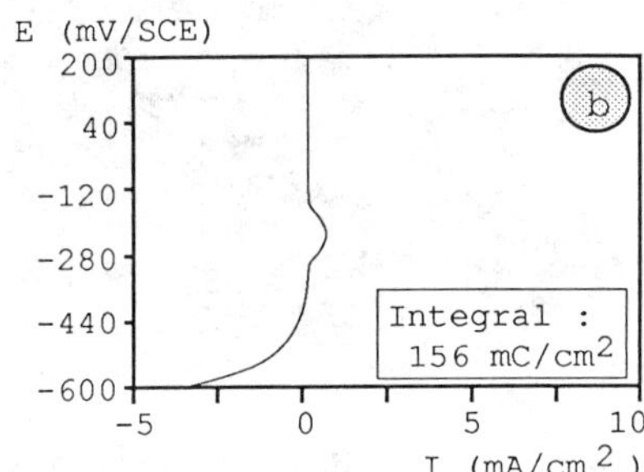

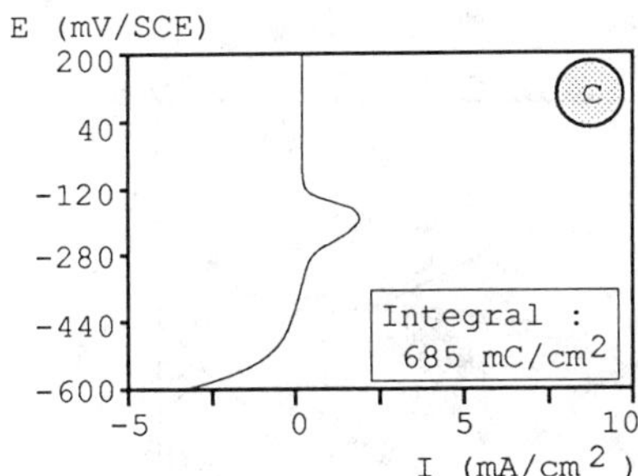

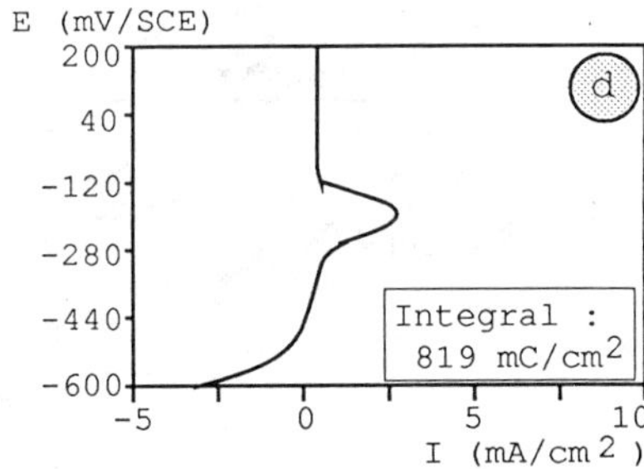

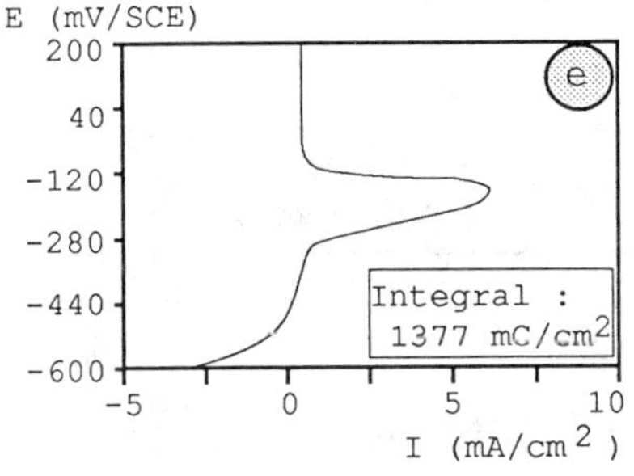

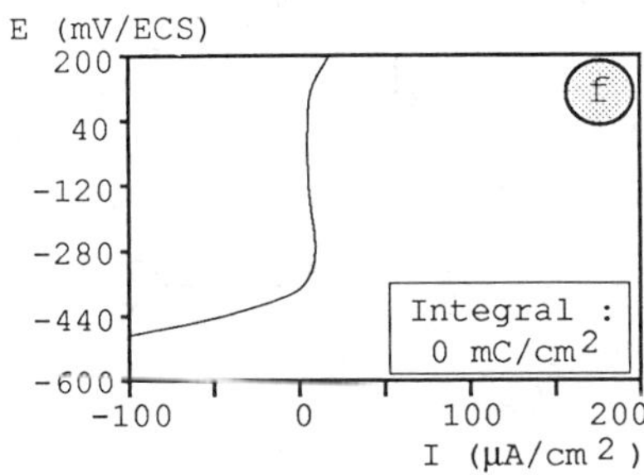

FIG. 5--EPR curves for UNS 31803 after the following heat-treatments: (a) Water quenched; (b) Water quenched + 800°C/10 min - air; (c) Water quenched + 800°C/30 min - air; (d) Water quenched + 800°C/1h-air; (e) Water quenched + 800°C/5h-air; (f) Water quenched + 600°C/1h-air

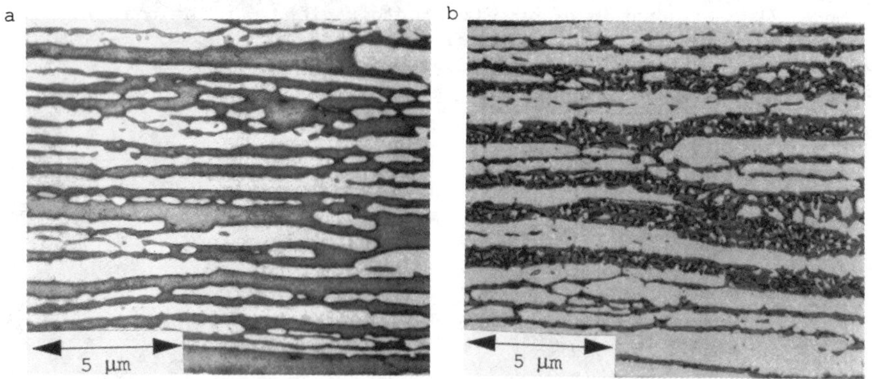

FIG. 6--(a) Micrograph on a water-quenched UNS 31803 plate; (b) Micrograph on the same plate after a 800 °C/5h-air heat treatment

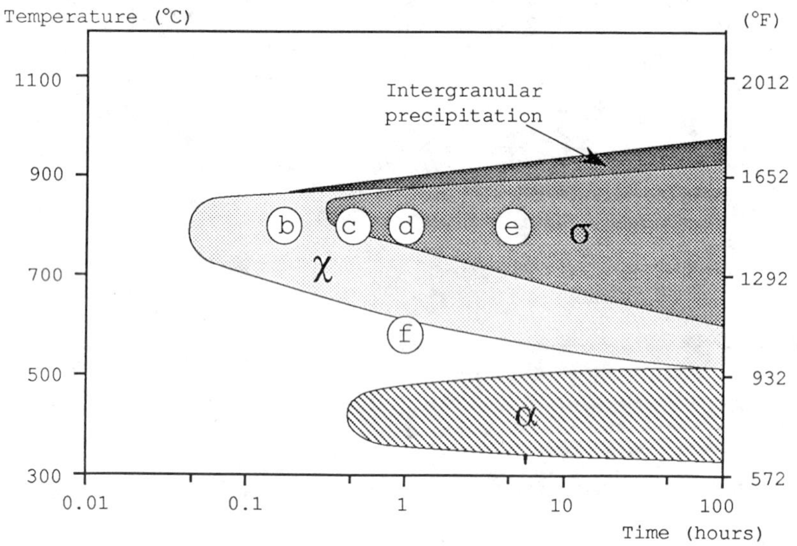

FIG. 7--Phase diagram of UNS S31803 [7]

ON-SITE MEASUREMENTS

One of the main advantages of the EPR method is that it is a non-destructive test. Thus, it allows on-site measurements. Indeed, the electrochemical method does not require that a sample is removed from the material. A specific electrochemical cell can be directly put on the surface of the steel to be tested, regardless of size.

Once calibrated, fabricators or steels producers can easily apply the test to evaluate the susceptibility of stainless steels plates or pipes to intergranular corrosion, or to detect eventual phase precipitation in their products. The EPR method could advantageously supplant the classical corrosion tests (Strauss, Streicher,etc). The rapidity of the test and its high sensitivity indicate the advantages of applying the method in the future.

This non-destructive method has the advantage of operations that are not possible with the classical tests although they would provide information on the quality of assembled vessels, for example :

1. Control of the sensitization after welding
2. Surface investigations with qualification of the pickling operations

Different equipments exist in this field of investigations. In practice, we have just conceived a portable equipment, easy to use on site (Fig. 8). This prototype consists of an electrochemical cell, with its three electrodes (working, reference and auxiliary), coupled with a portable computerized potentiostat. The cell is fixed on the steel to characterize, that constitutes the working electrode. Many different positions are possible, due to some articulations. This will make easier the development of the cell to work either on plates or pipes, elbows, welds, etc. The computerized potentiostat drives the experiment, and records the data instantaneously. At the end of the experiment, all the data are stored, and the computer calculates the integral of the reactivation peak. The degree of sensitization of the steel investigated is thus known in a few minutes only.

Only a few experiments have already been carried out with this new cell. The first results, obtained on flat plates, have shown a good accordance between laboratory tests and on-site tests. The method looks both sensitive and reliable. The future work will consist in testing different types of products. Some modifications will certainly be necessary to improve the prototype and to fit it to all shapes and sizes of products.

CONCLUSION

The paper has presented the development and some potential applications of an electrochemical method used to evaluate the degree of sensitization by phase or carbide precipitations of stainless steels and alloys. The main characteristics of the EPR test are its rapidity, its high sensibility, its reliability and the possibility of use as a non-destructive test. Therefore, a specific portable prototype has been conceived, to realize on-site measurements. After the first experiments, a good correlation has been observed between the results of the test and the real surface state of the materials. The development of the method should interest both steels producers, fabricators and final users.

FIG. 8--On-site measurements of the quality of pickling on stainless steels

REFERENCES

[1] Cihal, V., "Advances in Potentiodynamic Reactivation Method," Use of special steels, alloys and new materials in chemical process industries, Bulletin du Cercle d'Etudes des Métaux, Tome 16, May 1991, paper 17.

[2] Clarke, W.L., Cowan, R.L., and Walker, W.L., "Comparative Methods for Measuring Degree of Sensitization in Stainless Steels," Intergranular Corrosion of Stainless Alloys, ASTM STP 656, R.F. Steigerwald, Ed., American Society for Testing Materials, 1978, pp. 99-132.

[3] Majidi, A.P., and Streicher, M.A., "Nondestructive Electrochemical Tests for Detecting Sensitization in AISI 304 and 304L Stainless Steels," Electrochemical Techniques for Corrosion Engineering, R. Baboian, Eds., National Association of Corrosion Engineers, 1985

[4] Majidi, A.P., and Streicher, M.A., "Potentiodynamic Reactivation Method for Detecting Sensitization in AISI 304 and 304L Stainless Steels," Corrosion, Vol. 40, No. 8, 1984, pp. 393-408.

[5] Verneau, M., Charles, J., and Lojewski, C., "Technique de Contrôle In Situ de la Sensibilité à la Corrosion Intergranulaire d'Aciers Inoxydables et Alliages de Nickel," Proceedings of the European

Conference on Corrosion in Chemical and Parachemical Industries, Lyon, October 1991.

[6] Verneau, M., Lojewski, C., and Charles, J., "Modified EPR Test for Duplex Stainless Steels Surface Contamination and Microstructural Investigations," Duplex Stainless Steels'91, Editions de Physique, Beaune, October 1991, Vol. 2.

[7] Herbsleb, G., Schwaab, P., "Precipitation of Intermetallic Compounds, Nitrides and Carbides, in AF 22 Duplex Steel and their Influence on Corrosion Behavior in Acids," Proceedings of Duplex Stainless Steels, Ed. ASM, 1983, p.15.

Izumi Muto,[1] Eiji Sato,[1] and Satoshi Ito[1]

NEW ELECTROCHEMICAL METHOD TO EVALUATE THE ATMOSPHERIC CORROSION RESISTANCE OF STAINLESS STEELS

REFERENCE: Muto, I., Sato, E., and Ito, S., **"New Electrochemical Method to Evaluate the Atmospheric Corrosion Resistance of Stainless Steels,"** Application of Accelerated Corrosion Tests to Service Life Prediction of Materials, ASTM STP 1194, Gustavo Cragnolino and Narasi Sridhar, Eds., American Society for Testing and Materials, Philadelphia, 1994.

ABSTRACT: The quantitative evaluation of atmospheric corrosion resistance of stainless steels exposed to a marine environment was carried out by using a newly developed electrochemical method. This new method simulates field atmospheric conditions where most of stainless steels experience rust staining under a thin electrolyte layer. The incubation time to the breakdown of passive film under a thin electrolyte layer corresponds to the rating number of initial rust staining of the specimen exposed to the actual environment for 1 year, and depends on alloy compositions. Through the analysis of change in the rating number of the specimens exposed to the marine environment for 10 years, it was clarified that tendency of the change of rating number was directly related to exposure time, not to alloy compositions. The rating number of rust staining of stainless steels after several years can be predicted from the equation introduced through the results of both the accelerated test and the exposure test.

KEYWORDS: atmospheric corrosion, marine atmosphere, stainless steels, accelerated tests, electrochemical methods, rust staining

In recent years stainless steels have been utilized for architectural applications such as wall building side wall, where the initial surface appearance must be maintained. Then it is important to evaluate

[1]Steel Research Laboratories, Nippon Steel Corporation, 20-1 Shintomi, Futtsu, Chiba-ken 299-12, Japan.

of two aspects atmospheric corrosion resistance: rust staining resistance and pitting resistance. Atmospheric exposure testing remains the most reliable method for evaluating the resistance to rust staining and/or pitting. The results of exposure tests give us the base for estimation of long term performance of stainless steels [1]-[4].

The development of new accelerated tests [1][5]-[8] for evaluating quantitative atmospheric corrosion resistance is still under intensive study, where the rust staining resistance has been mostly examined from the point of appearance of corroded specimens. Through these evaluation methods, the relative resistance to rust staining of stainless steels can be estimated, but it is difficult to predict the rating number of rust staining of specimens exposed to an actual environment.

In this paper, in order to predict the service lifetime to rust staining of stainless steels used in a marine environment, the relationship between the incubation time to passive film breakdown under a thin electrolyte layer by using a newly developed electrochemical method and the rating number of specimens exposed to an actual marine environment for 10 years was studied, and the rating number of rust staining of stainless steels exposed to an actual environment was predicted.

TABLE 1-- Chemical compositions (mass%) and surface finishes of stainless steels exposed to a marine environment.

Specimen	AISI Type/ Tradename(Finish)	C	Mn	Cr	Ni	Mo	Other
16Cr-9Mn-2Ni	YUS120[a] (2D)	0.130	9.02	16.28	2.36	...	...
18Cr-6Ni	YUS27A[a] (2B)	0.050	1.09	17.58	6.40	...	...
17Cr-7Ni	301 (2B)	0.100	1.20	17.49	7.57	...	...
18Cr-8Ni	304 (2D)	0.050	0.95	18.26	9.14	...	...
18Cr-8Ni	304 (2B)	0.050	0.91	18.03	9.17	...	...
17Cr-12Ni-2.5Mo	316 (2B)	0.060	1.02	17.00	12.60	2.21	Cu:0.30
18Cr-7Ni-2Cu-1Mo	YUS316C[a](2B)	0.050	1.03	17.80	7.35	0.97	Cu:1.80
25Cr-13Ni-0.8Mo	YUS170[a] (2B)	0.022	1.50	24.25	12.97	0.82	N:0.345
10.9Cr	409 (2D)	0.040	0.57	10.85	...	...	Ti:0.47
11Cr	YUS409D[a](2B)	0.009	0.33	11.00	...	...	Ti:0.28
12Cr	YUS410W[a](2B)	0.017	0.47	12.10	...	...	...
13Cr	410 (2B)	0.090	0.24	12.85	...	...	...
16.5Cr	YUS430D[a](2B)	0.007	0.78	16.53	...	...	Ti:0.35
17Cr	430 (2B)	0.058	0.17	16.60	...	...	...
19Cr-2Mo	YUS190[a] (2B)	0.005	0.12	19.00	Mo:1.78	Nb:0.29	Ti:0.12

[a]Tradename of Nippon Steel Corporation.

EXPERIMENTAL PROCEDURE

Exposure Test

Fifteen different grades of stainless steels were exposed to a marine atmosphere for 10 years. The chemical compositions and surface finishes of the specimens were listed in Table 1. The specimens were machined to 100mmX150mm in size from commercially produced stainless steel sheets, with two different surface finishes, No.2D and No.2B. No.2D surface was produced by light skin-pass using dull rolls after annealing and pickling. No.2B surface was produced by light skin-pass using polished rolls. The atmospheric exposure test site was located in 150m distance from the seashore at Ako-city (100km west of Osaka-city in Japan). The average concentration of airborne chloride is $1.48 g \cdot m^{-2} \cdot day^{-1}$. This exposure test was started in 1978. The specimens were exposed facing to a south direction at a 30-degree angle to the ground. The degree of rust staining was examined visually and evaluated on the basis of the standard scale given in Table 2, referred to the standard rating panels shown in Fig.1.

TABLE 2-- Relationship between rating number and appearance of specimens.

Rating number	Appearance of surface
6	No rust or staining
5-3	Increased degree of rust staining, separate and small spots of rust
2-1	Discoloration and/or increased amount of rust
0	Rust covering major part of surface

FIG. 1--Standard rating panels.

Accelerated test

Rust staining of stainless steels in the atmosphere has three steps: (1) the formation of the water drop containing chloride ion and/or sulfur dioxide (SO_2) on the surface, (2) increased concentration of corrosive ions and diffusion of dissolved oxygen through drying process, (3) local breakdown of passive film and subsequent dissolution of metal. The incubation time to local breakdown of passive film under a thin electrolyte environment corresponds to initiation of rust staining of stainless steels exposed to an atmospheric environment.

Fig.2 shows the schematic illustration of the newly developed accelerated test apparatus newly developed. The apparatus was similar to that previously used [8], except for the use of a sheet heater, which accelerates the formation of a thin electrolyte layer. The specimen was adhered on the sheet heater. The cotton cloth, both sides of which were immersed in test solution reservoir, was placed on the specimen. A thin electrolyte layer, simulating an actual atmospheric environment, was formed by the capillary phenomena through the cotton cloth. The incubation time to the breakdown of passive film was monitored by the change of electrode potential with a saturated calomel electrode.

The chemical compositions of the stainless steels with No.2B surface finish used in this accelerated test were given in Table 3. The specimens were covered with silicone rubber except the test area of 20mmX13mm. Synthetic seawater, corresponding to a marine environment, was employed as an electrolyte. The surface temperature of specimens was kept at 333K by the sheet heater.

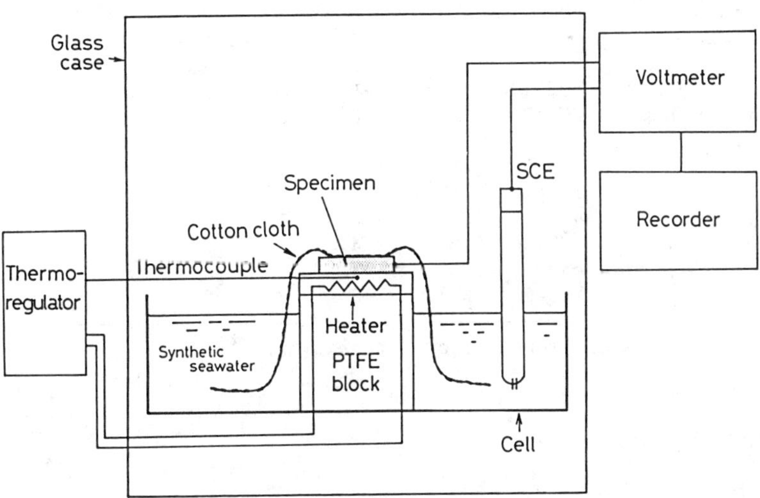

FIG 2--Schematic illustration of accelerated corrosion test apparatus.

TABLE 3-- Chemical compositions (mass%) of stainless steels for the accelerated test.

Specimen	AISI Type/ Tradename(Finish)	C	Mn	Cr	Ni	Mo	Other
11Cr	YUS409D[a](2B)	0.013	0.35	10.74	0.06	...	Ti:0.34
17Cr	430 (2B)	0.058	0.13	16.01	0.10	...	Al:0.10
19Cr-1Mo	—— (2B)	0.012	0.14	19.28	0.29	0.98	Cu:0.37
19Cr-2Mo	YUS190[a] (2B)	0.011	0.15	19.02	Mo:1.81	Nb:0.29	Ti:0.18
18Cr-8Ni	304 (2B)	0.069	0.15	18.29	8.74	...	...
17Cr-12Ni-2.5Mo	316 (2B)	0.072	0.70	17.62	11.30	2.19	Cu:0.29
25Cr-13Ni-0.8Mo	YUS170[a] (2B)	0.019	0.43	24.68	12.78	0.73	N:0.35

[a]Tradename of Nippon Steel Corporation

TABLE 4-- Rating number of specimens exposed to the marine environment.

Specimen	AISI Type/ Tradename(Finish)	Rating number			
		1	2	4	10 (year)
16Cr-9Mn-2Ni	YUS120[a] (2D)	5	2	1	1
18Cr-6Ni	YUS27A[a] (2B)	5	2	2	1
17Cr-7Ni	301 (2B)	5	2	2	1
18Cr-8Ni	304 (2D)	5	2	2	1
18Cr-8Ni	304 (2B)	4	2	2	1
17Cr-12Ni-2.5Mo	316 (2B)	5	3	2	2
18Cr-7Ni-2Cu-1Mo	YUS316C[a](2B)	5	2	2	1
25Cr-13Ni-0.8Mo	YUS170[a] (2B)	6	3	2	2
10.9Cr	409 (2D)	2	0	0	0
11Cr	YUS409D[a](2B)	2	0	0	0
12Cr	YUS410W[a](2B)	2	0	0	0
13Cr	410 (2B)	3	1	0	0
16.5Cr	YUS430D[a](2B)	4	2	2	1
17Cr	430 (2B)	4	1	0	0
19Cr-2Mo	YUS190[a] (2B)	6	3	2	2

[a]Tradename of Nippon Steel Corporation.

Results and Discussion

Exposure test

Table 4 shows the results of the long term exposure test in the marine environment. In order to know the relationship between the rating number of initial rust staining and alloy compositions, the rating numbers of the specimens after 1 year were examined visually. The dependence of the rating number on the alloy composition was analyzed by multi-regression method as shown in Fig.3. The rating number of stainless steels exposed for 1 year, R_{1year}, was given by the following equation:

$$R_{1year}=0.36X_1-1.64 \quad (1)$$

where X_1 is the alloy index, or the atmospheric corrosion resistance index depending on the alloy contents derived from multi-regression analysis. This index was written as follows:

$$X_1=[Cr]-0.08[Ni]+1.05[Mo]=[Cr]+[Mo] \quad (2)$$

Here [Cr], [Ni], and [Mo] indicate alloy content in mass% of chromium, nickel, and molybdenum respectively. In Equation (2), the coefficient of nickel content is almost zero. This alloy index after 1 year in the marine environment can be expressed as X_1=[Cr]+[Mo].

This result suggests that nickel has no influence on the resistance to initial rust staining of stainless steels. This alloy index X_1=[Cr]+[Mo] is also similar to the pitting index ([Cr]+3.3[Mo]) that was derived from the relationship between alloy compositions and pitting potential in a chloride solution [9].

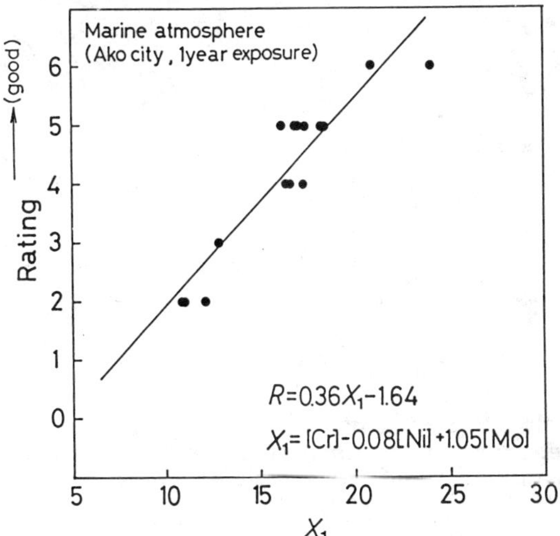

FIG. 3-- Effect of alloy compositions on the rating number after 1 year in the marine environment.

The change of rating number with time of four different grades of stainless steels is shown in Fig.4. These results suggest that most of stainless steels experienced severe corrosion within the first two years in this environment (Fig.4(a)). Fig.4(b) also shows the same tendency of the change in the rating number with the square of the inverse time. The change in rating number, ΔR, was expressed as follows:

$$\Delta R = 4t_1^{-2} - 4 \quad (3)$$

Here t_1 is exposure time in year.

Combination of the Equation (1) and (3) leads to the following equation, which gives the relation between the rating number, R, after several years and exposure time.

$$R = R_{1year} + \Delta R \quad (4)$$

$$= (0.36X_1 - 1.64) + (4t_1^{-2} - 4) \quad (5)$$

Here X_1 is the atmospheric corrosion resistance index as indicated in Equation (2). According to Equation (5), the resistance to initial rust staining resistance of stainless steels was related to chromium and molybdenum contents, but the change in rating number was directly related to exposure time.

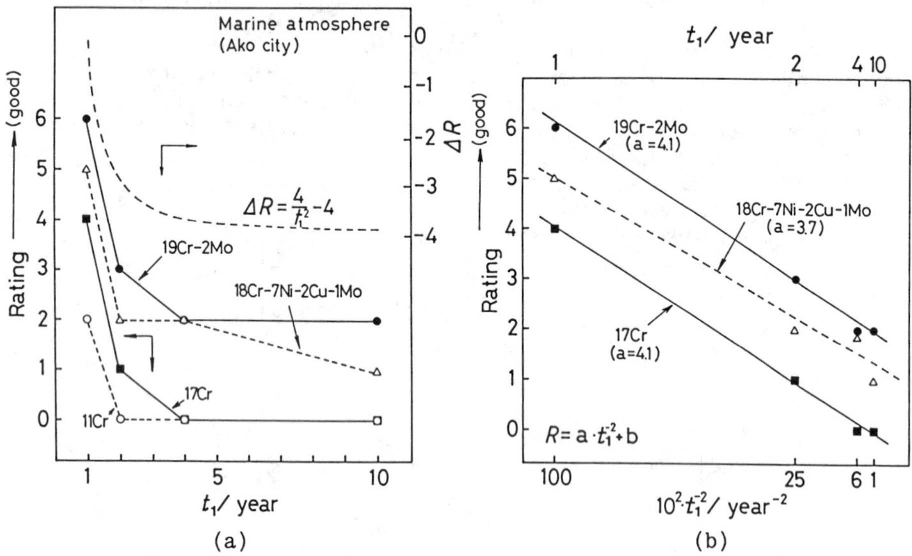

FIG. 4-- Atmospheric corrosion behavior of different grades of stainless steels. The dependence of the rating number on (a) exposure time and (b) the square if the inverse time.

Accelerated corrosion test

The atmospheric corrosion resistance index for stainless steels can be also derived from the newly developed accelerated test. Fig.5 shows the corrosion potential oscillations for stainless steels under the thin electrolyte layer of the accelerated test, which simulates the actual atmospheric conditions (Fig.2). One oscillation corresponds to local breakdown of passive film and repassivation process. All the specimens experienced potential oscillations and finally had the less noble potential around -0.4V, which indicates the formation of an irreversible macro pit on the surface. The similar pits were observed on the exposed specimens to the marine atmosphere. Therefore, the incubation time to pit initiation was employed as evaluation parameter for quantitative rust staining resistance of stainless steels in the marine environment.

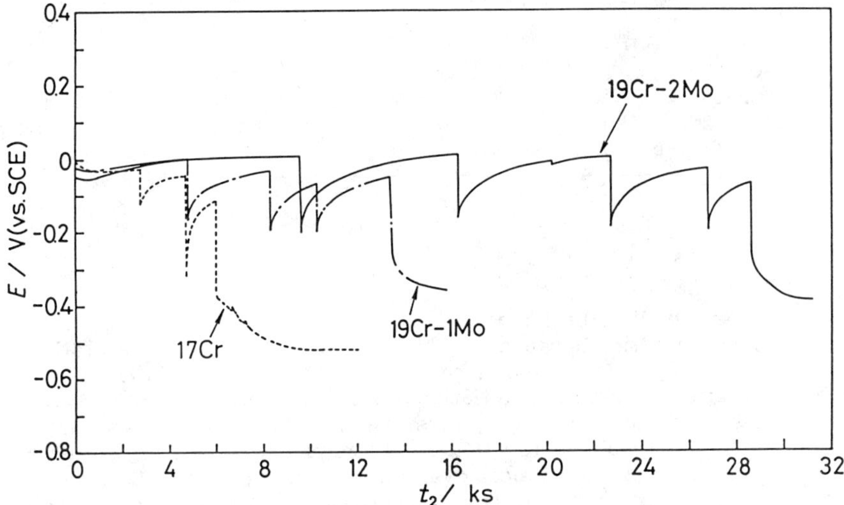

FIG. 5-- Potential-time curves for stainless steels obtained in the accelerated test.

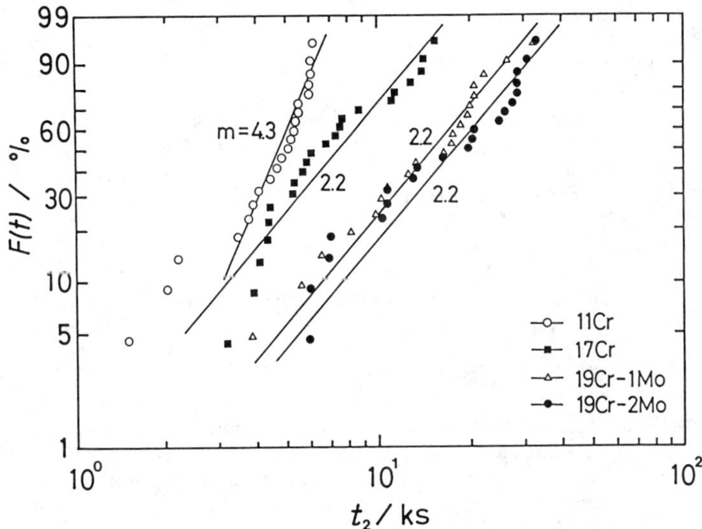

FIG. 6-- Weibull distributions of the incubation time to pit initiation.

TABLE 5-- Parameters of the Weibull distributions of the incubation time to pit initiation in the thin electrolyte layer.

Specimen	Shape parameter, m	Mean value, μ (ks)
11Cr	4.3	4.50
17Cr	2.2	6.78
19Cr-1Mo	2.2	13.80
19Cr-2Mo	2.2	16.26
18Cr-8Ni	3.6	4.92
17Cr-12Ni-2.5Mo	2.1	9.72
25Cr-13Ni-0.8Mo	2.5	24.12

The Weibull distribution was applied to determine the mean value of the incubation time, because the pit nucleation process is a statistic phenomenon [10]. Fig.6 shows the plot of the Weibull distributions of the incubation time to pit initiation under the thin electrolyte layer. These data follow straight lines, meaning that the incubation time to pit initiation obeys the Weibull distribution. These distributions indicate that the statistical evaluation of the time to pit initiation is useful to compare the corrosion resistance between the specimens due to overlapping of the distributions. The parameters of these distributions are given in Table 5. The slope of the straight line corresponds to the shape parameter, and these values listed in Table 5 are in the range of 2.1 to 4.3. This result suggests that the shapes of the distributions are almost the same. Therefore, the relative resistance to pit initiation under the thin electrolyte layer can be evaluated from the mean value of the distributions. Then, the mean value of the Weibull distribution, μ, is employed for the representative.

The relationship between the Weibull mean value, μ, of the incubation time and the chemical composition was determined by multi-regression analysis similar to that used for determination of Equation (1). The mean value, μ(ks), was given by the following equation:

$$\mu = 1.58X_2 - 16.70 \qquad (6)$$

where X_2 is the alloy index obtained by this accelerated test. The index X_2 is defined by Equation (7).

$$X_2 = [Cr] - 0.18[Ni] + 0.88[Mo] = [Cr] + [Mo] \qquad (7)$$

Fig.7 shows the relationship between X_2 obtained by multi-regression analysis and the mean value, μ, of the incubation time to pit initiation. The mean value, μ, has a linear relation with X_2 value, and this alloy index is almost the same as that obtained from the exposure test in Equation (2).

$$X_2 = X_1 \quad (= [Cr] + [Mo]) \qquad (8)$$

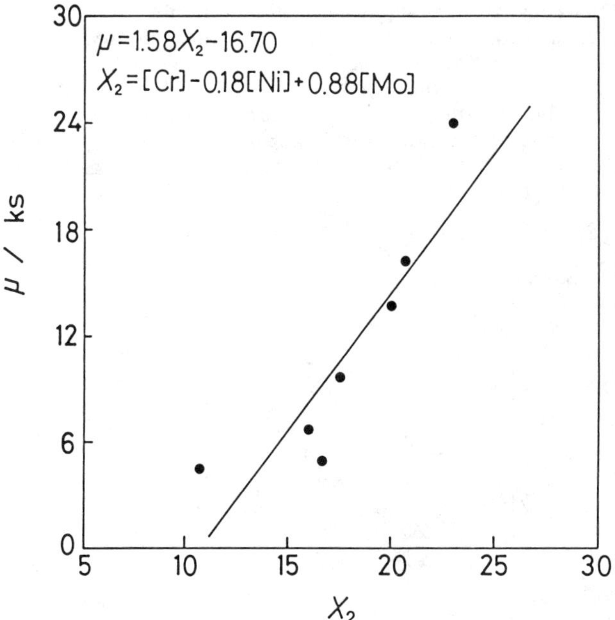

FIG. 7-- Effect of alloy compositions on the mean value of the incubation time to pitting in the accelerated test.

This result suggests that the accelerated test conditions approximately simulate the actual atmospheric environment, and the quantitative atmospheric corrosion resistance can be estimated from the mean value, μ, of the incubation time. The difference in the mean value, μ, indicates the quantitative difference in rust staining resistance of specimens. Furthermore, when $X_2=X_1$ is substituted in Equation (6), the following equations are derived.

$$\mu=1.58X_2-16.70=1.58X_1-16.70 \qquad (9)$$

$$X_1=\mu/1.58+10.6 \qquad (10)$$

Substituting the equation (10) for X_1 in the equation (5) yields the following equation, which gives the rating number, R, after several years in the marine environment.

$$R=(\mu/4.39+2.18)+(4t^{-2}-4) \qquad (11)$$

Here t is exposure time in years, and μ (ks) is the mean value of the incubation time to pitting in the accelerated test.

The rating number estimated by this accelerated test is more accurate than that calculated from the alloy index X_1=[Cr]+[Mo] by using

Equation (5), because this newly developed test also evaluates the influence of other alloy elements, inclusions, and surface finish on rust staining of specimens.

Fig.8 shows the relationship between the rating number of specimens exposed to the marine environment and that estimated by Equation (11), suggesting that the rating number and rust staining resistance of stainless steels in the marine environment can be evaluated by this newly developed test.

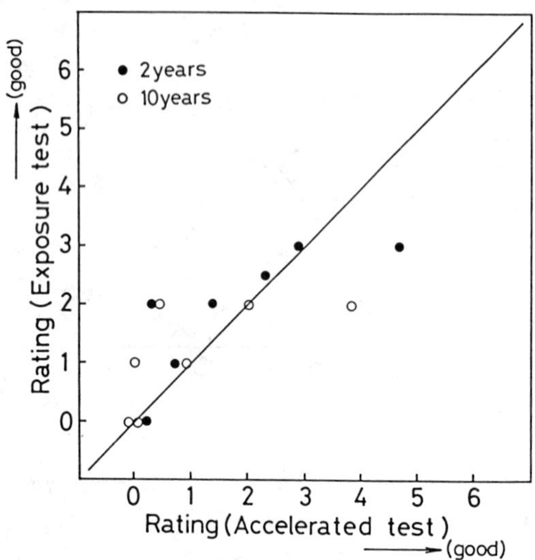

FIG. 8-- Relationship between the rating number of the specimens exposed to the marine environment and that estimated in the accelerated test.

CONCLUSIONS

1. The rating number of rust staining of stainless steels exposed to the marine environment was expressed as follows:

$$R=(0.36X_1-1.64)+(4t^{-2}-4)$$

$$X_1=[Cr]+[Mo]$$

where X_1 is the alloy index that indicates the relative corrosion resistance to rust staining, t is exposure time in years.

2. The rating number of rust staining of stainless steels was also evaluated in the accelerated test and was expressed as follows:

$$R=(\mu/4.39+2.18)+(4t^{-2}-4)$$

where μ (ks) is the mean value of the incubation time to pitting in the accelerated test, t is exposure time in years.

Good correlation was observed between the rating numbers of speci-ns in the exposure test and that estimated in the accelerated corro-n test.

FERENCES

Black, H. L. and Lherbier, L. W., "A Statistical Evaluation of Atmospheric, In-Service, and Accelerated Corrosion of Stainless Steel Automotive Trim Material," Metal Corrosion in the Atmosphere, ASTM STP 435, American Society for Testing and Materials, Philadelphia, 1968, pp.3-32.

Karlson, A. and Olsson, J., "Atmospheric Corrosion of Stainless Steels," 7th Scandinavian Corrosion Congress, 1975, pp.71-86.

Needham, N. G., Freeman, P. F., Wilkinson, J., and Chapman, J., "Atmospheric Corrosion Resistance of Stainless Steels," Stainless Steels '87, The Institute of Metals, York, 1988, pp.215-223.

Kearns, J. R., Johnson, M. J., and Pavlik, P. J., "The Corrosion of Stainless Steels in the Atmosphere," Degradation of Metals in the Atmosphere, ASTM STP 965, American Society for Testing and Materials, Philadelphia, 1988, pp.35-51.

Baker, E. A. and Lee, T. S., "Long-Term Atmospheric Corrosion Behavior of Various Grades of Stainless Steel," Degradation of Metals in the Atmosphere, ASTM STP 965, American Society for Testing and Materials, Philadelphia, 1988, pp.35-51.

Bates, J. F. and Phelps, E. H., "An Appraisal of Evaluation Tests for Stainless Steel Automotive Trim," Advances in the Technology of Stainless Steels and Related Alloys, ASTM STP 369, American Society for Testing and Materials, Philadelphia, 1963, pp.200-208.

Bush, G. F., Garwood, W. J., and Tiffany, B. E., "Accelerated Tests As a Method of Predicting Service Corrosion of Exterior Automotive Trim," Advances in the Technology of Stainless Steels and Related Alloys, ASTM STP 369, American Society for Testing and Materials, Philadelphia, 1963, pp.209-222.

Ito, S., Yabumoto, M., Omata, H., and Murata, T., "Atmospheric Corrosion of Stainless Steels," Passivity of Metals and Semiconductors, Elsevier Science Publishers B.V., Amsterdam, 1983, pp.637-642.

Speidel, M. O., "Corrosion Science of Stainless Steels, " Proceedings of International Conference on Stainless Steels (Stainless Steels '91), Iron and Steel Institute of Japan, 1991, pp.25-35.

Shibata, T., "Evaluation of Corrosion Failure by Extreme Value Statistics," ISIJ International, Vol. 31, No. 2, 1991, pp.115-121.

Author Index

Subject Index

L

M

N

O

P

Q

R

S

T

V

W

Y

Z